Instrumental
Methods
of Chemical
Analysis

Instrumental
Methods
of Chemical
Analysis

Galen W. Ewing

Professor of Chemistry
Seton Hall University

Fourth Edition

McGraw-Hill Book Company

New York • St. Louis • San Francisco • Auckland • Düsseldorf
Johannesburg • Kuala Lumpur • London • Mexico
Montreal • New Delhi • Panama • Paris • São Paulo
Singapore • Sydney • Tokyo • Toronto

544
E w 5 i
9 7 0 9 0
May 1976

Instrumental Methods of Chemical Analysis

1234567890   KPKP   798765

This book was set in Linotype No. 21 by Bi-Comp,
Incorporated. The editors were Robert H. Summersgill and
Anne T. Vinnicombe; the cover was designed by Joseph
Gillians; the production supervisor was Sam Ratkewitch.
The new drawings were done by ANCO Technical Services.
Kingsport Press, Inc., was printer and binder.

Library of Congress Cataloging in Publication Data

Ewing, Galen Wood, date
  Instrumental methods of chemical analysis.

  Includes bibliographies and indexes.
  1. Chemistry, Analytic. 2. Chemical apparatus.
I. Title.
QD75.2.E94 1975   543'.08   74-23870
ISBN 0-07-019853-5

Contents

Preface

As in previous editions, the general objective in this book is to survey the theory and practice of modern analytical instrumentation. Emphasis is placed on the possibilities and limitations inherent in the various methods.

The text is planned for use in upper-level undergraduate or first-year graduate classes. To be most effective, this course should follow work in elementary quantitative analysis and a year of physics; it may follow or run concurrently with physical chemistry.

It is always a difficult matter to decide what to include and what to omit. The words "analytical" and "instrumental" are not amenable to objective definition. With respect to the former, H. A. Laitinen has written: "The vital point here is that if the research is aimed at methods of solution of a measurement problem, it is properly classified as analytical chemistry, whereas the interpretation of the results of the measurements infringes upon other fields of chemistry." [Editoral, *Anal. Chem.*, **38**:1441 (1966).] I have attempted to include just enough interpretive material to suggest the areas in which a method can be useful.

With respect to the term *instrumental*, I have tried to be led more by usefulness to the chemistry student than by a strict definition of the term. Thus such an important instrument as the analytical balance is not discussed, whereas paper chromatography is.

The major changes in the fourth edition are the addition of chapters on electron spectroscopy and automated analysis. The treatment of magnetic resonance and mass spectroscopy have been considerably expanded. The former chapters on flames, solvent extraction, and electrical separations have been deleted, and their content integrated elsewhere.

The treatment of electronics, as in the previous edition, is placed at the end and cross-referenced from the earlier chapters as needed. It can be treated as a separate block at the start of an instrumentation course by those instructors preferring such a sequence. It is my feeling that this arrangement interferes least with the logical flow of ideas concerning instrumental analysis, while still providing access to this essential background material.

I have tried to convert uniformly to the SI system of units, as described in Chapter 1, though there are a few exceptions. My use of the British spelling "metre," for the unit, not the measuring instrument, can be considered a bit of crusading.

Mention of the products of individual manufacturers does not necessarily imply that I consider them superior to competing items. The aim is to describe instruments typical of their class or possessing some special features of interest, not to write a complete catalog of analytical apparatus.

I wish to express my sincere appreciation to my colleagues and students, past and present, who have offered advice and pointed out errors. Particular thanks go to the following individuals who have done critical readings of various portions of the manuscript: Drs. N. L. Alpert, J. Jordan, L. C. Love, C. A. Lucchesi, M. Margoshes, T. D. Schroeder, and H. Veening. I am greatly indebted also to the personnel of instrument companies and distributors, too numerous to list, without whose cooperation the book could not be a success.

GALEN W. EWING

Instrumental
Methods
of Chemical
Analysis

Chapter 1

Introduction

Analytical chemistry may be defined as the science and art of determining the composition of materials in terms of the elements or compounds contained. Historically, the development of analytical methods has followed closely the introduction of new measuring instruments. The first quantitative analyses were gravimetric, made possible by the invention of a precise balance. It was soon found that carefully calibrated glassware made possible considerable saving of time through the volumetric measurement of gravimetrically standardized solutions.

In the closing decades of the nineteenth century, the invention of the spectroscope brought with it an analytical approach that proved to be extremely fruitful. At first it could be applied only qualitatively; gravimetric and volumetric methods remained for many years the only quantitative procedures available for nearly all analyses. Gradually a few colorimetric and nephelometric methods were introduced. Then it was found that electrical measurements could be used to advantage to detect end points in titrations. Since about 1930, the rapid development of electronics has resulted in a major revolution in analytical instrumentation. Today the chemist, whether he calls himself an analytical specialist or not, must have a working knowledge of a dozen or so instrumental methods virtually unknown a generation ago.

1

Nearly any physical property characteristic of a particular element or compound can be made the basis of a method for its analysis. Thus, the absorption of light, the conductivity of a solution, or the ionizability of a gas can serve as an analytical tool. A whole series of related techniques depends on the varying electrical properties of different elements, as evidenced by their redox potentials. The phenomena of artificial radioactivity have led to several analytical methods of great significance. It is the purpose of this book to investigate the possibilities of many of these modern instrumental methods of analysis.

PHYSICAL PROPERTIES USEFUL IN ANALYSIS

The following is a list of physical properties that have been found applicable to chemical analysis. The list is not exhaustive, but it certainly includes all those properties that have been extensively investigated, as well as some not yet fully exploited.

EXTENSIVE PROPERTIES
1. Mass
2. Volume (of a liquid or gas)

MECHANICAL PROPERTIES
1. Specific gravity (or density)
2. Surface tension
3. Viscosity
4. Velocity of sound

PROPERTIES INVOLVING INTERACTION WITH RADIANT ENERGY
1. Absorption of radiation
2. Scattering of radiation
3. Raman effect
4. Emission of radiation
5. Refractive index and refractive dispersion
6. Rotation of the plane of polarized light and rotatory dispersion
7. Circular dichroism
8. Fluorescence and phosphorescence
9. Diffraction phenomena
10. Nuclear and electron magnetic resonance

ELECTRICAL PROPERTIES
1. Half-cell potentials
2. Current-voltage characteristics
3. Electrical conductivity
4. Dielectric constant

THERMAL PROPERTIES

1. Transition temperatures
2. Heats of reaction
3. Thermal conductivity (of a gas)

NUCLEAR PROPERTIES

1. Radioactivity
2. Neutron cross section
3. Isotopic mass

METHODS OF SEPARATION PRIOR TO ANALYSIS

It would be desirable to discover analytical methods that are *specific* for each element or radical or class of compounds. Unfortunately only a few methods are completely specific, and it is therefore frequently necessary to perform quantitative separations with the objective of either isolating the desired constituent in a measurable form or removing interfering substances. Some methods of separation are:

1. Precipitation
2. Electrodeposition
3. Distillation
4. Solvent extraction or sublation
5. Partition chromatography
6. Adsorption chromatography
7. Ion exchange
8. Electrophoresis
9. Dialysis

ELECTRONICS

The fundamental task to be performed by an instrument is to translate chemical information into a form directly observable by an operator. It does this by means of a *transducer,* a component capable of transferring energy from one domain (chemical) to another (usually electrical). Then it is up to the electronics of the instrument to process the electrical signal (i.e., isolate it, amplify it, compare it with a standard, modify it functionally) and finally to read it out on a meter or an automatic recorder.

The great majority of laboratory instruments involve electronic devices. It is characteristic of analytical instruments that they are designed to be as sensitive as possible, and hence must be able to measure precisely the smallest signal that can be produced by the primary measuring element. Amplification is almost always required, and the most versatile way of achieving this amplification is through electronics.

In many methods it is necessary to apply some sort of stimulus to the chemical system (a beam of radiation, for example), and this stimulus is often produced or at least measured and regulated by additional electronic devices.

Because of these close relations with chemical instruments, the fundamentals of electronics form an integral part of any treatment of instrumentation. Fortunately, modern electronics has developed in the direction of modularization. A variety of amplifiers and logic elements is available as low-cost plug-in units that can be used as building blocks for the construction of most of the electronic equipment described in this book. Commercial equipment frequently does not make use of these building blocks as separate physical entities, but the principles involved are the same.

A short treatment of those aspects of electronics pertinent to our main subject is given in the concluding chapters of this book. This may be studied separately if desired, or used as resource material to aid in a better understanding of the instruments as they are taken up.

TITRATION

Titration is defined as the measurement of an unknown constituent by establishment of the exactly equivalent amount of some standard reagent. Physical measurements are involved in two ways: in the detection of the equivalence point, and in the measurement of the quantity of reagent consumed. Usually, and unless otherwise specified, the quantity of reagent is measured volumetrically with a buret. The chief exception is *coulometric titration*, in which the reagent is generated electrolytically on the spot as required, and its quantity determined by electrical measurements.

Many of the analytical methods described in this book can be used to follow the course of a titration, often with reduced demands on the instrumentation. Such applications are discussed toward the ends of the respective chapters.

NOTE ON UNITS

Worldwide agreement on units and their symbols is of prime importance to the future of science. A determined effort in this direction has been mounted by a number of international conferences with the cooperation of the National Bureau of Standards in the United States and corresponding offices in other countries. This effort has resulted in the *Système International d'Unités* (International System of Units, abbreviated SI). The system is described in a booklet published by the National Bureau of Standards.*

* C. H. Page and P. Vigoureux (eds.), *The International System of Units (SI)*, 1972. Available from the Superintendent of Documents, U.S. Government Printing Office, Washington, D.C. 20402; SD Catalog No. C 13.10:330/2; price 30 cents.

In the present text, SI units are used throughout.   Where other units are widely employed, this fact will be pointed out.   The chief difference so far this book is concerned is the abandoning of such well-known units as the angstrom, micron, gauss, and torr.   Another point of departure from American custom is the spelling of metre.   This form is preferred as obviating confusion between a unit (e.g., micrometre, 10^{-6} m) and an instrument (micrometer, a measuring tool).

The standard prefixes for multiples and submultiples of units are:

Factor	Prefix	Symbol
10^{12}	Tera	T
10^{9}	Giga	G
10^{6}	Mega	M
10^{3}	Kilo	k
10^{2}	Hecto	h
10^{1}	Deka	da
10^{-1}	Deci	d
10^{-2}	Centi	c
10^{-3}	Milli	m
10^{-6}	Micro	μ
10^{-9}	Nano	n
10^{-12}	Pico	p
10^{-15}	Femto	f
10^{-18}	Atto	a

BIBLIOGRAPHY

The student who wishes to pursue in greater depth any of the topics mentioned in this book has many avenues to which to turn.   There are of course the general sources, such as *Chemical Abstracts*, applicable to all branches of chemistry.

In the analytical field there is a great proliferation of journals of primary interest.   *Analytical Chemistry, Analytica Chimica Acta, Talanta, The Analyst* (including *Analytical Abstracts*), the *Zeitschrift für analytische Chemie*, and *Analytical Letters* attempt general analytical coverage. In specific fields are the *Journal of Electroanalytical Chemistry*, the *Journal of Chromatography*, the *Journal of Gas Chromatography*, *Spectrochimica Acta, Analytical Biochemistry*, and many others.   With emphasis on instruments per se, one finds the *Review of Scientific Instruments*, the *Journal of Scientific Instruments*, and *Chemical Instrumentation*.   The *Journal of Chemical Education* runs a monthly column on topics in chemical instrumentation, in addition to many articles of analytical interest.

On the theoretical side, the "Treatise on Analytical Chemistry," edited by I. M. Kolthoff and P. J. Elving (Wiley-Interscience, New York), is invaluable, especially Part I. Many of the volumes of the series "Physical Methods of Chemistry," edited by A. Weissberger and B. W. Rossiter (Wiley-Interscience) present a wealth of information on analytical instrumentation. Also not to be overlooked is the series "Advances in Analytical Chemistry and Instrumentation," also published by Wiley-Interscience.

The *Annual Reviews* issue of *Analytical Chemistry,* published each April, contains critical reviews in all fields of analysis; in even years the reviews are classed by the analytical principles involved, and in odd years by field of application. The *CRC Critical Reviews in Analytical Chemistry* is another source of thorough treatment of selected topics.

An immense amount of useful information, with succinct reviews of theoretical principles, has been collected under the editorship of L. Meites in the "Handbook of Analytical Chemistry," published by McGraw-Hill, New York, 1963.

Chapter 2

Introduction to Optical Methods

A major class of analytical methods is based on the interaction of radiant energy with matter. In the present chapter we shall review some of the pertinent properties both of radiation and of matter, and then discuss those features of optical instrumentation that apply to all or several spectral regions. In subsequent chapters each major spectral range (visible, ultraviolet, infrared, x-ray, microwave) will be considered separately, with respect to theory, instrumentation, and chemical applications.

THE NATURE OF RADIANT ENERGY

An investigation into the properties of radiant electromagnetic energy reveals an essential duality in our understanding of its nature. In some respects its properties are those of a wave, while in others it is apparent that radiation consists of a series of discrete packets of energy (*photons*). The photon concept is almost always required in the rigorous treatment of the interactions of radiation with matter, although the wave picture may be used to give approximately correct results when large numbers of photons are involved.

Radiant energy can be described in terms of a number of properties or parameters. The *frequency* ν is the number of oscillations per second described by the electromagnetic wave; the usual unit of frequency is the *hertz* (1 Hz = 1 cycle per second). The *velocity* c of propagation is very nearly 2.9979×10^8 m · s⁻¹ for radiation traveling through a vacuum, and somewhat less for passage through a transparent medium.

The *wavelength* λ is the distance between adjacent crests of the wave in a beam of radiation. It is given by the ratio of the velocity to the frequency. The units of wavelength are the *micrometre* (1 μm = 10^{-6} m; formerly called the *micron*, μ), and the *nanometre* (1 nm = 10^{-9} m; formerly the *millimicron*, mμ). The *angstrom* (1 Å = 10^{-10} m) is widely used in spectroscopy. Another quantity which is often convenient is the *wave number* $\bar{\nu}$, the number of waves per centimetre.* The unit is the *reciprocal centimetre*, (cm⁻¹), sometimes called the kaiser.

The velocity, wavelength, frequency, and wave number are related by the expression

$$\nu = \frac{c}{\lambda} = c\bar{\nu} \qquad (2\text{-}1)$$

The energy content E of a photon is directly proportional to the frequency:

$$E = h\nu = \frac{hc}{\lambda} = hc\bar{\nu} \qquad (2\text{-}2)$$

where h is Planck's constant, very close to 6.6256×10^{-34} J · s. Thus there is an inverse relationship between energy content and wavelength, but a direct relation between energy and frequency or wave number. It is for this reason that the presentation of spectra in terms of frequency or wave number rather than wavelength is gaining in favor.

It is convenient, particularly with nuclear radiations and x-rays, to characterize the radiation by the energy content of its photons in *electron volts* (eV); 1 eV = 1.6020×10^{-19} J, corresponding to frequency $\nu = 2.4186 \times 10^{14}$ Hz or to the (in vacuo) wavelength $\lambda = 1.2395 \times 10^{-6}$ m. The multiples kilo-electron volt (keV) and mega-electron volt (MeV) are often encountered.

A beam of radiation carries energy which is emitted from a source and propagated through a medium or series of media to a receptor where

---

* It is unfortunate that the symbol $\bar{\nu}$ has been chosen to indicate the wave number, because of its likely confusion with ν for frequency; indeed, in certain areas of physics it is customary to use these symbols in exactly the opposite sense. Expressions such as "a frequency of 1600 wave numbers," though often found in the literature, are not correct. Frequency may be *proportional* to wave number, but cannot be equal to it, as the dimensions are unlike. Furthermore, a wave number is not a unit, so "1600 wave numbers" is no more correct than describing this page as "6 distances wide."

it is absorbed. On its way from source to ultimate absorber, the beam may undergo partial absorption by the media through which it passes, it may be changed in direction by reflection, refraction, or diffraction, or it may become partially or wholly polarized.

Since energy per unit time is power, one is often interested in the *radiant power* of the beam, a quantity often loosely referred to as intensity. *Intensity* more correctly refers to the power emitted by the source per unit solid angle in a particular direction. A photoelectric cell gives a response related to the total power incident on its sensitive surface. A photographic plate, on the other hand, integrates the power over the time of exposure to the beam, and hence its response (silver deposit) is a function of the total incident energy (rather than power) per unit area. In both photoelectric cells and photographic plates, as well as in the human eye, the sensitivity is a more or less complicated function of the wavelength.

SPECTRAL REGIONS

The spectrum of radiant energy is conveniently broken down into several regions, as shown in Table 2-1. The limits of these regions are determined by the practical limits of appropriate experimental methods of production and detection of radiations. The figures in the table are not in themselves especially significant, and should be considered only as rough boundaries.

The differentiation of spectral regions has additional significance for the chemist in that the interactions of the radiations with chemical systems follow different mechanisms and provide different kinds of information.

Table 2-1   Regions of the electromagnetic spectrum*

| Designation | Wavelength limits | | Frequency limits, Hz | Wave number limits, cm^{-1} |
	Usual units	Metres		
X-rays	10^{-2}–10^2 Å	10^{-12}–10^{-8}	10^{20}–10^{15}	
Far ultraviolet	10–200 nm	10^{-8}–2×10^{-7}	10^{16}–10^{15}	
Near ultraviolet	200–400 nm	2×10^{-7}–4.0×10^{-7}	10^{15}–7.5×10^{14}	
Visible	400–750 nm	4.0×10^{-7}–7.5×10^{-7}	7.5×10^{14}–4.0×10^{14}	25,000–13,000
Near infrared†	0.75–2.5 μm	7.5×10^{-7}–2.5×10^{-6}	4.0×10^{14}–1.2×10^{14}	13,000–4000
Mid infrared†	2.5–50 μm	2.5×10^{-6}–5.0×10^{-5}	1.2×10^{14}–6.0×10^{12}	4000–200
Far infrared†	50–1000 μm	5.0×10^{-5}–1×10^{-3}	6×10^{12}–10^{11}	200–10
Microwaves	0.1–100 cm	1×10^{-3}–1	10^{11}–10^8	10–10^{-2}
Radio waves	1–1000 m	1–10^3	10^8–10^5	

* Where a numerical factor is omitted, it is because the precision of delineation of the region does not warrant a greater number of significant figures.
†The limits for the subdivisions of the infrared follow the recommendations of the Triple Commission for Spectroscopy; *J. Opt. Soc. Am.*, **52**:476 (1962).

The most important atomic or molecular transitions pertinent to the successive regions are:

X-ray	*K*- and *L*-shell electrons
Far ultraviolet	Middle-shell electrons
Near ultraviolet⎫	
Visible ⎬	Valence electrons
Near and mid infrared	Molecular vibrations
Far infrared	Molecular rotations and
	low-lying vibrations
Microwave	Molecular rotations

INTERACTIONS WITH MATTER

Electromagnetic radiation originates in the deceleration of electrically charged particles, and can be absorbed by the reverse process, contributing its energy to accelerate particles. Hence an understanding of the interactions between matter and radiation can only be built on a knowledge of the electronic structure of atoms and molecules.

Figure 2-1 shows a few of the energy levels for the neutral sodium atom that apply to the outer (valence) electron. Under ordinary conditions essentially all the atoms in a body of sodium vapor are in the ground state, that is, their valence electrons lie in the 3*s* level. If irradiated with light containing the wavelengths 589.00 and 589.59 nm, the outer electrons of many of the atoms will absorb photons and be transferred to the 3*p* levels. (The two very close 3*p* levels differ only in their spin characteristics.) The excited electron has a strong tendency to return to its normal (3*s*) state, and in so doing emits a photon. This emitted photon possesses a very definite amount of energy, dictated by the spacing of the energy levels. In the present example, the emitted radiation constitutes the familiar yellow light of the sodium flame or lamp. This simple case, in which the outer electron is raised by one energy level and then returns, is known as *resonance absorption* and *radiation*. The important analytical technique of *atomic absorption* is based on this phenomenon.

If the electron is given more than enough energy to produce resonance, it may be raised to some higher level than 3*p*, such as 4*p* or 5*p*. In that case it will not drop back to 3*s* by a single process, but will pause at intermediate levels, like a ball rolling down steps. This situation no longer fits the definition of resonance radiation, but is more complex. For one thing, not all conceivable transitions are actually possible—some are "forbidden" by the selection rules of quantum mechanics.

With a highly energetic source, many electrons (not only the outermost) in any element can be excited to varying degrees, and the resulting radiation may contain up to several thousand discrete and reproducible

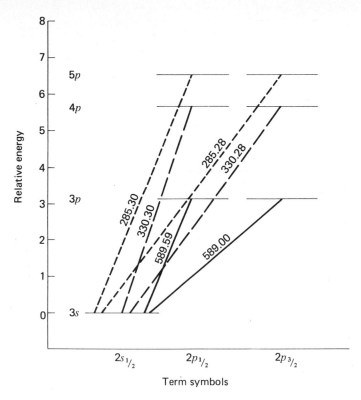

Fig. 2-1   Partial energy-level diagram for the valence electron in sodium.

wavelengths, mostly in the ultraviolet and visible regions. This is the basis of the analytical method of *emission spectroscopy*.

If the source of excitation provides even more energy, an inner electron can be torn entirely away from its atom. An electron from some higher level will then drop in to fill the vacancy. Since the energy change corresponding to this inner orbital transition is much greater than in the case of excited outer electrons, the photons radiated will be of much greater frequency and correspondingly shorter wavelength. This describes the emission of x-rays from atoms subjected to bombardment, for example, by a beam of fast-moving electrons.

MOLECULAR SPECTRA

In a typical molecule, as contrasted with an atom, the first few energy levels might show relations such as those of Fig. 2-2. The molecule to which this diagram applies has a singlet ground state designated S_0, which

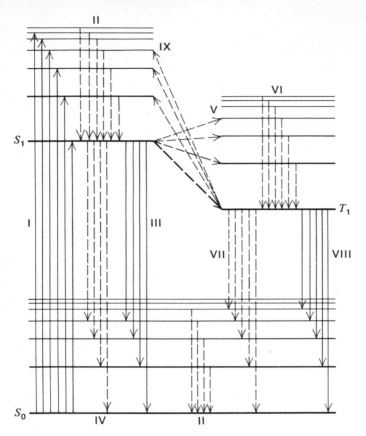

Fig. 2-2 Energy-level diagram of a typical organic molecule including only ground singlet, first excited singlet, and its corresponding triplet state. Solid lines indicate radiational transitions, and dashed lines indicate nonradiational transitions. Process I: absorption. Process II: vibrational deactivation. Process III: fluorescence. Process IV: quenching of excited singlet state. Process V: intersystem crossing to triplet state. Process VI: vibrational deactivation in the triplet system. Process VII: quenching of triplet state. Process VIII: phosphorescence. Process IX: intersystem crossing to excited singlet state. [*Academic* (1).]

represents its normal, unexcited condition. Two series of excited states exist, the *singlet* series, S_1, S_2, . . . , and the *triplet* series, T_1, T_2, ... These two series refer to a difference in the net electronic spin of atoms in the various levels. A triplet level always has less energy than the corresponding singlet. It is difficult to effect a change in electron spin, so the absorption of radiant energy can only raise an atom from S_0 to a high S level or sublevel. Triplet states can only be reached by an indirect process. Similarly, it is difficult for a molecule in a triplet state to drop back to the ground level.

Each electronic level (*S* or *T*) has associated with it a series of *vibrational sublevels,* corresponding to the energy required to excite various modes of vibration within the molecule. The closely spaced lines related to each vibrational sublevel correspond to the energy of rotation of atoms or groups of atoms within the molecule. Excellent discussions of this whole field can be found in references 1–3, among other works.

MOLECULAR ABSORPTION SPECTRA

Transitions within molecules are usually studied by the selective absorption of radiation passing through them, and less commonly by emission processes such as fluorescence and phosphorescence. Transitions between electronic levels are found in the ultraviolet and visible regions; those between vibrational levels, but within the same electronic level, are in the near and mid infrared, and those between neighboring rotational levels, in the far-infrared and microwave regions.

Electronic transitions involve jumps to and from the various sublevels, so that ultraviolet absorption spectra consist of the summations of absorptions at many closely spaced frequencies. The absorption contributions of individual transitions are too broad to be resolved under the usual conditions of observation, and hence the spectra appear as wide bands often with some residual fine structure. If spectra are taken in the gas phase or at low temperature (i.e., cooled with liquid nitrogen), the fine structure is sharpened and increased in detail. Cooling is usually not required for analytical work.

Absorption spectra are readily measured in each spectral region and are of great utility in analytical studies, as will become evident in subsequent chapters.

FLUORESCENCE

The energy gained by a molecule on the absorption of a photon does not remain in that molecule, but is lost by any of several mechanisms. Of considerable importance in solution chemistry is the case in which part of the energy is converted to heat, lowering the net energy of the molecule to the lowest vibrational and rotational level within the same electronic (singlet) level. The remainder of the energy is then radiated, returning the molecule to its ground state. This is the phenomenon of *fluorescence,* processes II and III in Fig. 2-2. The emitted radiation has less energy per photon than the exciting radiation, and hence a longer wavelength. Many organic and some inorganic compounds, when irradiated with ultraviolet light, fluoresce in the visible region. Fluorescence also occurs in atomic spectra, both in the ultraviolet-visible and the x-ray regions.

PHOSPHORESCENCE

In some molecules it is possible for a nonradiative transition to take place from an excited singlet state to the corresponding triplet level, from which the remaining energy is radiated as the molecule reverts to its ground state (processes II, V, and VI in Fig. 2-2). Since the transition from the triplet to the ground state has a low probability, the radiation may last for a measurable time interval after the exciting radiation is turned off. This persistent radiation is called *phosphorescence,* and contrasts with fluorescence which has no measurable persistence.

RAMAN SPECTRA

A phenomenon which bears some relation to fluorescence is the *Raman effect.* Here also, radiation is emitted from the sample with a change in wavelength from the exciting incident radiation. But whereas to excite fluorescence the primary radiation must be absorbed by the sample, the incident radiation in order to produce the Raman effect must *not* be appreciably absorbed. The shift in wavelength in the Raman effect is caused by the extraction of energy from the photons of incident radiation to raise molecules to higher vibrational states. The emergent photons can thus be thought of as the same ones which entered, but with less energy. Further discussion will be postponed to a later chapter.

REFRACTION

We now turn from atomic and molecular phenomena to "bulk" phenomena concerning matter in its interaction with radiation.

The *index of refraction* of an optical medium is an important bulk property. It is defined as the ratio of the velocity of radiation of a particular frequency in a vacuum to that in the medium. The variation of refractive index of a substance with wavelength is called its *refractive dispersion,* or simply its *dispersion.* The dispersion of a substance throughout the electromagnetic spectrum is intimately related to the degree to which radiation is absorbed. In regions of high transparency, the refractive index decreases with increasing wavelength (not linearly); in regions of high absorbance, the index rises abruptly with wavelength, but is usually difficult to measure precisely. The shape of the dispersion curve in regions of transparency is an important property, particularly with solids, because it is largely this curve that dictates the design of lenses and prisms.

POLARIZATION AND OPTICAL ACTIVITY

Another property sometimes shown by matter is the ability to polarize light. A beam of radiation can be thought of as a bundle of waves with vibratory

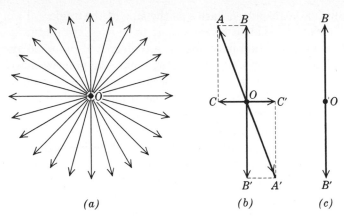

(a) (b) (c)

Fig. 2-3 Vibration vectors in ordinary and plane-polarized electromagnetic radiation.

motions distributed over a family of planes, all of which include the line of propagation. Figure 2-3a shows a cross section of such a ray which is proceeding in a direction perpendicular to the plane of the paper. If this beam of light is passed through a *polarizer*, each separate wave of the bundle, for example, that vibrating along the vector **AOA′** (Fig. 2-3b), is resolved into its components **BOB′** and **COC′** in the directions of the orthogonal x and y axes characteristic of the polarizer. The polarizing material has the property of eliminating one of these component vibrations (say **COC′**) and passing the other (**BOB′**). Thus the emergent beam consists of vibrations in one plane only (Fig. 2-3c) and is said to be *plane-polarized*.

A second polarizer (called the *analyzer*) placed in the beam will similarly pass only that component of the light vibrating parallel to its axis. Since the beam is already polarized, that means that in one position essentially all the radiation will come through, but turning the analyzer through a 90° angle will reduce the power to 0. This is illustrated in Fig. 2-4; radiation from a lamp, rendered parallel by a lens, passes through a polarizer A, which has its axis oriented vertically. The analyzer B, also with a vertical axis, has no further effect on the beam, but C, with its axis oriented horizontally, cuts the light to 0. If C is rotated in its own plane,

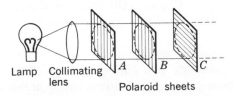

Lamp Collimating lens A B C Polaroid sheets

Fig. 2-4 Plane polarization of radiant energy.

the power of the radiation transmitted varies as the sine of the angle. Two polarizers placed in series are said to be *crossed* if their axes are mutually perpendicular. A beam of radiation may possess any degree of plane polar- ization from 0 (complete symmetry) to 100 percent (complete polarization).

Polarization is of importance in chemistry because of the ability exhib- ited by some crystals and liquids to rotate the plane of polarized light pass- ing through them. This property is known as *optical activity.* Its varia- tion with wavelength is called *optical rotatory dispersion,* and is related to regions of absorption in much the same manner as refractive dispersion.

A number of transparent crystalline materials show a phenomenon known as *double refraction* or *birefringence,* which is evidenced by the fact that a beam of light passing into the crystal is split into two beams of equal power which diverge from each other at a small angle. The two beams are found to be plane-polarized at right angles to each other. This effect is of great value in the identification and study of crystals.

A number of optical devices can serve as polarizers. One type consists of a pair of prisms made of a birefringent crystal such as quartz or calcite, cut with particular reference to the optical axes. Each dual prism is able to resolve a beam of unpolarized radiation into x- and y-polarized compo- nents. The prism may be designed to allow one component to pass nearly undeviated, while directing the other away from the optical axis of the sys- tem. Such prism assemblies usually go by the names of their inventors; best known is the Nicol, but sometimes superior for instrument use is the Glan-Thompson or the Rochon. For details see any optics text.

Another kind of polarizer depends on the combined effects of myriad submicroscopic crystals embedded in a film of plastic material. A stress is applied during manufacture so that all the tiny crystals line up with their axes parallel. *Polaroid* is the best-known example; it is much less expensive than a crystal prism, especially where a large area is required, but cannot be expected to be optically perfect.

The concepts of circularly and elliptically polarized radiation are con- sidered in Chap. 9.

PRACTICAL SOURCES OF RADIATION

From a purely physical standpoint it is convenient to classify sources ac- cording to whether they produce continuous or discontinuous spectra.

Continuous sources (sometimes called *white* sources) emit radiations over a wide band of wavelengths. They are utilized in the study of absorp- tion spectra, and as illuminants in such other fields as microscopy and turbidimetry.

The most familiar sources of continuous radiation are incandescent. Any substance at a temperature above absolute zero emits radiation. The theory of this thermal emission has been thoroughly worked out in terms

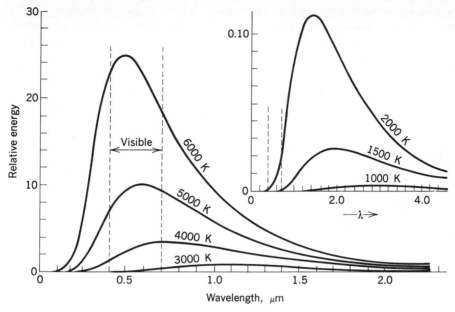

Fig. 2-5 Blackbody radiation as a function of temperature. [*McGraw-Hill* (4).]

of an ideal emitter called a *blackbody*. Figure 2-5 shows the manner in which blackbody radiation is distributed as a function of wavelength for various temperatures (4). Note that the wavelength corresponding to maximum energy moves toward higher energies and shorter wavelengths as the temperature is raised. This means that incandescent sources are very practical in the infrared and visible, but must be operated at inconveniently high temperatures for appreciable ultraviolet coverage.

Actual materials may vary to a considerable degree from the blackbody curves, in that they may give less emission in some wavelength regions for a given temperature. These are sometimes called *gray bodies*.

A continuous spectrum can also be obtained from an electrical discharge in a relatively high-pressure gas. A discharge through a gas typically produces a *line spectrum*. At low pressures each line approaches a single wavelength, but as the pressure is increased the lines broaden in proportion until, at sufficiently high pressures, neighboring lines coalesce, and a continuous spectrum results. The pressure required for a given degree of broadening depends in a complex manner on the molecular weight of the gas. Continuous spectra can be obtained from discharges in various gases, for example, in *xenon* at several atmospheres and in *mercury* vapor at pressures that may go higher than 100 atm. The very useful continuous discharges in *hydrogen* and *deuterium* in the neighborhood of 1 to 2 kPa* arise because

* One kPa (kilopascal) $\approx$ 7.5 torr; 1 Pa = 1 N $\cdot$ m^{-2} (newton per square metre).

of the complete dissociation of these gases at the voltages applied. Electrons are continually dropping from varying "infinite" distances to any of the atomic or molecular energy levels, resulting in a continuous spectrum with a definite cutoff limit toward longer wavelengths, and with considerable superimposed fine structure (5).

The useful ranges of representative silica-window lamps are (in nanometres): xenon, 250 to 1000; high-pressure mercury, 280 to 1400; hydrogen or deuterium, 160 to 365; tungsten-iodine, 380 to 1200.

LINE SOURCES

Sources producing discrete wavelengths are required for some instrumental applications. In the ultraviolet and visible regions a spectrum composed of a series of individual line images of an entrance slit is often useful. Such a line spectrum can be obtained easily from an arc discharge in a gas containing excited monatomic neutral or ionic species such as a metallic vapor or a noble gas. Excitation may be either thermal, as in a flame, or electrical, as in an arc or spark. Elements can be found which give lines spaced throughout most of the ultraviolet and visible regions. Characteristic infrared emission lines can be obtained from heated polyatomic gases.

LASERS (6)

A laser is a source of monochromatic radiation available principally in the visible and infrared regions. The earliest example, reported in 1960 and still one of the best, consists of a carefully ground rod of ruby (Al_2O_3 with Cr_2O_3 as a minor constituent) with parallel ends. A mirror is placed at one end so that all light approaching from the interior of the crystal is reflected back. The mirror at the other end is coated with a thin layer of silver so that only a fraction (typically 80 to 90 percent) of the incident light is reflected, the remainder escaping. When the rod is subject to an intense flash of light, as from a xenon discharge lamp (Fig. 2-6), nearly all the chromium atoms become excited, and most of them immediately drop into a metastable energy level (cf. discussion of phosphorescence above). Then the first electrons to return to the ground state from the metastable level radiate photons of the corresponding wavelength, 694.3 nm. Some of this light is directed parallel to the axis of the rod and is reflected back and forth many times. Laser action results because the presence of this radiant energy at exactly the required frequency *stimulates* emission from the remainder of the metastable chromium atoms, so that the radiant flux builds up rapidly. At every reflection from the partially silvered end some light escapes, forming the output of the device. The action is so efficient

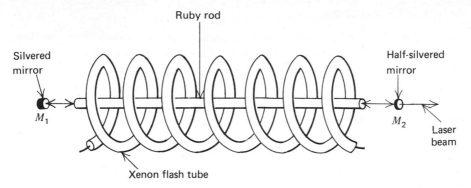

Fig. 2-6 Ruby laser, schematic.

that a large pulse of monochromatic light is emitted within a period of perhaps 0.5 ms. The power in each pulse may reach the megawatt level.

Lasers can be made with a variety of other active materials. Among solids, a glass matrix containing a few percent neodymium or another rare earth is important, as is yttrium aluminum garnet (YAG). A number of gases can be made to lase by an electrical discharge directly through the gas; notable in this area are helium-neon, argon, nitrogen, and carbon dioxide. Gas lasers can give continuous as well as pulsed outputs.

All the lasers mentioned produce only certain discrete wavelengths that cannot be varied. There is another type capable of being tuned over a considerable wavelength range, the *dye laser* (7–9). The active substance is a fluorescent organic compound such as fluorescein or rhodamine 6G in solution. Laser emission can be made to occur anywhere within the fluorescent emission spectrum, and can be tuned by means of a prism or diffraction grating. The details need not concern us here, but it is likely that this laser, which is the subject of intensive research, will become of great significance as a light source in high-resolution spectrophotometry.

Light from a laser has several unique properties (6). It is highly *monochromatic*. The light emitted is *coherent,* which means that the waves originating from all the atoms or molecules of the emitting substance are in phase with each other (not true of conventional light sources). Partly as a consequence of the coherence, the collimated beam of laser radiation has very little tendency to spread out (lose collimation) as it propagates. This permits a large amount of energy to be concentrated on a small target, even though at a considerable distance.

The importance of lasers for analytical purposes lies in the high degree of monochromaticity and the high power levels which can be achieved. Lasers find application as a source of localized heating, as an exciter for Raman spectroscopy, and as an illuminator for precise interferometry.

WAVELENGTH SELECTION

In the study of absorption spectra it is usually necessary to utilize a rather narrow band of wavelengths. In some instances a line source or laser can provide the narrow band required, but more often it is necessary to start with radiation from a continuous source and select a band of wavelengths from it. This results in greater flexibility than a line source can provide, as the chosen band can be taken at any desired location within the range covered by the source.

There are two basic methods of wavelength selection: (*1*) the use of filters, and (*2*) geometrical dispersion by means of a prism or diffraction grating.

A *filter* is a device which will transmit radiations of some wavelengths but absorb wholly or partially other wavelengths. Filters employed in the visible region are usually of colored glass. A great number of such filters are available, more or less evenly spaced throughout the visible region.

Filters are also made which function on the interference principle. Figure 2-7 represents a section through such an *interference filter*. This device is made by depositing on a glass plate a semitransparent film of silver, on the silver a very thin layer of transparent material such as magnesium fluoride, and then on the magnesium fluoride another film of silver. Each silver film reflects about half and transmits the other half of any radiation that strikes it. Part of the incident beam is reflected repeatedly by the silver layers, but at each reflection some is transmitted outward. The several emergent rays to the right reinforce each other for radiation, such that the distance between the silver films is an exact multiple of half the wavelength ($n\lambda/2$, where $n = 1, 2, 3 \ldots$). For all other wavelengths, the beams interfere destructively, so that essentially no energy can pass through. In commercial interference filters the thin layers are covered with a second glass plate for protection. The wavelength band isolated by this

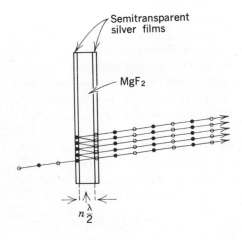

Fig. 2-7 An interference filter, schematic. The open circles represent crests and the black ones troughs in the wave of radiation; shown for $n = 1$.

Semitransparent
silver films

MgF$_2$

$n\dfrac{\lambda}{2}$

filter is much narrower and the peak transmittance much greater than in the corresponding colored glass filter. Interference filters transmit radiation of multiple *orders*, i.e., values of n. The unwanted orders can be removed by an absorbing layer used in conjunction with the interference filter. For the visible range, the second and higher orders lie in the ultraviolet, hence are eliminated by the glass plates.

Filters of one type or another are available which, alone or in combination, permit selection of wavelength bands in almost any region of the spectrum, from x-rays through the infrared.

MONOCHROMATORS

A monochromator is an instrument that can isolate a selected narrow band of wavelengths anywhere within a comparatively wide spectral range. It can be adjusted by a manual or automatic control to any desired wavelength position.

The monochromator consists of a dispersing element (a prism or a diffraction grating) together with two narrow slits to serve as entrance and exit ports for the radiation. The entrance slit defines a narrow beam of radiation which falls on the dispersing element. This deflects the beam through an angle depending on the wavelength, and thus "fans out" the beam. This is shown for the visible region in Fig. 2-8. An exit slit can be positioned so as to pass a narrow band of wavelengths at any point in the spectrum. (A practical monochromator generally also requires lenses or mirrors, incidental to its major function.)

A *spectrograph* is an instrument similar to a monochromator but with the exit slit omitted. A photographic film or plate is mounted so that successive wavelengths are focused at adjacent points. Thus an entire spectral region can be photographed simultaneously.

A *spectrophotometer* is an instrument consisting of a source of continuous radiation, a monochromator, and a detector such as a photoelectric cell, suitable for observing and measuring an absorption spectrum.

Except for the source and the camera or photoelectric detector, instru-

Fig. 2-8   Dispersion of white light by (*a*) a prism and (*b*) a transmission grating.

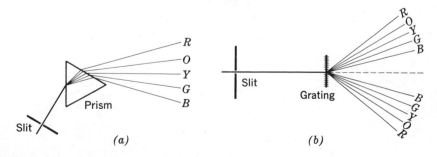

(*a*)   (*b*)

mental features are the same for monochromator, spectrograph, and spectrophotometer, so it is convenient to discuss them together.

DISPERSION BY PRISMS

Prisms are suitable dispersing elements from the mid-ultraviolet through the mid-infrared regions, but are not generally applicable elsewhere. In principle, any transparent medium may be used to fashion a prism, but its usefulness is determined by its *dispersion*, that is, the slope of its refractive index–wavelength curve, and its region of transparency.

In the ultraviolet, the only generally useful solid materials are silica and alumina.* *Silica* can be used either as quartz or in the vitreous form often called *fused quartz*. *Alumina*, in the form of artificial sapphire, is more expensive than silica and offers little advantage as a prism material. These materials transmit freely from somewhat below 200 nm in the ultraviolet up to about 4 μm in the infrared.

In the visible region silica is inferior to optical glass with respect to dispersion. For infrared work beyond about 3 μm, prisms are constructed of NaCl, KBr, or CsBr. These and other infrared-transmitting materials are considered further in Chap. 5.

The simplest prism monochromator is based on a 60° prism and two lenses, as in Fig. 2-9. Radiation enters through slit S_1, is rendered parallel by a collimating lens, and falls at an oblique angle on one face of the prism. The dispersed radiation emerging from the prism is focused by a second lens, so that the desired wavelength is centered on the exit slit S_2.

* Several ionic crystals, such as NaCl and KBr, are transparent in the ultraviolet and visible, but are seldom used in these regions. One disadvantage is that intense ultraviolet irradiation tends to produce color centers in the salt crystals, thereby reducing the transparency.

Fig. 2-9 Simple 60° prism monochromator or spectrograph. For spectrographic service, the plate holder must be placed so that the focus of each wavelength will be intercepted, i.e., points P_1 and P_2, in addition to S_2, the location of the exit slit when used as a monochromator.

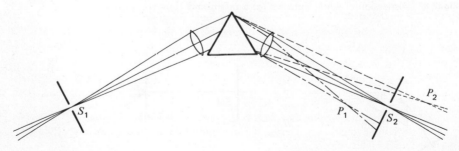

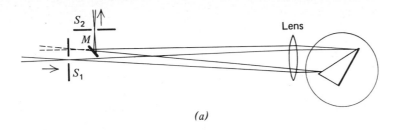

(a)

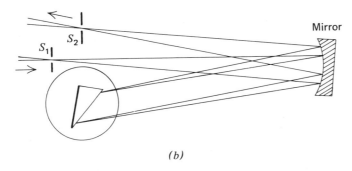

(b)

Fig. 2-10 Littrow-mounted prism monochromators. (a) With lens, exit slit offset; (b) with concave mirror. The dispersing prism, with mirrored back, is mounted on a table that can be turned to select wavelengths.

This instrument can be made more compact and economical by folding it around its center line, resulting in the *Littrow* design, as in Fig. 2-10a. The 60° prism is replaced by one with a 30° angle (for the same dispersive power) with its back surface mirrored. Now a single element can serve both to collimate the incoming radiation and to focus the dispersed beams.*

For many applications it is desirable to substitute a concave mirror for the lens (Fig. 2-10b). This has the great advantage that one component (the mirror) will serve equally well over the whole optical range that can be reflected by a metallic surface: the ultraviolet, visible, and infrared. For one thing, the radiation passes through no solid medium which would limit its range. Also, a mirror does not show the chromatic aberration characteristic of a lens; that is, it focuses all wavelengths at the same point.

Various other types of prisms are employed occasionally when particular features are desired, such as constant deviation, zero deviation, or auto-

---

* The Littrow prism was originally introduced as a means of compensating the birefringence of crystalline quartz by passage of the radiation in both directions. This is now principally of historical interest.

collimation. Some of these are very ingenious. For details, see any text-book on spectroscopy.

DISPERSION BY GRATINGS

A beam of monochromatic radiation, in passing through a transparent plate which has a large number of very fine parallel lines ruled on it, is split into a number of beams. One of these proceeds straight through, as though the plate were unruled. The other beams are deviated from this forward direction, as in Fig. 2-8b, through angles which depend on the spacing of the ruled lines and on the wavelength of the radiation. This can be explained by the assumption that each clear portion between the lines, when illuminated from behind, acts as though it were itself a source of the radiation which emanates from it in all forward directions (Huygens' principle). However, the rays coming from these numerous secondary sources will be destroyed by interference in most directions. Only at those angles where the geometry is just right will the beams reinforce each other. Figure 2-11 shows one of the possible deviated beams. The angle of deviation is θ, the difference in the lengths of the paths taken by beamlets from successive transparent areas is a, and the distance between centers of adjacent lines (the grating space) is d. Thus $a = d \sin \theta$. The many beamlets can only reinforce each other when the difference in path length is equal to an integral number of wavelengths of the radiation. This gives the fundamental relation called the *grating equation:*

$$n\lambda = d \sin \theta \tag{2-3}$$

where n is any integer, 0, 1, 2, 3, . . . , called the *order*, and λ is the wavelength.

It follows from Eq. (2-3) that, if a beam of polychromatic radiation is passed through the grating, it will be fanned out into a series of spectra

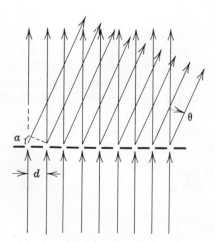

Fig. 2-11 Diffraction at a plane grating.

located symmetrically on each side of the normal to the gyrating. On each side there will be a spectrum corresponding to each of the first few values of n. The equation further shows that for a particular angle θ there will be several wavelengths for which the value of $n\lambda$ is the same. For example, a grating with 2000 lines per centimetre (grating space $d = 1/2000 = 5 \times 10^{-4}$ cm) will deflect through an angle $\theta = 6.00°$ radiation of those wavelengths, given by

$$\lambda = \frac{d \sin \theta}{n} = \frac{(5 \times 10^{-4})(\sin 6.00°)}{n}$$

$$= \frac{(5 \times 10^{-4})(0.1045)}{n} = \frac{0.5225 \times 10^{-4}}{n} \text{ cm}$$

$$= \frac{522.5}{n} \text{ nm}$$

The actual wavelengths corresponding to successive orders at this angle will be:

Order, n	1	2	3	4	...
Wavelength λ, nm	522.5	261.2	174.2	130.6	...

This relation is shown diagrammatically in Fig. 2-12, which gives selected wavelengths for the first four orders on one side of the normal.

The fact that successive orders of spectra overlap might seem to be

Fig. 2-12   Overlapping orders in a grating spectrum.

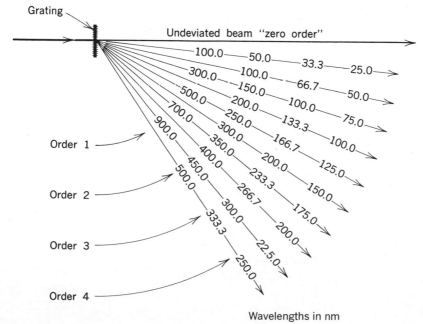

Wavelengths in nm

a great drawback, but in practice it gives little trouble. If the spectrum is to be observed visually, the question will not arise, since the visible regions (400 to 750 nm) of the various orders do not overlap. If the spectrum is to be recorded photographically, the spectral sensitivity of the plate will limit the degree of overlap, but some may still be encountered. Overlapping may be reduced or eliminated by placing ahead of the grating an auxiliary prism of small deviation, called a *foreprism* or *Order-Sorter*, or by the use of absorbing filters to remove one region of the spectrum while allowing another to pass.

The grating discussed above is of the type known as a plane *transmission grating*. In practical instruments other than small, hand-held ones, *reflection gratings* are more common. In these the lines are engraved on the surface of a mirror, which may be either a polished metal slab or a glass plate on which has been deposited a thin metallic film.

It is possible to rule a grating in such a way as to throw a maximum fraction of the radiant energy into those wavelengths that are diffracted at a selected angle. This is accomplished by ruling with a specially shaped diamond point held at a specified angle. The resulting grating is called an *echelette*, and is said to have been given a *blaze* at a particular angle. Figure 2-13 shows the geometry of a portion of an echelette reflection grating. The wider faces of the grooves make an angle ϕ with the surface of the grating. A ray incident at angle α will be reflected from the groove face at angle β such that $\alpha + \phi = \beta - \phi$. The rays reflected from successive grooves then undergo interference as already described. Because of the efficiency of specular reflection at the metal surface, much more energy will be diffracted at this angle (β) than at any other, for a given value

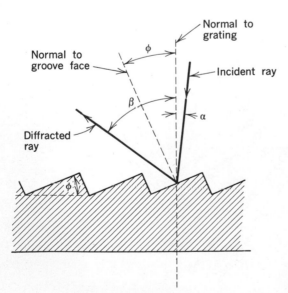

Fig. 2-13 The geometry of a blazed reflection grating (echelette).

of α. The energy will be only slightly less at angles close to β, so the grating can be used to advantage for a considerable wavelength span in a given order. A grating blazed for a particular wavelength in the first order will also be blazed for half that wavelength in the second order, one-third of it in the third order, and so on. Very little energy will be found at the symmetric position on the other side of the normal to the grating, and very little in the "zero order."

In general best results are obtained with a grating in which the spacing between lines is of the same order of magnitude as the wavelength region to be dispersed. For special purposes, other grating spaces may be found useful. An *echelle*, for example, is a grating with step-shaped rulings a few hundred times wider than the average wavelength to be studied. It must be used at an order n of 100 or more, which produces difficult problems of overlapping orders, but is capable of tremendous dispersion.

The manufacture of precision gratings is very exacting work. It is performed with an extremely precise and delicate machine, called a *ruling engine*, that scribes the fine parallel lines with a diamond point. Most spectrographs and practically all grating spectrophotometers use *replica gratings*, made by casting a plastic material on an original grating and then stripping it off and mounting it on a rigid support. The art of replication has reached the point where gratings made this way are of very nearly as good quality as the originals.

There are several ways in which a plane diffraction grating can be mounted in a monochromator or a spectrograph. One of these is the Littrow mounting. A diagram of it is analogous to Fig. 2-10a or b with the prism replaced by the grating on the same rotatable table. An example may be found in Fig. 5-4b.

The other common mounting for a plane grating was invented by Ebert in 1889, but was little used until resurrected and improved by Fastie (10) (1952). In this design (Fig. 2-14) a single, large, spherical mirror serves

Fig. 2-14 Ebert mounting for a plane reflection grating. The wavelength is selected by turning the grating around a vertical axis at its center.

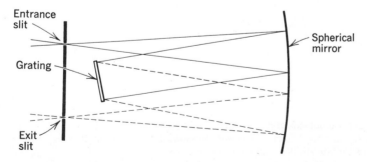

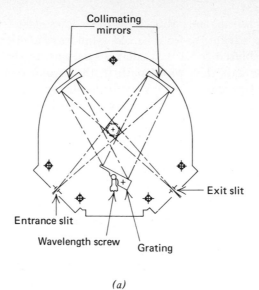

(a)

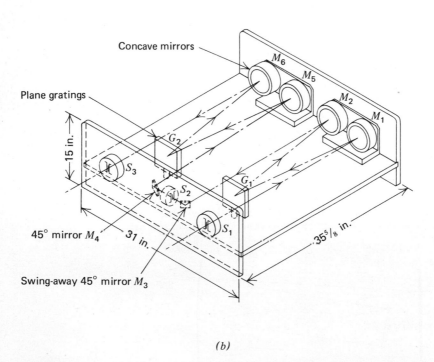

(b)

Fig. 2-15 Czerny-Turner monochromators. (a) Modified to provide a 90° angle between in and out radiation. (*Farrand Optical Company.*) (b) Double monochromator with intermediate slit S_2. (*Spex Industries.*)

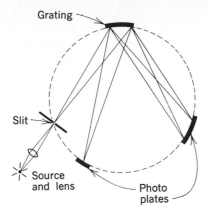

Grating

Slit

Source
and lens

Photo
plates

Fig. 2-16 Plan of a concave grating spectrograph, showing the principle of the Rowland
circle.

for both collimation and focusing, with symmetrically placed slits. Wavelength selection is effected by rotating the grating. Czerny and Turner (11)
(1930) suggested using two small spherical mirrors mounted symmetrically,
to save the expense of the large Ebert mirror, much of which was unused,
and most current instruments with this geometry incorporate the best features of the Czerny-Turner and Ebert designs. Figure 2-15 shows two commercial examples.

There is little choice between the modified Littrow and Ebert mountings. The Littrow is slightly more compact and saves one mirror, but the
two slits must be close together, usually one above the other, and this tends
to cramp the design. The Ebert is somewhat freer of aberrations, because
of the symmetry.

Another class of instruments makes use of a reflection grating in which
the lines are ruled on a concave spherical surface. Rowland, in 1882, discovered the following principle of design which bears his name: If a circle
(the *Rowland circle*) is drawn tangent to the concave grating at its center,
but with a diameter equal to the radius of curvature of the grating, then
the diffracted images of the entrance slit will lie on the circle if the slit
itself is on the circle. This will apply to all wavelengths of all orders of
diffraction, and is illustrated in Fig. 2-16. Several mechanical designs
have been devised to make use of this principle, ranging from very
large spectrographs (35 ft or more in diameter) to small portable
spectrophotometers.

Instruments based on the Rowland circle suffer from the inherent defect
of *astigmatism*. This means that, although the image of the entrance slit
is very sharply focused along the circle, so that its wavelength can be measured with precision, its height is not sharply defined. This makes quantitative measurements difficult, as some radiation from the entrance slit is
masked off at the top and bottom of the exit slit.

DISPERSION CURVES

The dispersion of a spectrograph is usually defined as the derivative $d\lambda/dx$, where x is distance measured along the focal plane, i.e., on the surface of a developed photoplate. It may be specified in nanometres per millimetre. In a monochromator or spectrophotometer the corresponding quantity is the *effective bandwidth*, in nanometres (or micrometres) per millimetre of slit width. This is convenient, especially in instruments in which the slit width can be varied.

A grating instrument produces a *normal* spectrum, that is, one that is spread out uniformly on a wavelength scale. The dispersion or bandwidth is then uniform over the whole spectrum. A prism, on the other hand, gives an unequally spaced spectrum with the longer wavelengths crowded together as compared with the shorter. The bandwidth for a given width of slit is no longer constant, but differs from one wavelength to another and from one instrument design to another.

Figure 2-17 shows the bandwidth curves for a number of commercial monochromators and spectrographs or spectrophotometers. Notice that, other variables being equal, the spectral purity is better the smaller the ratio of bandwidth to slit, that is, the lower down on the graph one can operate. Note also that, for grating instruments, the curve is merely a horizontal straight line. Relative position on this graph tells only part of the story, of course. A grating instrument with a bandwidth of 0.002 μm · mm^{-1} may give an actual bandwidth of much less than 0.002 μm, if the intensity of the lamp, the sensitivity of the detector, and other variables are such as to permit a slit narrower than 1 mm.

A very narrow slit produces an interference pattern (Fig. 2-18) consisting of a large central maximum and a series of much smaller maxima (usually negligible) symmetrically located on both sides. The width of the central maximum (measured between adjacent minima) is $2f\lambda/d$, where d is the slit width, f is the focal length of the focusing lens or mirror, and λ is the wavelength. As the entrance slit is opened up the spot of light appearing on the focal plane, initially diffuse and of very low intensity, will become narrower until it eventually passes through a minimum width; it then broadens, the intensity steadily increasing as more energy enters the slit. The minimum position corresponds to an optimum slit width given by

$$d_{\mathrm{opt}} = \frac{2f\lambda}{w} \tag{2-4}$$

where w is the diameter of the lens or mirror. For a typical case, where $w = 2$ cm, $f = 40$ cm, and $\lambda = 500$ nm, the optimum slit width d_{opt} becomes 20 μm.

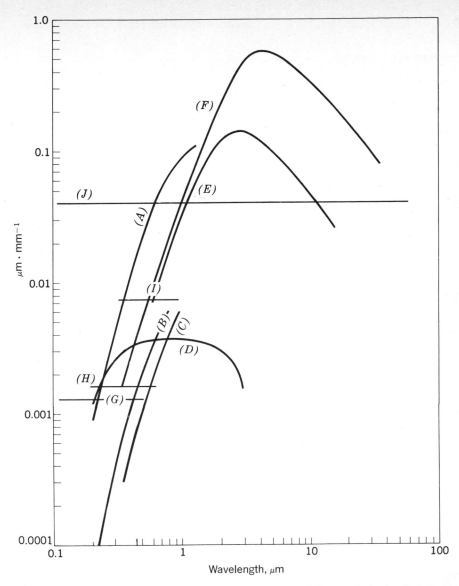

Fig. 2-17 Effective bandwidth curves and useful ranges for several spectrophotometers and spectrographs. The curves refer to the following: (*A*) Beckman Model DU, quartz Littrow; (*B*) Bausch & Lomb large quartz Littrow spectrograph; (*C*) same, with glass optics; (*D*) Cary Model 14, combining a quartz prism and 600 line per millimetre grating monochromators; (*E*) Beckman IR-4 with NaCl prism, extended; (*F*) same, with CsBr prism, extended; curves (*E*) and (*F*) were prepared from Beckman's literature for the regions to the right of the maxima and extended to shorter wavelengths by normalized comparison with data from Harshaw Chemical Company; this extended region does not correspond with any commercial instrument; (*G*) McPherson Model 218 vacuum monochromator with 2400 line per millimetre grating; (*H*) Bausch & Lomb Spectronic-505 with 1200 line per millimetre grating; (*I*) Bausch & Lomb Spectronic-20, 600 line per millimetre grating; (*J*) McPherson, same as (*G*), with 75 line per millimetre grating.

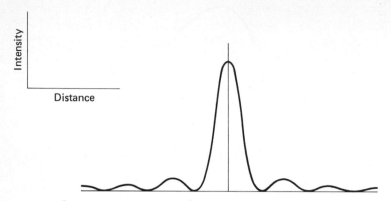

Fig. 2-18   Diffraction pattern from a single slit.

This optimum slit width will give the instrument the greatest *resolving power* (ability to separate close wavelengths) of which it is capable.   This parameter is arbitrarily defined by the criterion proposed many years ago by Lord Rayleigh: Two wavelengths differing by $\Delta\lambda$ are said to be resolved when the central diffraction maximum of one coincides with the first minimum of the other.   The resolving power is then

$$R = \frac{\lambda}{\Delta\lambda} \qquad\qquad (2\text{-}5)$$

where λ is the mean of the two wavelengths.

A monochromator is often operated with wider slits than indicated by Eq. (2-4), if the source is of low intensity, in order to gather enough energy to measure with precision; the resolution suffers, however.

Fig. 2-19   Band of wavelengths emerging from a narrow slit.

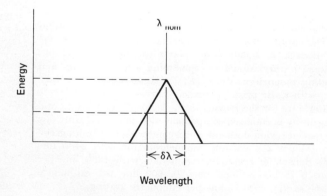

Almost invariably monochromators are provided with entrance and exit slits of equal width; if they are variable, a single control operates both, maintaining their equality. The wavelength of the radiation emerging from the exit slit includes contributions from both sides of the central or nominal wavelength (Fig. 2-19). The band $\delta\lambda$ between the half-height points is the effective bandwidth, as plotted in Fig. 2-19.

PROBLEMS

2-1 In Fig. 2-1, determine the wave number in reciprocal centimetres and the energy in electron volts for each of the transitions shown.

2-2 The legal definition of the metre in the United States (12) is 1,650,763.73 wavelengths in vacuo of the $2p_{10}$ to $5d_5$ transition in isotopically pure ^{86}Kr. Compute the wave number in reciprocal centimetres; the wavelength in angstroms, in nanometres, and in micrometres; the frequency in hertz; and the energy per photon in electron volts for this radiation. Take the velocity of light in vacuo as $(2.99792458 \pm 0.004$ ppm$) \times 10^8$ m·s^{-1}, and Planck's constant as $(6.626176 \pm 5.4$ ppm$) \times 10^{-34}$ J·s. Give each answer with the appropriate uncertainty limits. (If you do not have access to a calculator that can handle enough digits, round off the numbers as necessary and explain the extent to which this degrades your calculations.)

2-3 Calculate figures for two more columns in Table 2-1, giving the energy content of photons in the respective ranges in (a) electron volts and (b) calories per mole.

2-4 A transmission grating for a spectrograph has 15,000 lines per inch. Calculate the wavelengths that will be diffracted at an angle θ of 25° for the first two orders. At what angles will radiation of 2537 Å appear in each of the first two orders?

REFERENCES

1. Udenfriend, S.: "Fluorescence Assay in Biology and Medicine," vol. II, p. 45, Academic, New York, 1969.
2. Barrow, G. M.: "Introduction to Molecular Spectroscopy," McGraw-Hill, New York, 1962.
3. Jaffé, H. H., and M. Orchin: "Theory and Applications of Ultraviolet Spectroscopy," Wiley, New York, 1962.
4. Jenkins, F. A., and H. E. White: "Fundamentals of Optics," 3d ed., McGraw-Hill, New York, 1957.
5. Penner, S. S.: "Quantitative Molecular Spectroscopy and Gas Emissivities," chap. 3, Addison-Wesley, Reading, Mass., 1959.
6. Lengyel, B. A.: "Introduction to Laser Physics," Wiley, New York, 1966.
7. Mass, M., T. F. Deutsch, and M. J. Weber: in A. K. Levine and A. J. DeMarie (eds.), "Lasers," vol. 3, Dekker, New York, 1971.
8. Smith, R. G.: *Anal. Chem.,* **41**(10):75A (1969).
9. Webb, J. P.: *Anal. Chem.,* **44**(6):31A (1972).
10. Fastie, W. G.: *J. Opt. Soc. Am.,* **42**:641 (1952).
11. Czerny, M., and A. F. Turner: *Z. Phys.,* **61**:792 (1930).
12. *Natl. Bur. Stand. Tech. News Bull.,* February and October, 1963.

Chapter 3

The Absorption of Radiation: Ultraviolet and Visible

If a beam of white light passes through a glass cell (cuvet) filled with liquid, the emergent radiation is less powerful than that entering. The diminution in power is generally of different extent for different colors. The loss is due in part to reflections at the surfaces and to scattering by any suspended particles present, but in clear liquids is primarily accounted for by the *absorption* of radiant energy by the liquid.

If the energy absorbed is greater for some visible wavelengths than for others, the emergent beam will appear colored. Table 3-1 gives the wavelength bands designated by familiar color names, along with their complements. These wavelengths are taken from a study originating at the National Bureau of Standards (1), and are inherently somewhat arbitrary. The apparent color of the solution is always the *complement* of the color absorbed. Thus a solution absorbing in the blue region will appear yellow, one that absorbs green will appear purple, etc.

In referring to color, we are for the moment restricting the discussion to the visible region of the spectrum, but most of the concepts and analytical methods will apply with no change in principle in both the ultraviolet and infrared ranges.

To the analytical chemist, the importance of colored solutions lies

Table 3-1 Colors of visible radiation (1)

Approximate wavelength range, nm	Color	Complement
400–465	Violet	Yellow-green
465–482	Blue	Yellow
482–487	Greenish blue	Orange
487–493	Blue-green	Red-orange
493–498	Bluish green	Red
498–530	Green	Red-purple
530–559	Yellowish green	Reddish purple
559–571	Yellow-green	Purple
571–576	Greenish yellow	Violet
576–580	Yellow	Blue
580–587	Yellowish orange	Blue
587–597	Orange	Greenish blue
597–617	Reddish orange	Blue-green
617–780	Red	Blue-green

in the fact that the radiation absorbed is characteristic of the material doing the absorbing. A solution containing the hydrated cupric ion absorbs yellow and is transparent to blue, so copper may be determined by measuring the degree of absorption of yellow light under standardized conditions. Any soluble colored material may be determined quantitatively in this way. In addition, a substance that is colorless or only faintly colored may often be determined by adding a reagent which will convert it to an intensely colored compound. Thus the addition of ammonia to a copper solution produces a much more intense color than that of the hydrated cupric ion itself, and therefore provides a more sensitive analytical method.

The general term for chemical analysis through measurement of absorption of radiation is *absorptiometry*. The term *colorimetry* should be applied only in relation to the visible spectral region. *Spectrophotometry* is a division of absorptiometry that refers specifically to the use of the spectrophotometer.

MATHEMATICAL THEORY

The quantitative treatment of the absorption of radiant energy by matter depends on the general principle known as *Beer's law*. Consider a glass container with plane parallel faces traversed by monochromatic radiation. Losses by reflection at the surfaces and absorption by the glass will be neglected. Suppose that the container is filled with an absorbing species dissolved in a nonabsorbing solvent. The radiation diminishes in power the further it penetrates into the liquid and the greater the concentration of solute. More generally stated, the diminution in power is proportional to the number of absorbing molecules in the path of the beam. The quantitative statement of this relation is Beer's law:* *Successive increments in the number of identical absorbing molecules in the path of a beam of monochromatic radiation absorb equal fractions of the radiant energy traversing them.* In terms of the calculus, this may be stated as

$$\frac{dP}{dn} = -kP \tag{3-1}$$

where dP is the power absorbed at power level P by an increment dn in the number of absorbing molecules; k is a constant of proportionality. Rearrangement followed by integration between limits gives

$$\int_{P_0}^{P} \frac{dP}{P} = -k \int_{0}^{N} dn$$

$$\ln \frac{P}{P_0} = -kN \tag{3-2}$$

P_0 in this equation represents the level of radiant power as it enters the cell, and N is the number of absorbing molecules traversed at a point where the power is reduced to P, for a beam of unit cross section. For a beam of cross-sectional area s square centimetres the right member of Eq. (3-2) must be multiplied by s:

$$\ln \frac{P}{P_0} = -k'Ns \tag{3-3}$$

The quantity Ns is a measure of the number of particles effective in absorbing radiation. A more convenient measure is the product of concentration c and length of path b, so we can write

$$\ln \frac{P}{P_0} = -k''bc \tag{3-4}$$

The constant k'' now takes the units of mass per unit area of cross section. In other words, for a given cross section the absorption is proportional to the mass of absorbing material in the path.

For convenience, we shall replace k'' by another constant a, which includes the factor for conversion of natural to common logarithms:

$$\log \frac{P_0}{P} = A = abc \tag{3-5}$$

---

* This relation is sometimes known as the Beer-Lambert or the Bouguer-Beer law, as contributions to its development were made by Lambert and by Bouguer (as well as by others). It is frequently stated in terms of intensities I rather than powers P. For informative discussions of the law, see references 2 and 3.

Note that the ratio P_0/P has been inverted to remove the negative sign. The quantity log (P_0/P) is so important that it has a special symbol A and is called the *absorbance*. The shortest statement of Beer's law is thus $A = abc$.

Since the transmitted power P can vary between the limits zero and P_0, the logarithm of the ratio, in theory, can vary from 0 to infinity. In practice, however, absorbances greater than 2 or 3 are seldom usable, and the range which will give adequate analytical precision is even more limited, the exact permissible values being determined in part by the type of measuring instrument employed.

ABSORPTIVITY

The constant a of Eq. (3-5) is called the *absorptivity*. It is characteristic of a particular combination of solute and solvent for a given wavelength. Its units are dependent on those chosen for b and c (b is customarily in centimetres), and the symbol varies accordingly, as indicated in Table 3-2. Other symbols and names that have been widely used in the past are included for reference. The present notation is that proposed by the Joint Committee on Nomenclature in Applied Spectroscopy in its report published in 1952 (2, 3).

It must be noted carefully that absorptivity is a property of a substance (an intensive property), while absorbance is a property of a particular sample (an extensive property) and will therefore vary with the concentration and dimensions of the container.

Percent transmittance (percent $T = 100P/P_0$) is a convenient quan-

Table 3-2   Units and symbols for use with Beer's law

Accepted symbol	Definition*	Accepted name	Symbol	Name
			Obsolete or alternate	
T	P/P_0	Transmittance	......	Transmission
A	$\log P_0/P$	Absorbance	D, E	Optical density, extinction
a	A/bc	Absorptivity	k	Extinction coefficient, absorbancy index
ϵ	AM/bc	Molar absorptivity	a_M	Molar (molecular) extinction coefficient, molar absorbancy index
b	....	Length of path	l, d	

* The definitions of P and P_0 are given in the text (Chap. 2). The units of c are grams per liter; of b, centimetres; M is the molecular weight. A symbol formerly much used is $E_{1cm}^{1\%}$, which may be defined as A/bc', where c' is the concentration in percent by weight and $b = 1$ cm.

tity if the transmitted radiation is of more interest than the chemical nature of the absorbing material. Filters for colorimetry and photography are commonly rated in terms of percent transmittance. The absorbance A or the absorptivity a is useful as a measure of the degree of absorption of radiation. The symbol a is used if the nature of the absorbing material, and hence its molecular weight, is not known. The molar absorptivity ϵ is preferable if it is desired to compare quantitatively the absorption of various substances of known molecular weight.

Beer's law indicates that the absorptivity is a constant independent of concentration, length of path, and intensity of incident radiation. The law provides no hint of the effect of temperature, the nature of the solvent, or the wavelength. In practice, the temperature is found to have only secondary effects, unless varied over an unusually wide range. The concentration will vary slightly with change in temperature, because of the volume change. Also, if the absorbing solute is in a state of equilibrium with other species or with undissolved solute (as in a saturated solution), more or less variation with temperature is to be expected. On the other hand, some substances show quite different absorption if cooled to liquid-nitrogen temperature. For much practical analytical work, temperature effects may be disregarded, especially when the absorption of an unknown is directly compared with a standard at the same temperature.

The effect of changing the solvent on the absorption of a given solute cannot be predicted in any general way. The analyst is frequently limited to a particular solvent or class of solvents in which the material is soluble, so that the question may not arise. A further restriction applies particularly to work in the ultraviolet, where many common solvents are no longer transparent. Water, alcohol, ether, and saturated hydrocarbons are satisfactory, but aromatic compounds, chloroform, carbon tetrachloride, carbon disulfide, acetone, and many others are not usable except in the very near ultraviolet. Table 3-3 gives the approximate limits of ultraviolet transmission of a number of useful solvents.

Even at constant temperature and in a specified solvent, it is sometimes found that the absorptivity may not be truly constant but may deviate toward either greater or smaller values. If the absorbance A is plotted against concentration, a straight line through the origin should result, according to the prediction of Eq. (3-5) (curve 1, Fig. 3-1). Deviations from the law are designated positive or negative, according to whether the observed curve is concave upward or downward.

It must be realized that conformity to Beer's law is not necessary in order for an absorbing system to be useful for quantitative analysis. Once a curve corresponding to Fig. 3-1 is established for the material under specified conditions, it may be used as a calibration curve. The concentration of an unknown may then be read off from the curve as soon as its absorbance is found by observation.

Table 3-3   Ultraviolet transmission limits of common solvents*

180–195 nm	*210–220 nm*	*265–275 nm*
Sulfuric acid (96%)	n-Butyl alcohol	Carbon tetrachloride
Water	Isopropyl alcohol	Dimethyl sulfoxide
Acetonitrile	Cyclohexane	Dimethyl formamide
200–210 nm	Ethyl ether	Acetic acid
Cyclopentane	*245–260 nm*	*280–290 nm*
n-Hexane	Chloroform	Benzene
Glycerol	Ethyl acetate	Toluene
2,2,4-Trimethylpentane	Methyl formate	m-Xylene
Methanol		*Above 300 nm*
		Pyridine
		Acetone
		Carbon disulfide

* Transmission limits taken arbitrarily at the point where $A = 0.50$ for $b = 10$ mm; within each group, solvents are arranged in approximate order of increasing wavelength limit.   Data supplied by Matheson Coleman & Bell, Cincinnati, Ohio.

In general, Beer's law may be expected to hold rather closely for radiation of any given wavelength, but the absorbance will change as the wavelength is varied.   The width of the band of wavelengths employed may also affect the apparent value of the absorptivity.   A specific example will help to make this clear.   Figure 3-2 shows the absorption spectrum of the permanganate ion in water solution.   Reference to Table 3-1 shows that a substance absorbing as this does in the range 480 to 570 nm should

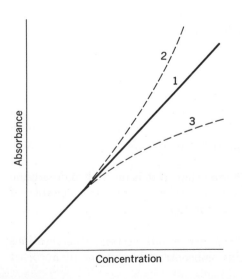

Fig. 3-1   Beer's law (1) obeyed, (2) positive deviation, and (3) negative deviation.

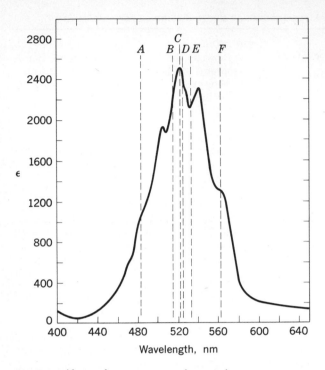

Fig. 3-2 Absorption spectrum of potassium permanganate in water solution, as determined on a Beckman DU spectrophotometer.

appear red-purple, which of course we know to be true. If an absorption measurement were made on this solution, using radiation passed by a filter of green glass with transmission limits approximately at the wavelengths marked A and F, the effects of the detailed peaks and valleys of the absorption curve would be averaged out, so that the value of the molar absorptivity so determined might be about 1700 to 1800. However, if by some means the wavelength range were limited to the region B to E, the value found would be the average of the true values within this space, perhaps 2300. If the width of the wavelength band were still further reduced to the region C to D, the molar absorptivity would approach its true value at this wavelength, 2500.

From this line of reasoning, it follows that, if it is desired to determine true absorption curves, it is necessary to use an instrument capable of isolating very narrow wavelength bands of light. Such an instrument is the spectrophotometer.

The derivation of Beer's law given here omits certain considerations which become important only in highly concentrated solutions of absorber or when the radiation is of unusually great power. These conditions are seldom encountered in analytical work.

DEVIATIONS FROM THE ABSORPTION LAW

The shape of an absorption curve may sometimes change with changes in concentration of the solution and, unless precautions are observed, apparent failure of Beer's law will result. This phenomenon may be due to interaction of the solute molecules with each other or with the solvent, or may be due to instrumental factors.

An example of chemically caused apparent deviation is the change in color of a solution of $K_2Cr_2O_7$ from orange to yellow on dilution with water. The absorption curves for $K_2Cr_2O_7$ and K_2CrO_4 (equal concentrations in terms of milligrams of $Cr(VI)$ per milliliter) are shown in Fig. 3-3. The

Fig. 3-3 Absorption spectra of (1) K_2CrO_4 in 0.05 N KOH and (2) $K_2Cr_2O_7$ in 3.5 N H_2SO_4. Both correspond to 0.01071 mg of Cr per ml; path length = 10.0 mm. Absorbance plotted on a logarithmic scale for convenience.

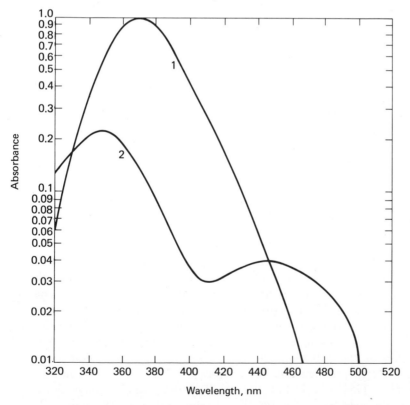

pertinent equilibrium is

$$Cr_2O_7{}^{2-} + H_2O \rightleftharpoons 2H^+ + 2CrO_4{}^{2-}$$
(Orange)   (Yellow)

As the dichromate solution is diluted, the absorbance will gradually change from curve 2 to curve 1, with respect to shape, though diminished in actual absorbance values.

Negative deviation can always be expected when the illumination is not monochromatic. This can be made evident by considering an extreme case. Take, for example, a series of solutions of potassium chromate at some constant high pH. At zero concentration, all colors are transmitted equally. At high concentrations, nearly all the blue will be absorbed, but light of wavelengths above about 500 nm will be transmitted as freely as before. At one-half this concentration, a large fraction of the blue light will come through, along with all the higher wavelengths. A moment's thought will show that the total absorbance over all visible wavelengths cannot possibly be linear with concentration. A mathematical proof of this statement is readily derived.

It is interesting to compare monochromatic light from a laser with that from a narrow-slit monochromator. A research report (4) describes the absorbance of a dye with a sharp maximum at 635 nm, as examined against a water reference and also differentially against a standard solution. The spectrophotometer required a slit width of 1.48 mm for a differential analysis with a tungsten lamp, as compared with 0.08 mm with a laser. Absorbance-concentration curves were found to be linear as far as they could be followed, to about 1.4 absorbance, with the laser, but marked negative deviation was observed with the tungsten lamp. This was caused in part by the wider wavelength band passed, and in part by excessive stray light effects.

PHOTOMETRIC ACCURACY

Even for a system that shows no deviation from Beer's law, the concentration range over which photometric analyses are useful is limited at both high and low values. At high concentrations of absorbing material, so little radiant energy will penetrate that the sensitivity of the photometer becomes inadequate. At low concentrations, on the other hand, the error inherent in reading the galvanometer or recorder will become large compared to the quantity being measured. In many photoelectric instruments, the deflection of the galvanometer or the setting of a balancing potentiometer is directly proportional to the power of the radiation falling on the photocell. This means that the smallest detectable change in power ΔP will be constant regardless of the absolute value of the power itself.

For greatest accuracy in the measurement of absorbance A, however, the increment ΔA, corresponding to the power change ΔP, must be as small a fraction as possible of the actual absorbance A; in other words, the quantity $\Delta A/A$ must be minimized. To determine the transmittance for which $\Delta A/A$ is a minimum, it is necessary to differentiate Beer's law twice and to set the second differential equal to 0. It is convenient to rewrite the law in the form

$$A = \log P_0 - \log P \qquad (3\text{-}6)$$

Then

$$dA = 0 - (\log e)\frac{1}{P}\,dP$$

from which

$$\frac{1}{A}\,dA = -\frac{0.4343}{AP}\,dP = -\frac{0.4343}{A}\frac{1}{P_0 10^{-A}}\,dP$$

Replacing differentials by finite increments,

$$\frac{\Delta A}{A} = -\frac{0.4343\,\Delta P}{P_0}\frac{1}{A\,10^{-A}}$$

Differentiating again (remember that ΔP is a constant),

$$\frac{d(\Delta A/A)}{dA} = -\frac{0.4343\,\Delta P}{P_0}\left(\frac{10^A \ln 10}{A} - \frac{10^A}{A^2}\right) \qquad (3\text{-}7)$$

The condition for the minimum value of $\Delta A/A$ is that the right-hand member of Eq. (3-7) be 0. Hence the factor within the parentheses must be 0, or

$$\frac{10^A \ln 10}{A} = \frac{10^A}{A^2}$$

from which

$$A = \frac{1}{\ln 10} = 0.4343 \qquad (3\text{-}8)$$

This means that the optimum value for the absorbance is 0.4343, which corresponds to a transmittance of 36.8 percent. The relative error in an analysis resulting from a 1 percent error in the photometric measurement for varying transmittance is shown graphically in Fig. 3-4. The

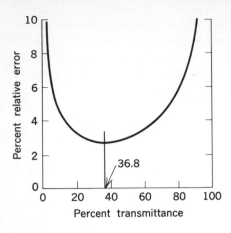

Fig. 3-4 Relative error as a function of transmittance.

situation can perhaps be visualized more readily, especially for those who follow the calculus with difficulty, with the aid of Fig. 3-5, in which Beer's law is plotted in the form $P/P_0 = 10^{-abc}$. An arbitrary value of $\Delta T = 1$ percent is plotted in three positions, as 10, 37, and 90 percent T. The corresponding uncertainty in percent concentration is largest in absolute value at the 10 percent point, which results in poor precision. At the other extreme, 90 percent T, the uncertainty is much smaller but represents a large fraction of the total concentration, which again gives poor precision. It is evident that there must be some intermediate point where the two tendencies become equal and the error is a minimum. This is the 37 percent point.

It is evident from Fig. 3-4 that, although the error is least at 37 percent T, it will not be much greater over a range of transmittance of about 15 to 65 percent (absorbance of 0.8 to 0.2).

The conventional method of plotting a calibration graph for a photometric analysis is either the exponential curve of Fig. 3-5 or the straight

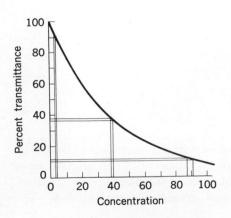

Fig. 3-5 Beer's law plotted as transmittance versus concentration.

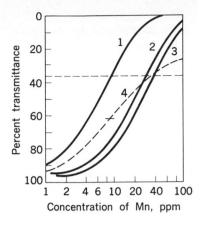

Fig. 3-6 Standard curves for permanganate. The solid curves are determined with a spectrophotometer at wavelengths (1) 526 nm, (2) 480 nm, and (3) 580 nm. The dashed curve (4) is from data taken with a filter photometer with a filter centering at 430 nm. Compare with Fig. 3-2. [*Analytical Chemistry* (6).]

line of Fig. 3-1. The latter has the advantage of showing the region over which Beer's law is followed, but it fails to give any indication of the relative precision at various levels of absorbance. Another method of plotting has been suggested (5, 6) that gives some added features; Fig. 3-6 shows the curve obtained by plotting percent transmittance against the logarithm of the concentration. If a sufficient range has been included, an S-shaped curve, known as a *Ringbom plot*, always results. If the system follows Beer's law, the point of inflection occurs at 37 percent transmittance; if not, the inflection is at some other value, but the general form of the curve is the same. The curve generally has a considerable region which is nearly straight. The extent of this straight portion indicates directly the optimum range of concentration for the particular photometric analysis. Furthermore, the precision of the analysis can be estimated from the slope of the curve since, the steeper it is, the more sensitive the test. It can be shown by differentiation that, if the absolute photometric error is 1 percent, the percent relative error in the analysis is given by $230/S$, where S is the slope, taken as the transmittance change in percent (read from the ordinate scale) corresponding to a tenfold change in concentration. The relative error in the determination of permanganate by curve 1 of Fig. 3-6 is shown by an application of this relation to be approximately 2.8 percent for every 1 percent absolute photometric error. If the error in reading the photometer (reproducibility) is 0.2 percent (a reasonable value with modern instruments), then the relative analysis error is about 0.6 percent. A similar analysis by means of curve 4 would be much less precise. The precision with curve 2 or 3 is about the same as with curve 1, but the range of usable concentrations is shifted to larger values. A detailed comparison of Figs. 3-2 and 3-6 will show the reason for this.

The range of concentration suitable for analysis with adequate precision, as indicated by the straight portion of a Ringbom plot, may be too short for application to the range of unknowns likely to be encountered. There are several methods by which this useful range can be extended in the direction of higher concentrations. The most obvious is simply quantitative dilution of the sample to bring it within the required limits. This approach, if carried too far, may defeat its own purpose, as the cumulative volumetric errors partially offset the gain in photometric precision. Similarly, a more dilute solution can be prepared by taking a smaller sample, the precision being limited only by the sensitivity of the balance available. These methods often suffice to permit the use of sensitive absorptiometric procedures with samples rich in the desired constituent.

Reference to Fig. 3-6 shows that the analyses may also be carried to higher concentrations by a different choice of wavelength. The useful range for permanganate is about 6 to 60 ppm of manganese at 580 nm, as compared with 2 to 20 ppm at 526 nm.

INSTRUMENTATION

Instruments for measurement of the selective absorption of radiation by solutions are known as *colorimeters, absorptiometers,* or *spectrophotometers.* The term colorimeter is restricted to the simpler visual and photoelectric devices for the visible region. The term absorptiometer includes the class of colorimeters, but can be applied to other spectral regions as well. Spectrophotometers vary from simple absorptiometers only in that a much narrower band of wavelengths is employed, as produced by a monochromator. These classes of instruments vary only in degree, and the distinctions are not sharply drawn.

COLORIMETERS

Before photoelectric instruments were generally available, colorimetric analyses were carried out by simple visual techniques. Many of these methods are still prevalent, as the apparatus is less expensive and the precision is adequate for many purposes. An absolute accuracy of ±5 percent may be expected, though it may often be improved by careful attention to details.

The apparatus required for visual comparison methods may be quite simple. One common comparison cell is the *Nessler tube,* a cylindrical glass tube of perhaps 30-cm length, with the lower end closed by a plane window and with an etched mark at a fixed height from the bottom. A series of standard solutions is placed in Nessler tubes to the exact height of the mark. The unknown, prepared under the same conditions, is placed in another tube and compared with the standards by looking downward through the solutions toward a uniform diffuse light source. The unknown can thus be bracketed between two standards, one slightly deeper and one slightly lighter, and the concentration estimated accordingly.

Another visual method uses the *Duboscq comparator,* an instrument that allows variation of the depth of liquid through which the light must pass to reach the eye. The depth is adjusted until the visual intensity through sample and standard are equal, when the concentration can be found from the relation

$$bc = b'c' \tag{3-9}$$

where primes refer to the standard.

FILTER PHOTOMETERS

An example of a photoelectric colorimeter is shown in Fig. 3-7. Two light beams from an incandescent lamp are passed through a glass filter to a pair of matched photovoltaic cells, one beam traversing the sample and the other bypassing it. (A photovoltaic cell, described in more detail in Chap. 26, produces a current proportional to the power in the light striking it.) Each of the photocells is shunted by a low-value resistor, that across the reference cell being provided with a sliding contact (a potentiometer). In use, the contact is first set at the upper end of its resistor, marked 100 on the scale, and the movable jaws adjusted to zero the galvanometer. This ensures equal illumination on the two photocells. The solvent is then replaced by the colored test solution. Since this solution absorbs more light than does the solvent, the current generated by the working cell is decreased, producing a smaller voltage drop across its resistor. The potentiometer must then be moved down to pick off an equal voltage from the reference cell, to return the galvanometer to 0. The potentiometer scale reads percent transmission directly. Comparable instruments have been produced by many manufacturers and have proved very useful, especially in clinical laboratories. The best precision to be expected is ±3 to 4 percent.

An example of a modern filter photometer is the Du Pont 400 shown in Fig. 3-8. This instrument consists of a series of building-block or modular components that can be assembled in various configurations, of which

Fig. 3-7 Schematic representation of the Klett-Summerson photoelectric colorimeter. (*Klett Manufacturing Company.*)

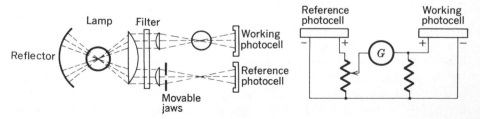

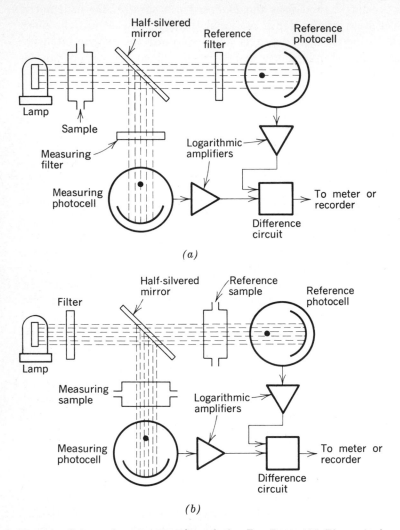

Fig. 3-8 Schematic representation of the Du Pont 400 Photometric Analyzer, shown in two configurations. (*E. I. Du Pont de Nemours & Company.*)

two are shown in the figure. In (*a*) the radiation passes through the sample and then is split into two beams. The beams are filtered separately so that the indicator will give the ratio of radiant powers in two wavelength bands. The *reference* beam is usually set at a wavelength not absorbed by the substance analyzed for, while the *measuring* beam consists of wavelengths that are absorbed by this material. In the configuration shown in (*b*), the radiation is passed through a single filter before splitting into

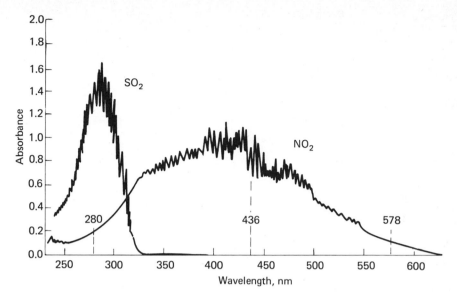

Fig. 3-9 Absorption spectra of SO_2 and NO_2, showing filter wavelengths. (*E. I. Du Pont de Nemours & Company.*)

two beams. One beam then traverses a reference solution, and the other the sample to be analyzed. The readout will then indicate the ratio of the concentration of absorbing substance in the unknown to that in the standard. The electronic system gives an output proportional to the ratio of powers falling on the two vacuum phototubes.

The Du Pont photometer is intended primarily for continuous monitoring of flowing streams of liquid or gaseous samples. An example of its application is the analysis of products of combustion for oxides of sulfur and nitrogen. The configuration of Fig. 3-8a is used, with a mercury-vapor–lamp source. Figure 3-9 shows the absorption spectra of SO_2 and NO_2, and indicates the mercury wavelengths utilized in the analysis. The reference beam uses only the yellow doublet at 578 nm, where the desired oxides do not absorb appreciably. For SO_2 determinations, the measuring beam is so filtered that only the complex of mercury lines near 280 nm is observed, and for NO_2 the blue line at 436 nm. Provision is made for in situ oxidation of NO to NO_2, so that before and after measurements at 436 nm will give information on both oxides.

SPECTROPHOTOMETERS

Instruments in this class are divided between single and double-beam types. Single-beam spectrophotometers require interchange of sample and refer-

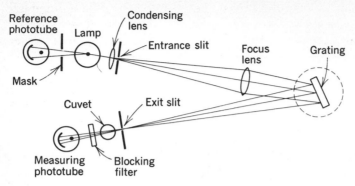

Fig. 3-10 Optical diagram of the Bausch & Lomb Spectronic-20 spectrophotometer. The grating is turned from a panel knob by a cam-and-bar linkage. The blocking filter is used only with the infrared phototube, to eliminate overlapping orders. In one version of the instrument the reference phototube and mask are omitted. (*Bausch & Lomb, Inc.*)

ence solutions for each wavelength, and are therefore better suited to manual than automatic operation.

A widely used instrument primarily for the visible range (340 to 625 nm, capable of extension to 950 nm by change of phototube) is the Bausch & Lomb Spectronic-20 (Fig. 3-10). A 600 line per millimetre plane reflection grating is mounted on a rotatable table. Note that this is *not* a Littrow mounting, as the lens is not traversed twice. The fact that the incident radiation is slightly convergent rather than collimated means that the degree of monochromaticity is less than the maximum obtainable with the same grating and represents a compromise between precision and expense. The wavelength band isolated is 20 nm in width.

The detector is a gas-filled phototube, the signal from which is passed through a special amplifier, compensated to reduce nonlinearity and drift. Both transmittance and absorbance are indicated on the large panel meter. There are only three controls: wavelength selection, zero adjustment, and 100 percent adjustment. Spurious deflections caused by lamp fluctuations are effectively eliminated by a magnetic or electronic voltage regulator. The latter involves an auxiliary photocell to monitor the lamp directly. As the Spectronic-20 is a relatively inexpensive and easily operated instrument, it is often used for convenience where a filter photometer would serve equally well.

ULTRAVIOLET SPECTROPHOTOMETERS

Spectrophotometers for the ultraviolet and visible generally extend down to between 165 and 210 nm. The upper limit is never less than about 650 nm, and many extend to 1000 nm or even further. Until about 1963, very few instruments were usable much below 210 nm, a limit imposed principally by the absorption of natural quartz. The advent of optical-quality vitreous silica has opened up great possibilities in the observation of compounds containing isolated double bonds and certain other groups that absorb in this region.

The earliest ultraviolet spectrophotometer with photoelectric detection on the American market was the Beckman Model DU. This instrument, introduced in 1941, has contributed greatly to our knowledge of the ultraviolet region for both theoretical and analytical purposes. It has been modified somewhat over the years to increase its sensitivity and convenience of operation, but it remains essentially the same instrument, and is in very wide use today.

The optical system of the Beckman DU is shown in Fig. 3-11. It is seen to follow the classical Littrow pattern. There are interchangeable tungsten and hydrogen or deuterium lamps, and also interchangeable photo-tubes as required to cover the complete range. A photomultiplier is available as an option and increases the sensitivity by about tenfold from the ultraviolet limit up to about 625 nm.

Manual instruments comparable to the Beckman DU are available from many manufacturers. They are most useful for quantitative analysis at specified wavelengths, but are inconveniently time-consuming for plotting entire absorption spectra.

An important variation has been introduced by Gilford* (7, 8). No essential changes in the optical or mechanical features are involved, but a different photometric system is employed. The detector is a photomultiplier operated at constant *current*, rather than at constant *voltage*, as in conventional photometers. This is accomplished through a special feedback circuit that electronically adjusts the voltage applied to the photomultiplier in an inverse relation to the illumination. Operated in this fashion, the

* Gilford Instrument Laboratories, Oberlin, Ohio.

Fig. 3-11 Optical system of the Beckman DU spectrophotometer. (*Beckman Instruments, Inc.*)

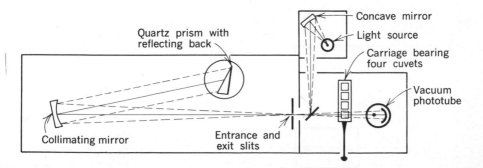

characteristic curve of the photomultiplier, plotted as anode potential against illumination, is very nearly logarithmic, and can be made precisely logarithmic by a simple electronic padding system. The advantage is that the output becomes direct-reading in absorbance, and accurately so (within 1 percent or better) up to an absorbance of 3 or even higher. Furthermore, the stability of the system, both long- and short-term, is unusually good, so that kinetic studies can be followed for hours or days without objectionable zero drift. This instrument has found its greatest usefulness in biochemical work.

DOUBLE-BEAM SPECTROPHOTOMETERS

A possible source of error in any type of absorptiometer lies in the fluctuations which are apt to appear in the intensity of the source of radiation. Short-term variations usually result from lack of regulation in the power line. In the long term, additional variations may be due to aging effects in the lamp or other components. Any change in intensity during the comparison of unknown and standard will of course result in error. The error can be reduced by use of a voltage-regulating transformer or, as mentioned in connection with the Spectronic-20, an electronic control for the lamp itself, provided with a photocell to monitor its intensity.

The most effective way to eliminate the effects of source variations is a double-beam design, a solution that has other advantages as well. Consider the radiant power relations for a pair of identical cuvets (Fig. 3-12). Let P_0' denote the power of the beam coming from the source (equal for the two cuvets). P_S and P_R, the powers transmitted through the sample and reference, are both necessarily smaller than P_0'. The respective absorbances are, by Beer's law,

$$A_S = \log \frac{P_0'}{P_S}$$

Fig. 3-12 Power relations in a double-beam photometer. P_s, the power from the source, is assumed equal on the two cells; P_0 is passed by the reference, and P by the sample solution.

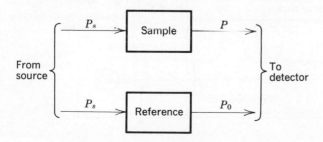

and

$$A_R = \log \frac{P_0'}{P_R}$$

Combining these gives

$$A_S - A_R = \log \frac{P_0'}{P_S} - \log \frac{P_0'}{P_R}$$

$$= \log \frac{P_R}{P_S} \qquad\qquad (3\text{-}10)$$

If the reference solution is identical to the sample except for the absorbing solute under test, it is permissible to replace P_R and P_S by P_0 and P, respectively, in the Beer's law statement,

$$A = \log \frac{P_0}{P}$$

where A is now the corrected absorbance of the sample. Note that P_0 is no longer the power of the incident beam, but rather represents that power diminished by all losses that are duplicated in the two cuvets. Specifically, since P_0' does not appear in the final equation, reasonable variations in the lamp intensity will not affect the measured absorbance. The contents of the reference cuvet should be as nearly as possible identical to the sample with respect to absorbing impurities and refractive index.

A double-beam spectrophotometer for the ultraviolet and visible regions normally uses some sort of beam splitter between the exit slit of the monochromator and the two cuvets. After passing through the cuvets, the beams may fall on the same detector or on two matched detectors. In all modern instruments, the radiation is *chopped* at some point by a rotating shutter or its equivalent; the chopper and beam splitter are often combined in a single component.

Chopping the radiation into a succession of pulses has several advantages. It makes it easier to design an electronic system to sort out signals pertaining to the two channels. It permits important discrimination against unwanted signals (noise) and unchopped stray radiation, in that the amplifier can be tuned to respond only to signals modulated at the chopping frequency.

As an example of a recording spectrophotometer for the ultraviolet, visible, and near-infrared regions, we will describe the *Cary Model 14*. The optical system (Fig. 3-13) includes two monochromators, the first with a vitreous silica prism and the second with a plane reflection grating (600

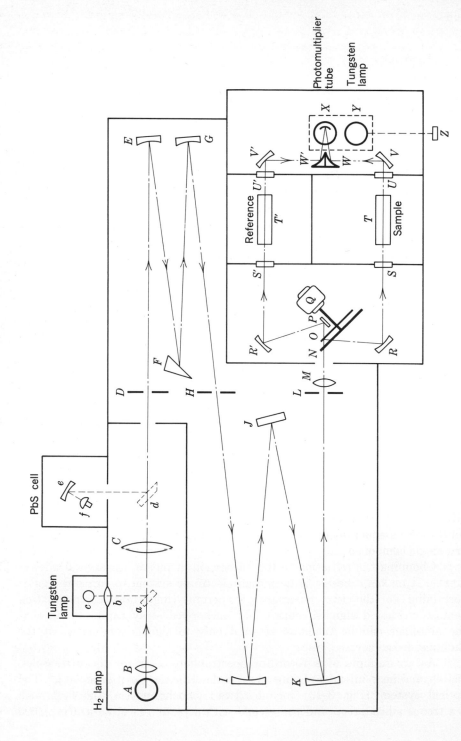

Fig. 3-13 Optical system of the Cary 14 spectrophotometer. (*Varian Associates.*)

lines per millimetre), both in the Czerny-Turner configuration. A combination of cams and bars, driven by a constant-speed motor, ensures that the two monochromators track each other accurately, and provides a scan that is linear in wavelength over the whole range, 186 nm to 2.65 μm.

The dispersion produced by the prism prevents the overlapping of spectral orders from the grating, and the presence of the grating causes the dispersion to be more nearly linear than would be the case with a second prism. The reciprocal dispersion curve is intermediate in shape between curves typical of prism and of grating instruments.

The operation of the instrument can be followed in Fig. 3-13. Radiation from the hydrogen lamp A or tungsten lamp c, selected by the movable mirror a, enters the double monochromator through the entrance slit D. It is dispersed by the 30° Littrow-type prism F and by the grating J; H is an intermediate slit, and L is the exit slit. All three slits are variable in width, and operated in unison by a servo system; slits D and L are always equal in width, while H is slightly wider to make allowances for slight residual inaccuracies in the optical system. Monochromatic radiation from the exit slit L enters the photometer, where it encounters a rotating chopper disk N and a semicircular mirror O, also rotating, that steers it alternately via mirrors R, V, and W, or mirrors P, R', V', and W', to the photomultiplier detector X. The rotating mirror alternates the radiation between the two beams at a frequency of 30 Hz. One beam passes through a reference cuvet T', the other through an identical cuvet T containing the sample.

The pulses of light from the two paths are out of phase with each other, so that the phototube receives radiation from only one beam at a time. Auxiliary photoswitches, operated by the same motor that chops the light beam, are connected electronically in such a way that the output from the photomultiplier is sampled synchronously and split into two electrical signals corresponding to the two optical beams. The signals are fed into a comparator, and the built-in strip-chart recorder plots the ratio, either directly (transmittance) or as its logarithm (absorbance).

For operation in the near infrared, the positions of source and detector are interchanged, so that the radiation traverses the optical system in the opposite direction. The changeover is accomplished by means of plunger Z, which not only moves the photomultiplier out of the way and the tungsten lamp Y into position but also swings a mirror d so as to deflect the beam to the lead sulfide detector f.

The reason for the inversion of the system is that the chopper itself radiates some infrared energy, as do all bodies not at absolute zero, but radiation from the chopper is modulated at the very frequency to which the electronics responds. This would cause a significant error if allowed to fall directly on the infrared detector, but not if it must first pass through the monochromator. In the ultraviolet it is desirable to disperse the radiation before incidence on the sample to avoid photochemical decomposition and to minimize fluorescence effects.

There are many recording double-beam spectrophotometers with only single-stage monochromators. These are naturally less expensive, but are limited in performance because of stray radiation effects (9, 10). *Stray radiation* is unwanted radiation which passes out of a monochromator through the exit slit but is not within the wavelength band for which the monochromator is set. Some of it may be higher-order diffraction from a grating, and some may be the result of reflections from optical surfaces of lenses or prisms. It can be reduced by careful placement of opaque screens but not eliminated entirely. The effects of stray radiation are most pronounced near the ends of the wavelength range of the monochromator, where it is necessary to open up the slits to obtain enough sensitivity. The residual stray radiation for a well-made single monochromator may be of the order of 0.1 percent of the total radiation through most of its range. A double monochromator of equal quality shows only 0.1 percent of 0.1 percent, or 1 ppm of stray radiation.

DUAL-WAVELENGTH SPECTROPHOTOMETERS

It is possible to design an instrument so that two beams of radiation of different wavelength pass through a single cuvet simultaneously. This is useful in several distinct ways: (*1*) It can be utilized without scanning to measure two components, recording their variation with time for kinetic studies or process control. (*2*) One wavelength can be fixed at a point of negligible absorbance while the other is scanned, thus permitting correction for variations in lamp intensity comparable to conventional dual-beam operation but with only a single cuvet. (*3*) It can be operated in the same configuration as (*2*) to give true absorption spectra in the presence of heavy turbidity (11); the effective path length is inherently uncertain in this case, because of the effect of suspended matter, and this arrangement guarantees that both beams follow the same path, whatever it is. Finally, (*4*) the two wavelengths can be scanned simultaneously, but with an offset of a few tens of nanometres, so that the resulting spectrogram will display the difference ΔA between the absorbances at the two wavelengths against the mean wavelength, a *derivative* curve.

The original design of a dual-wavelength spectrophotometer was described by Chance in 1951 (12); Fig. 3-14 shows a recent version (13). The radiation from the source is divided into two parts by a mask. The two parts trace their paths through the same Czerny-Turner monochromator, but are dispersed by separate gratings. The chopping disk allows the two to pass sequentially through the samples to a single photomultiplier.

There are two optional positions for sample cuvets. The *primary* position, almost in contact with the photomultiplier, is used whenever two

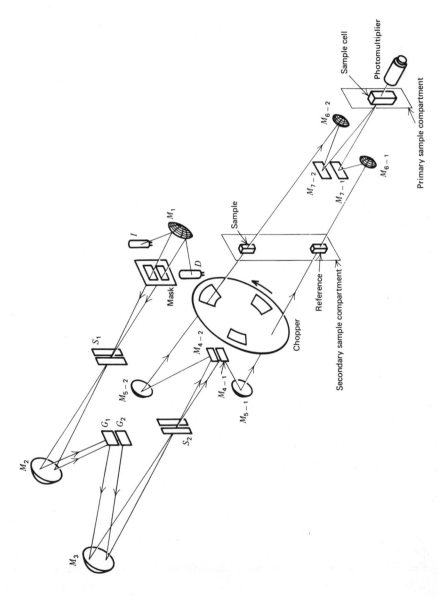

Fig. 3-14 Optical schematic, double-wavelength spectrophotometer, Perkin-Elmer Model 356, simplified. (*Perkin-Elmer Corporation.*)

wavelengths are to pass through a single cuvet; for turbid samples it is essential that the cuvet be close to the detector so that as much of the scattered radiation as possible will be collected. The *secondary* sample position is used for double-beam configurations in which two cuvets can be employed.

DERIVATIVE SPECTROGRAMS

The dual-wavelength spectrophotometer is not the only way to obtain derivative spectra. Another method is to install a miniature electric generator (a *tachometer*) directly connected to the pen drive motor (14). This produces a voltage proportional to the speed of the pen motor, hence to the slope of the spectrogram. A second recorder is necessary to plot the derivative curve.

Still another manner in which a derivative spectrum can be obtained is with an oscillating entrance slit (15, 16). A grating spectrophotometer is modified to accept a slit that can be moved laterally in simple harmonic motion with an amplitude corresponding to about 2 nm of wavelength. Hence the wavelength passing through the sample varies sinusoidally in time about a central value. This can be shown to give a second-derivative spectrum.

It will be demonstrated in Chap. 26 that a derivative of any continuous function, such as an absorption spectrum, can be taken by an operational amplifier circuit.

The advantage of derivative spectrophotometry (14) is shown in Fig. 3-15 to be greatly enhanced resolution, making available considerably more structural information. It must be pointed out that the maxima evident in this technique represent a mathematical analysis of the absorption curve, and do not correspond exactly to the actual frequencies of electronic transitions, although usually they are close.

SOURCES OF RADIATION

The tungsten filament incandescent lamp is unexcelled as a spectrophotometric source through the visible region and adjacent parts of the ultraviolet and infrared, approximately 320 nm to 3.5 μm. Its life at high temperatures can be extended by the inclusion of a small pressure of iodine vapor. This is the *tungsten-iodine* or *quartz-iodine* lamp, so called because it is provided with a vitreous silica envelope to permit operation at higher temperature. The iodine reacts with vaporized tungsten to form volatile WI_4, which pyrolyzes when it comes in contact with the hot filament, redepositing the tungsten atoms on the filament rather than allowing them to accumulate on the cooler walls of the bulb.

For the ultraviolet, hydrogen or deuterium discharge lamps are uni-

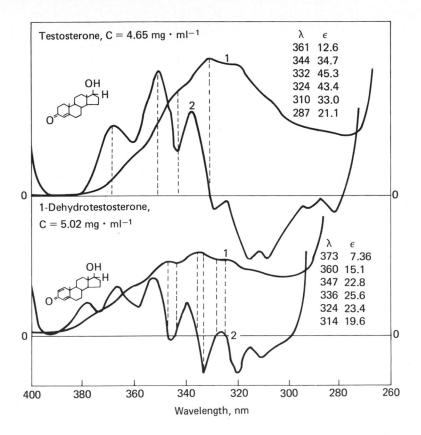

Fig. 3-15 Relationship between ultraviolet spectra (1) and their derivative spectra (2). (*Analytical Chemistry* (14).]

versally employed. They are useful from about 160 to 360 nm. The deuterium lamp gives somewhat greater intensity than its hydrogen counterpart, has a longer life expectancy, and costs more.

DETECTORS

The most widely used detector in the ultraviolet is the photomultiplier tube (17). This device utilizes the same type of photosensitive cathode as a simple phototube, namely, a thin layer of selected semiconductive material containing alkali metal, such as Cs_3Sb, K_2CsSb, Na_2KSb with a trace of Cs, or successively deposited Ag, O, and Cs. The active layer can be on the surface of a metallic support or in semitransparent form on the inner surface of the silica or glass envelope. Different cathode materials have different spectral response curves. Some tubes, such as

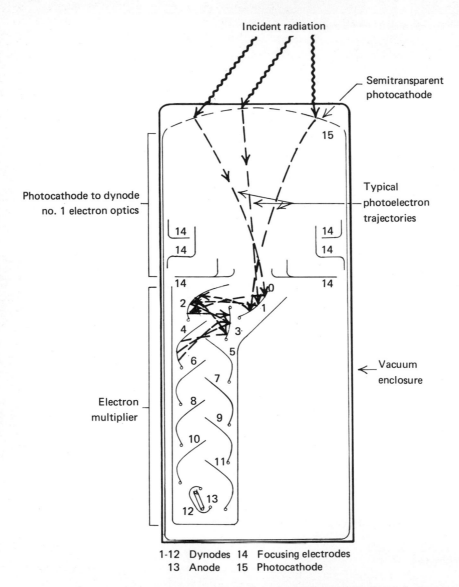

1-12 Dynodes 14 Focusing electrodes
13 Anode   15 Photocathode

Fig. 3-16 Schematic of a typical photomultiplier, showing some electron trajectories. (*RCA Corporation.*)

the widely used type 1P28, respond through the range 200 to 650 nm. Special tubes extend to as low as 150 nm, or as high as 1.18 μm.

The electrons ejected from the surface by incident radiation are focused by an electrostatic field onto the first of a series of dynodes, at successively more positive potentials. Each electron impacting on a dynode causes the ejection of perhaps four secondary electrons, which are focused on the next dynode, and so on down the series (see Fig. 3-16). If there are 12 dynodes, the amplification or electron multiplication will be 4^{12}, or nearly 17×10^6. These 17 million electrons, resulting from a single primary photoelectron, constitute a pulse that typically lasts for about 5 ns, so the resulting anode current is of the order of 1 mA.

Figure 3-17 shows an electronic circuit for a photomultiplier. The current from the anode (the 1 mA mentioned above) is fed into an *operational amplifier* (abbreviated "op amp" and symbolized by a triangle). The operational amplifier is an extremely valuable and versatile circuit component that will be encountered many times in this book. It is discussed thoroughly in Chap. 26. For present purposes the following three principles will suffice to understand its function in the photomultiplier circuit: (*1*) An operational amplifier must be provided with a feedback path from its output to the input marked — (here the resistor R forms this path); (*2*) no current can enter the amplifier inputs; (*3*) a properly connected operational amplifier will generate whatever voltage is needed at

Fig. 3-17 Typical photomultiplier connections, schematic. The resistors in the voltage divider chain are all equal, perhaps 50 kΩ each. The capacitors are needed only if fluctuating or pulsed radiation is being measured. The operational amplifier is discussed in the text.

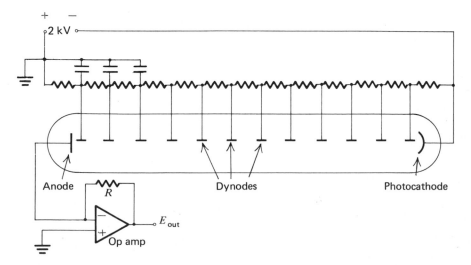

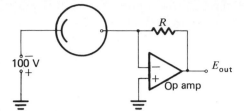

Fig. 3-18 Connections for a vacuum or gas-filled phototube.

its output to maintain (through the feedback loop) its two inputs at virtually equal potentials. In the present application, the current from the photomultiplier anode I_a, since it cannot enter the amplifier, must be exactly equaled by a current from the amplifier output through resistor R, so that the − input will be maintained at virtual ground (since the + input is grounded). Therefore the output voltage will be given by $E_{out} = I_a R$. This is an efficient and convenient method of measuring the tiny current. The output of the amplifier can be manipulated in any required way, such as comparing with a standard, and can be read out on a meter or recorder.

Where less sensitivity is required, a nonmultiplying phototube is often used for its relative simplicity and economy. As Fig. 3-18 shows, an operational amplifier can be employed in exactly the same way as with the photomultiplier. The resistor R is typically 1 to 10 MΩ which (by Ohm's law) makes E_{out} 1 to 10 V for a photocurrent of 1 μA. The spectral range of these tubes, determined by the cathode material and the transparency of the envelope, is the same as for photomultipliers. Tubes filled with a low pressure of argon have a sensitivity about 10 to 20 times greater than their vacuum counterparts, but are considered somewhat less stable and reliable for precision photometry.

A variety of solid-state photocells are available for applications in the visible and near-infrared regions. These include photovoltaic cells (Fig. 3-7), photoconductive diodes, and phototransistors. A photoconductive diode of PbS is the detector of choice for the near infrared and is incorporated in some ultraviolet-visible spectrophotometers, such as the Cary 14, to allow them to be used in this region. Details of semiconductor devices are given in Chap. 26.

PHOTON COUNTING (18)

As the intensity of the radiation diminishes, it is more and more difficult for the detector to distinguish it from the random noise that is always present. In this condition, the response can be greatly improved by observing incident photons individually. Each photon will produce a pulse of current in the photomultiplier, and these pulses can be counted electroni-

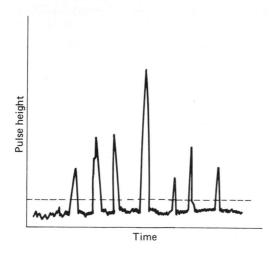

Fig. 3-19 A series of pulses rising above random noise. Only those pulses extending above some cutoff level (dotted line) are counted.

cally with greater accuracy than can be attained by averaging out the pulses and measuring a steady current. Figure 3-19 will show why this is so. Electronic counting methods are considered further in Chap. 23.

SPECTROPHOTOMETER OPERATION

There are a number of controls that must be provided in any spectrophotometer and others that apply only to certain classes. A listing of these will promote fuller understanding of these instruments.

Manual spectrophotometers require: (*1*) a wavelength (or wave number) control, usually a mechanical linkage to rotate the prism or grating; (*2*) a slit-width adjustment, also mechanical in nature; (*3*) a zero adjustment, often called a dark-current control, to ensure that the indicator reads zero transmittance with the detector in darkness; (*4*) a sensitivity control, to adjust the other extreme of the scale to 100 percent transmittance with solvent (or a standard) in the light path; and, if the meter is of the null type, (*5*) a balancing potentiometer with a calibrated scale, which forms the readout.

Recording spectrophotometers have all the foregoing controls except the last, and in addition: (*1*) a switch to start, stop, and reverse the scan drive; (*2*) a control to permit selection of scanning speeds (in some instruments this requires a change of gears); (*3*) recorder controls: paper speed, pen control switch, and frequently range controls to permit scale expansion or contraction.

Some of these adjustments are omitted in less expensive instruments, at the cost of reduced flexibility. On the other hand, more elaborate instruments have other additional controls to permit use in various modes of operation, such as to allow the operation of a double-beam spectrophotometer in a single-beam configuration.

There are a variety of attachments or accessories available for most spectrophotometers, to enable one to obtain reflectance or fluorescence spectra, special adaptors to accept unusually long-path absorption cells (for measurement of gases), and many others.

CHEMICAL APPLICATIONS

The absorption of radiant energy in the visible and ultraviolet spectral regions depends primarily on the number and arrangement of the electrons in the absorbing molecules or ions.

Among inorganic substances, selective absorption may be expected whenever an unfilled electronic energy level is covered or protected by

Table 3-4   Representative chromophores*

Compound	Chromophore	Solvent	$\lambda_{max},$ nm	log ϵ
Octene-3	C=C	Hexane	185	3.9
			230	0.3
Acetylene	C≡C	(Vapor)	173	3.8
Acetone	C=O	Hexane	188	2.9
			279	1.2
Diazoethyl acetate	N=N	Ethanol	252	3.9
			371	1.1
Butadiene	C=C—C=C	Hexane	217	4.3
Crotonaldehyde	C=C—C=O	Ethanol	217	4.2
			321	1.3
Dimethylglyoxime	N=C—C=N	Ethanol	226	4.2
Octatrienol	C=C—C=C—C=C	Ethanol	265	4.7
Decatetraenol	[—C=C—]₄	Ethanol	300	4.8
Vitamin A	[—C=C—]₅	Ethanol	328	3.7
Benzene		Hexane	198	3.9
			255	2.4
1,4-Benzoquinone		Hexane	245	5.2
			285	2.7
			435	1.2
Naphthalene		Ethanol	220	5.0
			275	3.7
			314	2.5
Diphenyl		Hexane	246	4.3

* Data collected from various sources; to be taken as illustrative only.

a completed energy level, usually formed by means of coordinate covalences with other atoms.

Consider copper as an example. The simple Cu^{2+} ion is never found in aqueous solution (though often written as such), because it has a great tendency to form coordinate bonds with any available molecules or ions that carry unshared pairs of electrons. Such unshared pairs are present in water, ammonia, cyanide ion, chloride ion, and many other entities. The structure of the cupric ion coordinated with any of these Lewis bases has only 17 electrons in the third (M) major energy level, whereas the fourth (N) level contains a stable octet. The several ions do not have identical colors, as the nature of the ligand has an effect on the energy of the electrons.

Selective absorption among organic molecules is again related to the deployment of electrons in the molecule. Completely saturated compounds show no selective absorption throughout the visible and the accessible ultraviolet regions. Compounds that contain a double bond absorb strongly in the far ultraviolet (195 nm for ethylene). Conjugated double bonds produce absorption at longer wavelengths. The more extensive the conjugated system, the longer will be the wavelengths at which absorption is observed. If the system is extended far enough, the absorption enters the visible region, and color results. Thus β-carotene, with 11 conjugated double bonds, absorbs strongly in the region 420 to 480 nm, and hence is yellow-green in appearance. The complete conjugated system in a compound is called its *chromophore*.

The wavelengths of the absorption maxima of a compound provide a means for identifying the chromophore it contains. The spectra are in general modified by the presence of various atomic groups when these are substituted for hydrogen atoms on the carbons of the chromophore. Such substituents usually have the effect of shifting the absorption bands toward longer wavelengths and changing their absorbance values. Substituents that produce these effects are known loosely as *auxochromes*.

In Table 3-4 are listed a number of organic compounds containing representative chromophores, together with their wavelengths of maximum absorption and approximate molar absorptivity values. Many of these illustrative compounds, especially those with larger chromophores, have many lesser maxima in the spectra. This table cannot be used in the precise identification of absorbing groups, the way the corresponding infrared table can. The wavelengths, but not the absorptivities, tend to change when the solvent is changed.

In aromatic compounds, the benzene ring is the simplest chromophore. Two or more rings in conjugation, as in either naphthalene or diphenyl, again shift the absorption toward the visible. Table 3-5 shows the effects of some auxochromes on the absorption of benzene (19).

The quinoid ring is much more effective as a chromophore than is

Table 3-5   Effect of auxochromes on the benzene chromophore (19)

Compound	Solvent	Ethylenic band		Benzenoid band	
		λ_{max}, nm	log ε	λ_{max}, nm	log ε
Benzene	Cyclohexane	198	3.90	255	2.36
Anilinium ion	Aq. acid	203	3.88	254	2.20
Chlorobenzene	Ethanol	210	3.88	257	2.23
Thiophenol	Hexane	236	4.00	269	2.85
Phenol	Water	210.5	3.79	270	3.16
Aniline	Water	230	3.93	280	3.16
Phenolate ion	Aq. base	235	3.97	287	3.42

the benzene ring.   An example contrasting the two types is phenolphthalein, which has the following structures in acidic and basic solutions, respectively:

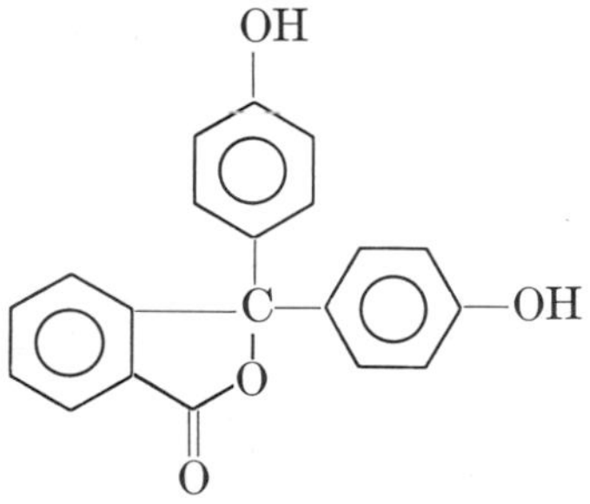

Colorless molecule   Red anion (in basic
(in acid solution)   solution)

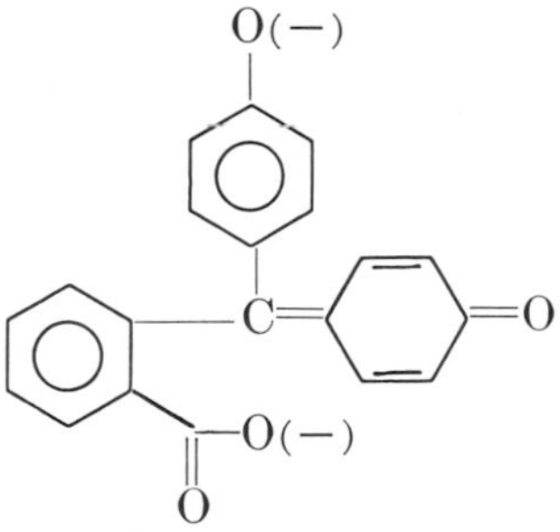

In the colorless form, conjugation does not extend outside the individual rings (except that one ring is conjugated with a carbonyl group).   In the red form, however, one ring has been converted to the corresponding quinone, which results in extending the conjugation to include the central carbon and, through it, the other two rings.   So we conclude that the entire anion constitutes a chromophore, whereas in acid solution, the molecule contains three separate and nearly identical lesser chromophores, the benzene rings.

QUALITATIVE ANALYSIS

Ultraviolet and visible absorption spectra provide a useful source of supporting evidence in the elucidation of structures of organic compounds. Selective absorption in itself is seldom sufficiently unique to serve as an identifying fingerprint for a particular structure, but it may serve to rule out certain alternatives. The lack of appreciable absorption in the region 270 to 280 nm is unequivocal evidence that a compound does not contain a benzene ring. If there is no absorption from about 210 nm to the visible, it is certain that the compound contains no conjugated unsaturation. If transparency extends down to 180 nm, then even isolated double bonds must be absent.

A number of useful correlations between wavelengths of absorption maxima and structure have been worked out. One of the most productive of these is due to Woodward (20), and is conveniently summarized by the Fiesers (21). For an α,β-unsaturated carbonyl compound, for example, a basic absorption can be assigned at 215 nm; additional conjugated double bonds will add 30 nm each; a saturated substituent on C-2 will add 10, on C-3, 12, and on C-4 or higher, 18; a 2-bromo substituent will add 23; etc. The measured absorption maximum can be compared with calculated values for various possible structures.

The variation of the absorption spectrum of an acid-base indicator as a function of pH provides an excellent method for determining the pK value of the indicator. In Fig. 3-20 are plotted the absorption curves for phenol red at a series of pH values. It is seen that with increasing pH the absorption at λ 610 nm increases, while the lesser absorption at λ 430 decreases. Note that the several curves cross very nearly at a common point at λ 495; this is called an *isoabsorptive point* or an *isosbestic point* and is characteristic of a system containing two chromophores which are interconvertible, so that the total quantity is constant.

If we plot the absorbance at λ 615 against pH, an S-shaped curve is obtained (Fig. 3-21). The horizontal portion to the left corresponds to the acidic form of the indicator, while the upper portion to the right corresponds to nearly complete conversion to the basic form. Since the pK is defined as the pH value for which one-half of the indicator is in the basic form and one-half in the acid form, this is determined by a point midway between the left and right horizontal segments, pH 7.0 in our example.

The dissociation constants of compounds with absorption in the ultraviolet rather then the visible can often be determined by a similar procedure. Examples are theobromine (22) (λ 240 nm), theophylline (22) (λ 240), and benzotriazole (23) (λ 274). It is instructive to plot the data in such cases in three dimensions (see Fig. 3-22). In this presentation, which has been called a *stereospectrogram,* the three axes refer to wavelength, pH, and absorbance (24). Note that the isoabsorptive point on graph F corresponds to a straight line parallel to the pH axis on the stereospectrogram.

The existence of an isoabsorptive point is evidence that a chemical equilibrium exists between two species. If instead of one absorption peak disappearing as another appears (the condition for an isoabsorptive point),

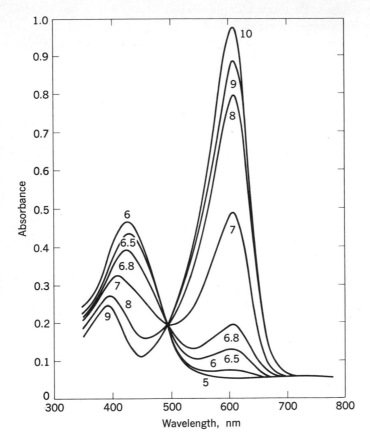

Fig. 3-20   Phenol red.   Absorbance curves at various pH values.

the wavelength of a maximum shifts gradually on change of pH, concentration, or other variable, then one can assume that the change is due to a physical interaction between the absorbing substance and its environment, or to the existence of a series of complexes of about the same stability.

DETERMINATION OF LIGAND/METAL RATIO IN A COMPLEX

Since organometallic complexes in general show selective absorption in the visible or ultraviolet, this property is widely employed to determine their composition as well as their stability constants. The stoichiometry of a stable complex can be determined by either of two related techniques: (1) the *mole-ratio* method introduced by Yoe and Jones (25) and (2) the method of *continuous variations* attributable to Job and modified by Vosburgh and Cooper (26).

In the mole-ratio method, the absorbances are measured for a series

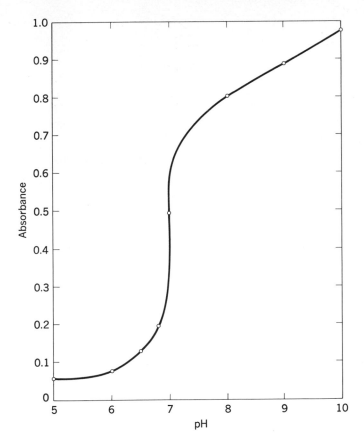

Fig. 3-21 Phenol red. Absorbance at 615 nm as a function of pH.

of solutions which contain varying amounts of one constituent with a constant amount of the other. A plot is prepared of absorbance as a function
of the ratio of moles of reagent to moles of metal ion. This is expected to
give a straight line from the origin to the point where equivalent amounts
of the constituents are present. The curve will then become horizontal,
because all of one constituent is used up and the addition of more of
the other constituent can produce no more of the absorbing complex. If
the constituent in excess itself absorbs at the same wavelength, the curve
after the equivalence point will show a slope that is positive, but of smaller
magnitude than that prior to equivalence. Figure 3-23 shows the result
of such an experiment utilizing the complex of diphenylcarbazone with
mercuric ions (**27**).

The method of continuous variations requires a series of solutions
of varying concentration of the two constituents wherein their *sum* is kept
constant. The difference between the measured absorbance and the absor-

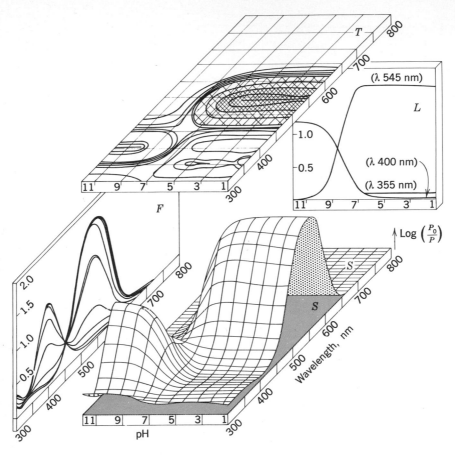

Fig. 3-22 Stereospectrogram of benzeneazodiphenylamine. The stereospectrogram S represents a three-dimensional plot of the absorbance (vertical axis) as a function of pH (left-right axis) and wavelength (oblique axis). F, T, and L are the three two-dimensional modes of representation of the same data. F corresponds to Fig. 3-20, and L to Fig. 3-21. In T, each line is an isoabsorbance line, and the graph can be read in the same manner as a topographical map. [*From the work of Jaffé, quoted by Archibald* (24).]

bance calculated for the mixed constituents on the assumption of no reaction between them is plotted against the mole fraction. The resulting curve will show a maximum (or minimum) at the mole fraction corresponding to that in the complex. An example is shown in Fig. 3-24 for the mercury-diphenylcarbazone complex (27).

The sharpness of the breaks in the curves of both of these methods of identification of a complex depends on the magnitude of the stability constant. The degree of curvature often provides a convenient means of

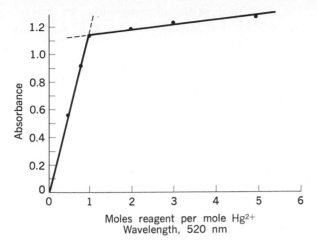

Fig. 3-23 Yoe-Jones plot for the mercury-diphenylcarbazone complex. [*Analytical Chemistry* (27).]

measuring the constant. The required calculations are too involved to develop here (28–30).

QUANTITATIVE ANALYSIS

Analysis for an absorbing substance can be carried out directly through Beer's law if no other absorbing material is present to interfere, or if the interfering material can be removed or corrected for. A good example of this approach is the measurement of ozone concentration in urban smog (31). A high-pressure mercury-arc lamp was set up on the roof of a building, and

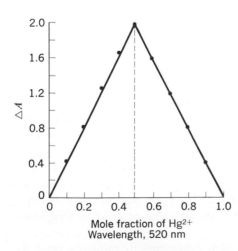

Fig. 3-24 Job plot for the mercury-diphenylthiocarbazone complex. The vertical coordinate ΔA represents the difference between the absorbance of the mixed solution and the sum of the absorbance that the reagents would have shown had they not reacted. [*Analytical Chemistry* (27).]

its radiation received with a simplified prism spectrophotometer located on another building several hundred feet distant. Since a double-beam system was impracticable, zero calibration was made at night, when atmospheric ozone was known to drop to negligible values. The ozone measurements were interfered with by other oxidants such as NO_2, but this effect was eliminated by determining the ratio of absorbances at the ozone maxima of λ 313 and 265 nm. Ozone was found to rise to about 22 parts per 10^8 at noon, as the average of 50 days.

Substances that do not show useful absorption can in many instances be determined spectrophotometrically following the addition of a reagent to produce an absorbing complex or other chromophore. One of the more important reagents is *dithizone* (diphenylthiocarbazone). This green compound, soluble in chloroform, reacts with cations of most of the transition metals to give red or violet complexes. The reagent can be made specific by adjustment of the pH. Details of the dithizone method are readily available (32, 33) and will not be repeated here. Another example of developed absorption is the determination of trace amounts of Hg(II) with the dye 4,4'-bis(dimethylamino)diphenylamine, known as Bindschedler's green, in citrate buffer (34). The complex, extracted into 1,2-dichloroethane, follows Beer's law from 8×10^{-7} to 4×10^{-6} M Hg(II). Only tin, out of 21 metals checked, interferes with the analysis.

ADDITIVITY OF ABSORBANCES

In our discussion of Beer's law it was pointed out that the absorbance is proportional to the number of particles that are effective in absorbing radiation at the specified wavelength. This is easily extended to cover the presence in the same solution of more than one absorbing species. Each species absorbs as though the others were not present. Hence we can write:

$$A = \sum_i A_i = b \sum_i a_i c_i \qquad (3\text{-}11)$$

which states that the absorbance is an additive property. This relation presumes, of course, that there is no chemical interaction between solutes.

This additivity can be useful in a number of ways. It permits subtraction from an observed absorbance of the contribution due to solvent or reagents, the familiar use of a blank. It also enables one to subtract from the spectrum of an unknown the absorbance due to a chromophore known to be present, in order to identify a second chromophore. For example, in Fig. 3-25, curve a is the observed absorption spectrum of the 4-nitrobenzoate of 7-dehydrocholesterol. To identify this ester, the known spectrum b of another ester of the same acid (cyclohexyl 4-nitrobenzoate) was subtracted point by point from curve a. The resulting curve c was found to be essentially identical to the spectrum determined on the free sterol itself (35).

The additivity of absorbance is also important in multiple analysis, the simultaneous determination of two or more absorbing substances in the same solution. The requirement for multiple analyses is merely that the absorption curves for the individual components do not approach too close to coincidence. Some partial overlapping is permissible but, the greater the overlap, the less precise the analysis. The relation can be understood by reference to Fig. 3-26. Curves 1 and 2 are the absorption spectra of the pure components, and curve 3 is the spectrum of the mixture. It is assumed that no other substances present absorb in this region.

Note that the maxima of curve 3 do not appear at exactly the same wavelengths as the corresponding maxima in curves 1 and 2. It will be seen that both substances contribute to the absorption at both wavelengths λ_u and λ_v. Let c_1 and c_2 represent the concentrations of the two components in the mixture, and let A_{1u} represent the absorbance of substance

Fig. 3-25 Ultraviolet absorption spectra of (a) 7-dehydrocholesteryl 4-nitrobenzoate, (b) cyclohexyl 4-nitrobenzoate, and (c) 7-dehydrocholesterol, determined by subtraction; solvent, n-hexane. [*Journal of the American Chemical Society* (35).]

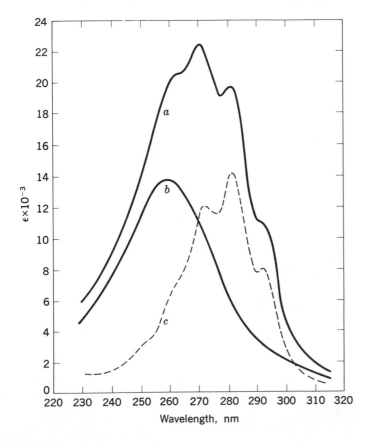

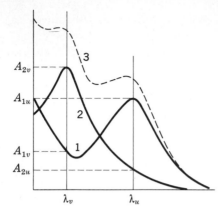

Fig. 3-26 Two-component analysis with a spectrophotometer, hypothetical example.

1 at wavelength λ_u, etc., as indicated in the figure. Then the absorbance of curve 3 at λ_v will be given by

$$A_v = A_{1v} + A_{2v} = a_{1v}bc_1 + a_{2v}bc_2$$

and at λ_u by

$$A_u = A_{1u} + A_{2u} = a_{1u}bc_1 + a_{2u}bc_2$$

Since A_v and A_u are determined by experimental observation on the mixture, and the a's may be obtained once and for all in advance, the above equations can be solved simultaneously for c_1 and c_2. For highest precision, a_{1v} and a_{2u} should be as low as possible, while a_{1u} and a_{2v} should be high. An example to show the possibilities of the method is the simultaneous determination of Mo, Ti, and V by means of their colored complexes with H_2O_2 (36). The standard curves are reproduced in Fig. 3-27.

DIFFERENTIAL SPECTROPHOTOMETRY

The standard procedure for absorptiometric analysis as previously described requires two preliminary adjustments: (*1*) the scale must be adjusted to read 0 with no light reaching the photocell (lamp turned off or shutter closed), and (*2*) another adjustment must assure that the scale reads 100 with pure solvent in the beam of radiation. (Here and in what follows we shall assume a linear transmittance scale marked off in 100 divisions. The discussion applies equally to a deflection or null type of photometric circuit.) To complete an analysis with a system known to follow Beer's law, one must make two readings of transmittance, one for the standard and one for the unknown. (Precision can be increased by multiple measurements of the standards.) It is desirable, particularly if some deviation from Beer's law is suspected, to have the concentrations of unknown and standard rather close together.

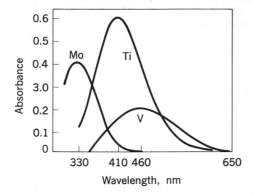

Metal	Absorbances, A		
	330 nm	410 nm	460 nm
Mo	0.416	0.048	0.002
Ti	0.130	0.608	0.410
V	0.000	0.148	0.200

Fig. 3-27 Comparison spectra of the products of reaction of hydrogen peroxide with Mo, Ti, and V. Concentration, 4 mg of metal per 100 ml. [*Analytical Chemistry* (36).]

This standard procedure leaves much to be desired with respect to precision, especially when operating near either end of the scale, i.e., with solutions either of very high or very low transmittance relative to the solvent. Consider, for example, an analysis in which the standard solution shows 10 percent transmittance and the unknown 7 percent. Then both of the readings to be compared utilize only 10 percent of the instrument's scale. A tenfold increase in precision can be obtained by effectively spreading out this first 10 percent to fill the whole scale by the simple expedient of setting the instrument to read 100 with the standard solution in the light beam instead of the solvent alone. This operation is shown schematically in Fig. 3-28, where the horizontal lines represent the scale of a photometer. The upper scale shows the position of the standard ($T = 10.0$ percent) and unknown ($T = 7.0$ percent) for the conventional method, and the lower scale shows that the transmittance of the unknown becomes 70.0 percent when the photometer is set at 100 for the standard.

Calculation of the concentration of the unknown involves no complications over the conventional method. Let us write Beer's law twice, designating the unknown by the subscript x and the standard by s (subscript 0 still refers to the solvent):

$$\log P_0 - \log P_s = abc_s \qquad\qquad (a)$$

$$\log P_0 - \log P_x = abc_x \qquad\qquad (b)$$

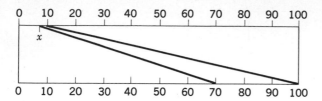

Fig. 3-28 Scale expansion for solutions of low transmittance.

Subtracting Eq. (*b*) from Eq. (*a*) gives

$$\log P_x - \log P_s = \log \frac{P_x}{P_s} = ab(c_s - c_x) \tag{3-12}$$

It is convenient to designate the quantity $\log (P_x/P_s)$ as the *relative absorbance*, and P_s/P_x as the *relative transmittance*. Equation (3-12) shows that, as long as Beer's law is followed, the relative absorbance is proportional to the difference in concentration between unknown and standard.

The literature during the 1950s contained many applications of this method, with standard deviations typically of the order of 2 or 3 percent, usually a more than tenfold increase in precision over conventional spectrophotometry. With the advent of atomic absorption spectrophotometry, however, this more laborious procedure became less popular. A critical review of the field has appeared recently (37).

TRACE ANALYSIS

An analogous situation exists with highly dilute solutions. Figure 3-29 diagrams such a case, for which a standard gives 90 percent transmittance and the unknown 93 percent. In conventional operation (upper scale), again only 10 percent of the scale is utilized. In this case, expansion of the scale is accomplished by setting the zero with a standard solution in the light path instead of an opaque shutter. The increase in precision for the example quoted is again tenfold.

Some types of photometers have no provision for offsetting the zero as far as may be required for this method. In general it can be accomplished with an instrument that is equipped with both sensitivity and dark-current controls. The zero point is adjusted by the dark-current control (a misnomer, since darkness is now replaced by a finite level of illumination), and the 100 point is set by use of the sensitivity control. In some instruments this may be rather inconvenient in that setting the full scale point may change the setting for zero, and vice versa, so that a successive approximation procedure must be applied.

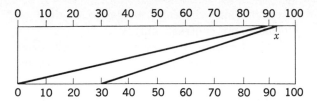

Fig. 3-29 Scale expansion for solutions of low absorbance.

For this method of photometry, it can be shown that the absorbance follows a power series:

$$A = -\log T_2 - \left(\frac{0.4343}{100}\right)\left(\frac{1}{T_2} - 1\right) R$$

$$+ \left(\frac{0.4343}{2 \times 10^4}\right)\left(\frac{1}{T_2} - 1\right)^2 R^2 - \cdots \quad (3\text{-}13)$$

where R designates the scale reading corresponding to absorbance A, and T_2 is the transmittance of the reference solution used to establish the zero of the scale. This is the equation of a straight line if the term in R^2 is neglected, and shows only slight curvature if that term is included. Higher terms are completely negligible.

"ULTIMATE PRECISION"

These two methods can be combined to give the most general modification of spectrometric analysis. Both ends of the scale are established with standard solutions, so that the unknowns are bracketed between them. Any region of transmittance, say 30 to 35 percent T, can be expanded to fill the entire scale. The working equation is the same as Eq. (3-13), except that $1/T_2$ within the second and fourth parentheses is replaced by T_1/T_2, where T_1 is the transmittance of the reference solution used to set the 100 point on the scale. Although capable of good precision, this method is rather inconvenient and is seldom used.

For further theory of these differential methods, see the paper by Ingle (38).

PHOTOMETRIC TITRATIONS

In conventional titrimetry the equivalence point in a reaction is detected by visual observation of a change in color, either inherent in one of the reactants (e.g., permanganate) or produced by an indicator. Under favorable conditions, precision within a few tenths of 1 percent is easily obtainable by operators with normal vision. Good results, however, are difficult or impossible to obtain in cases in which the color change is gradual or in which the colors of the two forms do not contrast sharply.

Such difficulties can often be overcome by carrying out the titration in a cuvet in a spectrophotometer or filter photometer. An optimum wavelength (or filter) is selected and the zero adjustments made in advance; then a photometric reading is taken following each incremental addition from a buret. Conventional spectrophotometers or colorimeters generally require some structural modifications to permit insertion of a titration vessel of convenient size, as well as a buret tip and some type of stirrer.

Numerous automatic or semiautomatic instruments that can carry out all the steps of a titration with a minimum of operator attention are commercially available. In some the results are plotted out by a pen recorder, while in others the stopcock of the buret is closed electrically at the end point. Such photometric titrators can be extremely convenient and economical, particularly for repetitive titrations.

The usual photometric titration curve is a plot of absorbance against volume of added reagent. If the absorbing substances (titrant, substance titrated, or both) follow Beer's law, then the titration curve will ideally consist of two straight lines, intersecting at the equivalence point. The intersection is likely to show some degree of curvature, as a result of incompleteness of reaction. This is usually of little consequence, because the segments of the curve more remote from equivalence are nearly straight and can be extrapolated to an intersection. Correction for dilution must not be overlooked.

An excellent example of a photometric titration is shown in Fig. 3-30, which gives the titration curve for a mixture of m- and p-nitrophenols titrated by sodium hydroxide (39). The absorbance was measured at 545 nm, a wavelength where the anions of both isomers absorb, but where the corresponding acids do not. The absorptivity of the m-isomer is greater than that of the p-isomer. The p-isomer is neutralized first, because it is the stronger of the two weak acids. The end points corresponding to the two straight-line intersections were in error by slightly over 1 percent in this particular experiment. It would be impossible to determine these two acids in the presence of each other by visual observation, with or without an indicator, as the color change corresponding to the first equivalance point would be very gradual. For a similar reason, the analysis would be impracticable also by potentiometric (pH-meter) techniques.

This method is excellent for a weak acid that absorbs at a wavelength different from that of its anion, or if either acid or anion does not absorb. Both the ionization constant and the concentration of the acid must be considered in determining how weak an acid can be titrated. Satisfactory results can be obtained (40) if CK_a, the product of the molar concentration and the acid ionization constant, is greater than about 10^{-12}.

Strong acids cannot be titrated in this way, because they are in the

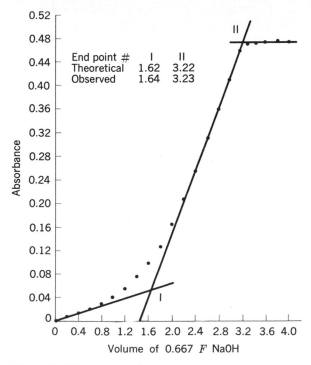

End point #	I	II
Theoretical	1.62	3.22
Observed	1.64	3.23

Fig. 3-30 Photometric titration of a mixture of 50 ml of 0.0219 F p-nitrophenol and 50 ml of 0.0213 F m-nitrophenol with 0.667 F NaOH; wavelength, 545 nm. [*Analytical Chemistry* (39).]

ionized state at all times. They may, however, be monitored photometrically by following the absorbance of an added indicator. The indicator is, as usual, a weak acid such that the free acid and its anion absorb at different wavelengths. The indicator is not selected on the same basis as for visual titrations, however. For visual work, it is desirable that the pK_a of the indicator coincide with the pH of the titration system at its equivalence point. For a photometric system, on the other hand, when measurements are made in the region of absorption of the indicator anion, it is desirable to choose an indicator which has a small enough pK_a value that it does not start to be neutralized until the stronger acid has essentially completely reacted. The equivalence point can then be found by the intersection of two straight-line segments, as at *a* in Fig. 3-31. Intersection *b* corresponds to complete titration of the indicator, so this system can be considered to be the titration of a mixture of a strong and a weak acid.

The titration of various metals by EDTA and similar complexogens

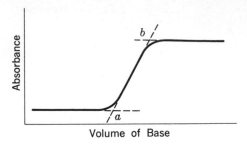

Fig. 3-31 Photometric titration curve; acid titrated by base, with added indicator.

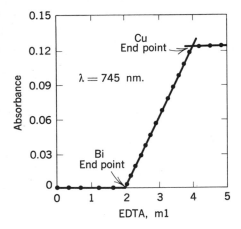

Fig. 3-32 Photometric titration of a bismuth-copper mixture with 0.1 F solution of EDTA. [*Analytical Chemistry* (41).]

is one of the most fruitful applications of photometric titration. Figure 3-32 shows an example in which bismuth and copper are determined successively in a single titration (41). Measurements were made at 745 nm, where the Cu-EDTA complex absorbs strongly but the Bi-EDTA complex does not.

Precipitation titrations represent another important application of photometric end-point detection, but discussion of them is postponed to Chap. 6.

PROBLEMS

3-1 A particular sample of a solution of a colored substance, known to follow Beer's law, shows 80 percent transmittance when measured in a cell 1.0 cm in length. (*a*) Calculate the percent transmittance for a solution of twice the concentration in the same cell. (*b*) What must be the length of the cell to give the same transmittance (80 percent) for a solution of twice the original concentration? (*c*) Calculate the percent transmittance of the original solution when it is contained in a cell 0.5 cm in length. (*d*) If the original concentration was 0.005 percent (weight/volume), what is the value of the absorptivity a?

3-2 In the preparation of a reference curve for an analysis with a photoelectric colorimeter, the following values were obtained:

Concentration, $mg \cdot l^{-1}$	P_0	P	Concentration, $mg \cdot l^{-1}$	P_0	P
0.00	98.0	98.0	6.00	100.0	27.9
1.00	97.0	77.2	7.00	99.0	23.4
2.00	100.0	63.5	8.00	98.2	20.3
3.00	99.5	50.0	9.00	100.0	18.1
4.00	100.0	41.3	10.00	100.0	16.4
5.00	100.0	33.5			

(*a*) Calculate absorbances, and plot against concentration. Do these data indicate a negative or positive deviation from Beer's law, or neither? (*b*) Plot percent transmittance against the logarithm of the concentration. Using the method of Ringbom, state the approximate range of concentration that will give adequate precision in this analysis.

3-3 Vitamin D_2 (calciferol), measured at wavelength 264 nm, its maximum, in alcohol, follows Beer's law over a wide range, with a molar absorptivity $\epsilon = 18,200$. (*a*) If the formula weight is 397, what is the value of a? (*b*) What range of concentration, expressed in percent, can be used for analysis if it is desirable to keep the absorbance A between the limits 0.4 to 0.9? Assume $b = 1$ cm.

3-4 On the basis of Fig. 3-3, predict the type of deviation, if any, from Beer's law at the following wavelengths in aqueous K_2CrO_4 solutions, as absorbance is plotted against total Cr(VI) in solution: 350 nm, 370 nm, 445 nm, 480 nm.

3-5 For each of the following situations, predict whether Beer's law would show an apparent negative deviation, a positive deviation, or practically none at all. (*a*) The absorbing substance is the undissociated form of a weak acid. (*b*) The absorbing entity is the cation in equilibrium with the weak acid. (*c*) A metal is being determined by means of a color-forming reagent, measured with a photoelectric photometer with the appropriate glass filter. (*d*) In the same system as (*c*), an insufficient amount of reagent is added to react completely with the 3 most concentrated samples of the 10 examined.

3-6 A silicate rock is to be analyzed for its chromium content. A sample is ground to a fine powder, and a 0.5000-g portion is weighed out for analysis. By suitable treatment the material is decomposed, the chromium being converted to Na_2CrO_4. The filtered solution is made up to 50.00 ml with 0.2 N H_2SO_4, following the addition of 2 ml of 0.25 percent diphenylcarbazide, a reagent that gives a red-violet color with Cr(VI). A standard solution is available containing 15.00 mg of pure $K_2Cr_2O_7$ per liter. A 5.00-ml aliquot of the standard is treated with 2 ml of the diphenylcarbazide solution and diluted to 50.00 ml with 0.2 N H_2SO_4. The absorbances of the two final solutions are determined in a filter photometer and found to be: standard, $A_s = 0.354$; unknown, $A_x = 0.272$.

(*a*) What is the amount of chromium in the rock, expressed in terms of percent Cr_2O_3? (*b*) Suppose the same solutions are to be compared directly with each other in the filter photometer, by the relative method. Describe how this would be done and how the calculation would be performed. Would any additional data be needed, and if so, what?

3-7 Set up simultaneous equations for the determination of Mo, Ti, and V mixtures by the peroxide method, using the data of Fig. 3-27 and the associated table. In one experiment, a test solution was treated with excess peroxide and perchloric acid, and diluted to 50.00 ml. The following absorbances were obtained:

λ, nm	330	410	460
A	0.248	0.857	0.718

Calculate the quantities of the three elements in milligrams present in the sample. Assume equal path length in all measurements.

3-8 Caffeine, $C_8H_{10}O_2N_4 \cdot H_2O$ (formula weight = 212.1), has been shown to have an average absorbance $A = 0.510$ for a concentration of 1.000 mg per 100 ml at 272 nm. A sample of 2.500 g of a particular brand of soluble coffee product was mixed with water to a volume of 500 ml, and a 25-ml aliquot was transferred to a flask containing 25 ml of 0.1 N H_2SO_4. This was subjected to the prescribed clarification treatment and made up to 500 ml. A portion of this treated solution showed an absorbance of 0.415 at 272 nm. (a) Calculate the molar absorptivity. (b) Calculate the number of grams of caffeine per pound of soluble coffee (1 lb = 453.6 g). Assume $b = 1$ cm.

3-9 According to Wetters and Uglam (42), the molar absorptivity of the nitrite ion is 23.3 at 355 nm, and the ratio of the absorptivity at 355 to that at 302 nm is 2.50. The molar absorptivity for the nitrate ion is negligible at 355 nm and 7.24 at 302 nm. A particular mixture gave $A_{302} = 1.010$ and $A_{355} = 0.730$. Calculate the molarities of both ions in the mixture. Assume $b = 1.0$ cm.

3-10 The solubility of $BaCrO_4$ is to be determined (at 30°C) through the color produced by diphenylcarbazide in a saturated solution. An excess of the solid $BaCrO_4$ is shaken with water in a constant-temperature bath for a long enough time to assure equilibrium. A 10-ml aliquot of the supernatant liquid is transferred to a 25-ml volumetric flask and treated with 1 ml of 5 N H_2SO_4 and 1 ml of a 0.25 percent solution of the reagent and diluted to the mark. The absorbance is then determined through a green filter. A standard solution containing 0.800 ppm of Cr(VI) is also measured. The results are: $A_{std} = 0.440$ and $A_x = 0.200$. Assuming Beer's law to hold, calculate the solubility of $BaCrO_4$ in grams per 100 g of water.

3-11 The following facts are abstracted from a published article (43). (a) Both As and Sb can be oxidized from the trivalent to the pentavalent state by Br_2, As more readily than Sb. (b) Sb(III) forms a complex with Cl$^-$ ion (in 6 N HCl) which absorbs in the ultraviolet at 326 nm, whereas Sb(V), As(III), and As(V) do not. (c) $KBrO_3$ and KBr dissolved together in water form a stable solution that will liberate Br_2 quantitatively upon being added to an acid solution, according to the equation $BrO_3^- + 5Br^- + 6H^+ \rightarrow 3Br_2 + 3H_2O$. (d) Free Br_2, in the presence of excess Br$^-$ ion, absorbs strongly in the ultraviolet, including the vicinity of 326 nm, though its maximum is at a shorter wavelength. Bromide alone shows no such absorption.

   On the basis of the above statements, show how trivalent As and Sb can be determined in mixed solution in HCl by titration with standard $KBrO_3$–KBr reagent.

The absorbance of the solution is to be determined at 326 nm after each addition of reagent. (*Hint:* A curve will result with two breaks corresponding to the two elements sought.)

3-12 Water has been determined by spectrophotometric titration in a solvent consisting of anhydrous acetic and sulfuric acids (44). The reagent is acetic anhydride, $(AcO)_2O$. The titration is followed by the absorption of radiation at 257 nm by the reagent. Sketch a titration curve and explain its shape.

3-13 A method has been reported (45) for the spectrophotometric determination of chlorate impurity in ammonium perchlorate, NH_4ClO_4, for use in rocketry. This is based on the reduction of chlorate to free chlorine: $ClO_3^- + 5Cl^- + 6H^+ \rightarrow 3Cl_2 + 3H_2O$. The chlorine then reacts with benzidine (I) to give a colored product (II) with a maximum absorption at 438 nm.

$$H_2N-\langle O \rangle - \langle O \rangle -NH_2 + Cl_2 \rightarrow$$

$$\text{(I)}$$

$$\left[H_2N = \langle\ \rangle = \langle\ \rangle = NH_2 \right]^{2+} + 2Cl^-$$

$$\text{(II)}$$

Experiments with standard $KClO_3$ solutions showed the following straight-line relation to hold under prescribed experimental conditions (in 10.0-mm cuvets):

$$A = (1.17 \times 10^3)C - 0.186$$

where C is the formal concentration of $KClO_3$. (*a*) Explain why (II) gives a colored solution while (I) does not. (*b*) What color is the solution of (II), and what color filter would be suitable for its determination? (*c*) Does this system obey Beer's law? (*d*) The term "−0.186" was ascribed to a reducing impurity in the reagents. Explain why this would be expected to give a subtractive term. (*e*) A sample of 6.000 g of commercial NH_4ClO_4 was suitably treated with reagent and diluted to 100.0 ml. A portion of this showed an absorbance of 0.450 in a 10.0-mm cuvet at 438 nm. Calculate the concentration of NH_4ClO_3 in the NH_4ClO_4 in terms of mole percent.

3-14 A 12-dynode photomultiplier is found to deliver an average anode current of 0.92 mA over a period of 4.2 ns, as a pulse resulting from a single ejected photoelectron. Calculate the multiplication factor for one dynode.

REFERENCES

1. Judd, D. B.: in M. G. Mellon (ed.), "Analytical Absorption Spectroscopy," chap. 9, Wiley, New York, 1950.
2. Hughes, H. K., et al.: *Anal. Chem.*, **24**:1349 (1952).
3. Liebhafsky, H. A., and H. G. Pfeiffer: *J. Chem. Educ.*, **30**:450 (1953).
4. Houle, M. J., and K. Grossaint: *Anal. Chem.*, **38**:768 (1966).
5. Ringbom, A.: *Z. anal. Chem.*, **115**:332 (1939).
6. Ayres, G. H.: *Anal. Chem.*, **21**:652 (1949).
7. Gilford, S. R., D. E. Gregg, O. W. Shadle, T. B. Ferguson, and L. A. Marzetta: *Rev. Sci. Instrum.*, **24**:696 (1953).

8. Wood, W. A., and S. R. Gilford: *Anal. Biochem.*, **2**:589 (1961).
9. Slavin, W.: *Anal. Chem.*, **35**:561 (1963).
10. Miranda, C., and P. Conte: *Appl. Spectrosc.*, **25**:557 (1971).
11. Cowles, J. C.: *J. Opt. Soc. Am.*, **55**:690 (1965).
12. Chance, B.: *Rev. Sci. Instrum.*, **22**:634 (1951); *Science*, **120**:767 (1954).
13. Porro, T. J.: *Anal. Chem.*, **44**(4):93A (1972).
14. Olson, E. C., and C. D. Alway: *Anal. Chem.*, **32**:370 (1960).
15. Hager, R. N., Jr., D. R. Clarkson, and J. Savory: *Anal. Chem.*, **42**:1813 (1970).
16. Williams, D. T., and R. N. Hager, Jr.: *Appl. Opt.*, **9**:1597 (1970).
17. "RCA Photomultiplier Manual," RCA Corporation, Harrison, N.J., 1970.
18. Malmstadt, H. V., M. L. Franklin, and G. Horlick: *Anal. Chem.*, **44**(8):63A (1972).
19. Silverstein, R. M., and G. C. Bassler: "Spectrometric Identification of Organic Compounds," 2d ed., p. 165, Wiley, New York, 1967.
20. Woodward, R. B.: *J. Am. Chem. Soc.*, **63**:1123 (1941); **64**:72, 76, (1942).
21. Fieser, L. F., and M. Fieser: "Steroids," p. 17 ff., Reinhold, New York, 1959.
22. Turner, A., Jr., and A. Osol: *J. Am. Pharm. Ass., Sci. Ed.*, **38**:158 (1949).
23. Fagel, J. E., Jr., and G. W. Ewing: *J. Am. Chem. Soc.*, **73**:4360 (1951).
24. Archibald, R. M.: *Chem. Eng. News*, **30**:4474 (1952).
25. Yoe, J. H., and A. L. Jones: *Ind. Eng. Chem., Anal. Ed.*, **16**:111 (1944).
26. Vosburgh, W. C., and G. R. Cooper: *J. Am. Chem. Soc.*, **63**:437 (1941).
27. Gerlach, J. L., and R. G. Frazier: *Anal. Chem.*, **30**:1142 (1958).
28. Momoki, K., J. Sekino, H. Sato, and N. Yamaguchi: *Anal. Chem.*, **41**:1286 (1969).
29. Likussar, W., and D. F. Boltz: *Anal. Chem.*, **43**:1265, 1273 (1971).
30. Lingane, P. J., and Z. Z. Hugus, Jr.: *Inorg. Chem.*, **9**:757 (1970).
31. Renzetti, N. A.: *Anal. Chem.*, **29**:869 (1957).
32. Sandell, E. B.: "Colorimetric Determination of Traces of Metals," 3d ed., p. 87 ff., Wiley-Interscience, New York, 1959.
33. Kolthoff, I. M., E. B. Sandell, E. J. Meehan, and S. Bruckenstein: "Quantitative Chemical Analysis," 4th ed., pp. 351, 1064, Macmillan, New York, 1969.
34. Tsubouchi, M.: *Anal. Chem.*, **42**:1087 (1970).
35. Huber, W., G. W. Ewing, and J. Kriger: *J. Am. Chem. Soc.*, **67**:609 (1945).
36. Weissler, A.: *Ind. Eng. Chem., Anal. Ed.*, **17**:695 (1945).
37. Svehla, G.: *Talanta*, **13**:641 (1966).
38. Ingle, J. D., Jr.: *Anal. Chem.*, **45**:861 (1973).
39. Goddu, R. F., and D. N. Hume: *Anal. Chem.*, **26**:1679, 1740 (1954).
40. Underwood, A. L.: In C. N. Reilley (ed.), "Advances in Analytical Chemistry and Instrumentation," vol. 3, p. 31 ff., Wiley-Interscience, New York, 1964.
41. Underwood, A. L.: *Anal. Chem.*, **26**:1322 (1954).
42. Wetters, J. H., and K. L. Uglam: *Anal. Chem.*, **42**:335 (1970).
43. Sweetser, P. B., and C. E. Bricker: *Anal. Chem.*, **24**:1107 (1952).
44. Bruckenstein, S.: *Anal. Chem.*, **31**:1757 (1959).
45. Burns, E. A.: *Anal. Chem.*, **32**:1800 (1960).

Chapter 4

Molecular Luminescence: Fluorometry and Phosphorimetry

It was pointed out in connection with Fig. 2-2 that both fluorescence and phosphorescence constitute possible mechanisms whereby electronically excited molecules can lose energy. During the process of excitation, most of the affected molecules acquire vibrational as well as electronic energy. The greatest tendency is for them to drop to lower vibrational states through collisions. If this radiationless process stops at an excited singlet electronic level, the molecules will be able to return directly from there to the ground state by the radiation of a photon (fluorescence); less commonly they may shift to a metastable triplet level before emitting radiation (phosphorescence).

In either case the molecule may end up in any of the vibrational states of the ground level, and so fluorescence and phosphorescence spectra generally consist of many lines, mostly in the visible region. In the presence of a solvent the lines are broadened and fuse together more or less, to give a structured spectrum of the same general appearance as an ultraviolet or visible absorption spectrum.

Fluorescence is more widely useful in analysis than is phosphorescence, so we will consider it first and in more detail.

In studies of fluorescence or phosphorescence, one is generally concerned with three types of spectra: *(1)* the absorption spectrum, *(2)* the excitation spectrum, and *(3)* the emission spectrum. The excitation spectrum is recorded by scanning the wavelength of the primary radiation while monitoring the emitted (luminescent) radiation. The curve so obtained is usually nearly identical to the conventional absorption spectrum. This is consistent with the scheme presented in Fig. 2-2, which shows that energy absorbed by excitation of electrons into higher levels is degraded as heat, emission only resulting from electrons leaving the *lowest* excited state. To ensure complete identity of absorption and excitation spectra, certain corrections must be applied; these will be discussed later in this chapter.

FLUOROMETRY

The shortwave portion of a fluorescence spectrum usually overlaps at least slightly with the long-wave end of the absorption spectrum that gives rise to the excitation. Figure 4-1 shows the reason for this. If all molecules are originally in the ground state, the least energy that can be absorbed in process I is equal in magnitude to the greatest energy transition

Fig. 4-1 Transitions involved in molecular fluorescence. This is a detailed view of a portion of Fig. 2-2. The transitions marked *(a)* are responsible for the overlap of excitation and fluorescence spectra.

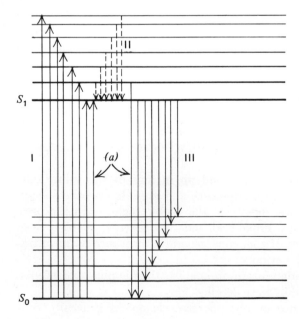

possible in fluorescence (process III) and, since the vibrational levels in the ground and excited states have nearly the same spacing, the fluorescence spectrum is roughly the *mirror image* of the absorption spectrum. Overlap can occur only if a significant proportion of molecules is originally in a vibrationally excited state, so that less energy will be needed to raise them to the lowest excited level S_1. Figure 4-2 shows a comparison of the absorption and fluorescent spectra of anthracene, in which the mirror image effect is evident. For further treatment of the theoretical aspects of fluorescence and phosphorescence, the reader is referred to the literature (1–3).

Since fluorescent radiation originates in the sample, it is emitted equally in all directions, and so can in principle be observed from any angle. In practice, three different geometries are used (Fig. 4-3). A 90° angle (*a*) is the most convenient from a design standpoint, and is selected for all less expensive fluorometers. Observation at a small angle (*b*) is advantageous if the solution is so concentrated that most of the absorption and hence most of the fluorescent generation takes place close to the irradiated surface. This configuration is also desirable if the solution absorbs appreciably at the wavelengths of fluorescence, because the emitted radiation need traverse only a minimal thickness of solution. The in-line construction of (*c*) has advantages if theoretical deductions (the quantum

Fig. 4-2 Spectra of anthracene. (*a*) Fluorescence emission; (*b*) excitation. Note that the mirror-image relationship does not extend to the absorption maximum at 250 nm. [*Research and Development* (4).]

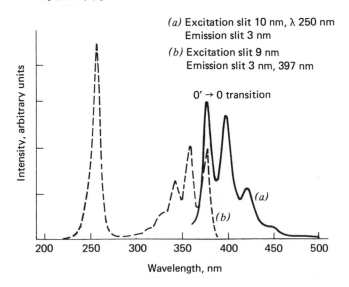

(a) Excitation slit 10 nm, λ 250 nm
Emission slit 3 nm

(b) Excitation slit 9 nm
Emission slit 3 nm, 397 nm

0′ → 0 transition

Intensity, arbitrary units

200   250   300   350   400   450   500

Wavelength, nm

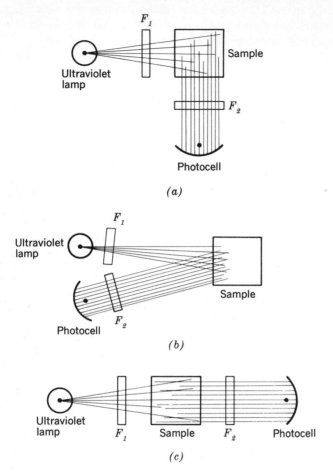

Fig. 4-3 Alternative fluorometer configurations. Observation (*a*) at 90°, (*b*) at a small angle, and (*c*) in a straight line. F_1 and F_2 are primary and secondary filters, either or both of which may be replaced by monochromators.

yield, for example) are to be drawn from the observations, as the equations involve fewer approximations (5).

The mathematical treatment of fluorescence is more complicated than the mathematics of simple absorption leading to Beer's law. For one thing, the amount of primary radiation absorbed varies exponentially throughout the body of the solution, in accordance with the absorption law. For another, the fluorescent light is always subject to at least slight absorption in the solution, and the thickness of solution through which it passes is not constant, because the radiation does not originate at a single point.

The complete equations are much too complex to include in the present discussion, but may be found in the literature.

The simplest case is that in which only a single absorbing and fluorescing species is present in a dilute solution. For this case, the governing equation is

$$F = P_0 K (1 - 10^{-A}) \qquad\qquad (4\text{-}1)$$

where F is the power of fluorescent light reaching the detector, and P_0 is the power of the incident ultraviolet radiation. K, a constant for a given system and instrument, is the factor for converting absorbed power to the fraction of unabsorbed fluorescence reaching the detector. A is the absorbance of the solution at the primary wavelength.

This expression can be transformed by application of an exponential series to a more useful form:

$$F = P_0 K \left[2.30A - \frac{(2.30A)^2}{2!} + \frac{(2.30A)^3}{3!} - \cdots \right] \qquad\qquad (4\text{-}2)$$

This means that at low concentrations, where the squared and higher terms become negligible, F is linear with A, and thus with the concentration C:

$$F = 2.30 P_0 K A = P_0 K' abc \qquad\qquad (4\text{-}3)$$

where $K' = 2.30K$, a is the absorptivity at the primary wavelength, and b is the path length in the primary direction. At higher concentrations, not only does Eq. (4-2) predict nonlinearity, but the factor K may no longer be constant. A full discussion of the origin of these nonlinearities may be found in a series of papers by Fletcher (5). For practical analytical purposes, their effect can be avoided only through use of a calibration curve prepared from known solutions. Figure 4-4 shows the fluorescence of phenol in water as a function of concentration; it is seen to be linear only to about 10 $\mu g \cdot ml^{-1}$ (6).

Two important conclusions can be drawn from Eq. (4-3). First, since this equation is valid only for small values of A (less than about 0.01, corresponding to approximately 98 percent transmission), it gives an indication of why fluorescence is most useful at a much lower concentration than is absorptiometry. In observing the fluorescence of a dilute solution, the photocell sees a faint light against a dark background, whereas in observing the absorbance of the same solution, it is necessary to measure the removal of a very small fraction of the radiation, which amounts to a precision measurement of the difference between two large numbers. Second, P_0 is greater for primary radiation selected by a filter than for that selected by a monochromator (for a given source), because of the

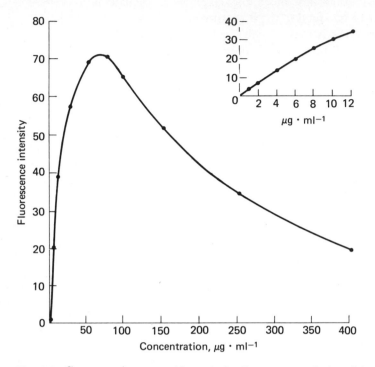

Fig. 4-4 Concentration quenching of the fluorescence of phenol in water. The large graph covers the concentration range 0 to 400 $\mu g \cdot ml^{-1}$, and the inset 0 to 12 $\mu g \cdot ml^{-1}$. Excitation, 295 nm; fluorescence, 330 nm; pH, 6.5. [*Journal of Clinical Pathology* (6).]

wider wavelength band allowed to pass. This accounts for the greater sensitivity of a filter fluorometer as compared to a spectrofluorometer.

FILTER FLUOROMETERS

The basic arrangement for a single-beam filter fluorometer is essentially that of Fig. 4-3a. The primary filter F_1 allows ultraviolet to pass, but cuts out the visible radiation produced by the lamp. The secondary filter F_2 transmits the visible fluorescent radiation, but absorbs any ultraviolet that might be scattered toward the photocell.

As in any single-beam photometer, the voltage supplies for both lamp and phototube must be stabilized to preserve constant sensitivity during an analysis. The dual-beam principle can be introduced in either of two ways: the reference photocell may monitor the ultraviolet lamp directly, or it may receive fluorescent light from a standard, sometimes called a *fluorescence generator* (Fig. 4-5). In both arrangements the effect of variations in brilliance of the lamp resulting from line-voltage fluctuations

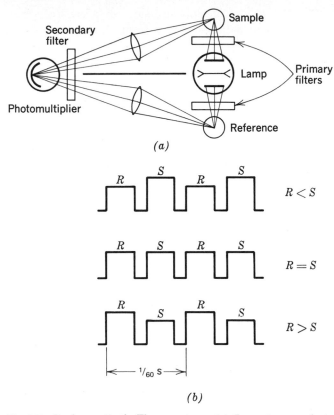

Fig. 4-5　Beckman Ratio Fluorometer. (*a*) Layout, somewhat simplified; (*b*) waveforms, where R and S represent, respectively, the powers in the reference and sample beams. (*Beckman Instruments, Inc.*)

is eliminated. The use of a fluorescent standard is to be preferred, however, as it also minimizes the effect of temperature changes and variations in phototube sensitivity and linearity to be expected at widely differing wavelengths.

It is most desirable to employ as a reference a standard solution of the substance sought, but sufficient purification to remove fluorescent impurities may be a major difficulty. Secondary standards of uranium glass are probably the best substitute; they can be obtained in a series of graded fluorescent power and are stable indefinitely. Acidic solutions of quinine sulfate are highly fluorescent and are often used as secondary standards. Quinine is, however, less than ideal for this purpose, and its use should be discouraged (7, 8).

An example of a double-beam filter fluorometer is the Beckman Ratio Fluorometer (Fig. 4-5). This instrument makes use of a specially designed mercury-vapor lamp with two anodes on opposite sides of a central structure that combines a symmetrical cathode with a light shield. The lamp operates on alternating current so connected that the two anodes receive the discharge, and hence produce the radiation, on alternate half-cycles of the exciting voltage. This has the effect of a chopper, so that the sample and standard receive identical radiation pulsed at 60 Hz and separated by a short dark interval. The two beams pass through primary filters, and the fluorescent radiation from the two cuvets converges through a common secondary filter onto a single photomultiplier which therefore sees a series of pulses (Fig. 4-5b). The electronic circuits sort out these pulses and utilize the magnitude of the reference pulses to adjust the dynode voltages on the photomultiplier so as to maintain a constant reference level, not necessarily 100 percent. Then the sample pulses will produce a meter deflection which can be made direct-reading in concentration units, as long as concentrations are within the linear region.

SPECTROFLUOROMETERS

These instruments fall into two classes: those consisting of a fluorescence attachment for a spectrophotometer, and those that are self-contained instruments, usually with two monochromators. The attachments provide for illumination of the sample at a 90° angle with ultraviolet from a mercury or xenon lamp via a filter or small, large-aperture monochromator. The fluorescent light is analyzed by the spectrophotometer, which may be scanned either manually or automatically. Even though the spectrophotometer may be of a double-beam type, it can only function in a single-beam mode when used as a fluorometer.

A spectrofluorometer designed as such should be expected to be more efficiently optimized for its intended function than an adapted spectrophotometer. Ideally such an instrument should be double-beam and ratio-recording, it should be capable of scanning either the primary or the secondary radiation at the choice of the operator, it should have an observation angle (as in Fig. 4-3) adjustable from 180° to as near 0° as practicable, it should have automatic control of the primary beam to maintain constant incident energy on the sample, and provision should be made for regulation of the temperature of both sample and standard.

Figure 4-6 shows the optical diagram of the Aminco-Bowman Spectrophotofluorometer, a single-beam instrument incorporating two Czerny-Turner grating monochromators with a 90° sample geometry. This is one of the more widely distributed instruments of its class.

Operation of any spectrofluorometer usually follows some such procedure as this: By rough preliminary observations, a suitable wavelength in the emission spectrum can be chosen, and the second monochromator set at this point. An excitation spectrum can then be plotted by scanning the first monochromator. Similarly an emission spectrum can be obtained by scanning the second monochromator with the first set at a suitable value.

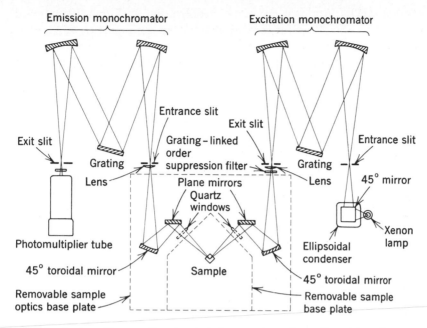

Fig. 4-6   Aminco-Bowman Spectrophotofluorometer, schematic.   (*American Instrument Company.*)

With most substances neither spectrum is found to vary greatly by a change in the wavelength selected for the *other* monochromator, as long as it is at a point of fairly high sensitivity.

Spectrofluorometers are useful for establishing conditions for analysis and for studying interferences, as well as for carrying out actual analyses. For the best precision in the latter function, results should be compared with standards run on the same instrument. It is found that supposedly comparable spectra measured on different instruments often give varying results, primarily as a result of the variations with wavelength of the output of the light sources, the absolute sensitivities of the detectors, and the efficiencies of the monochromators. The observed spectra represent some combination of the spectral properties of the substance itself with artifacts generated by the instrument. Such variations are particularly disturbing when it is desired to measure the quantum efficiency of the fluorescence process.

It is possible to calibrate a spectrofluorometer by painstaking comparisons involving a standard light source and thermocouple detector or precalibrated photomultiplier tube. The corrections so derived may be of considerable magnitude—a factor varying from less than unity to as high as 50 or more (9, 10). Much of the tedium of such methods can be eliminated by the use of a fluorescent material as a standard of comparison. The aluminum chelate of the azo dye Pontachrome blue-black R has been recommended (11). Application of the calibration techniques described in these references should go far toward eliminating the reporting of "uncorrected" fluorescence spectra.

Some spectrofluorometers have available attachments to permit on-line correction of fluorescence spectra. One instrument, the Turner Model 210, incorporates a true double-beam system, and also gives corrected spectra directly.

QUENCHING

This is the name given to any reduction in the intensity of fluorescence (or phosphorescence) due to specific effects of constituents of the solution itself. Quenching may occur as the result of excessive absorption of either primary or fluorescent radiation by the solution; this is called *concentration quenching,* or the *inner-filter* effect. If the effect is produced by the fluorescent substance itself, it is known as *self-quenching;* the decline in fluorescence of phenol with concentration, shown in Fig. 4-4, is an example.

Quenching can also be caused by nonradiative loss of energy from the excited molecules. The quenching agent may facilitate conversion of the molecules from the excited singlet to a triplet level from which emission cannot occur. The quenching of many aromatic compounds by dissolved oxygen is thought to follow this mechanism. Electron transfer from or to the excited molecule can cause fluorescence quenching in some systems, such as the fluorescence of methylene blue, which is quenched by ferrous ion.

Chemical quenching is a term sometimes applied to the reduction in emission due to actual changes in the chemical nature of the fluorescent substance. A common observation is pH sensitivity. Aniline, for example, shows blue fluorescence in the range pH 5 to 13, when excited at 290 nm. At lower pH, aniline exists as the anilinium cation, and in highly alkaline media as the anion; neither ion is fluorescent.

The amount of fluorescence of many compounds is dependent on the temperature to a greater degree than is usual in absorption spectroscopy. A hazard in fluorometry is the ease of contamination with fluorescent substances in amounts undetectable by most other means. Even water and aqueous solutions can extract fluorescent substances from rubber and plastic laboratory ware, even from Bakelite bottle caps (12). Significant amounts of quenchers can sometimes be picked up from similar sources; traces of chromium from chromic acid cleaning mixtures may offer a case in point.

APPLICATIONS

In its applications, fluorometry is noted for its great sensitivity, often 10^2 to 10^4 times greater than comparable absorptiometric techniques. It is also more selective, since fewer compounds fluoresce than merely absorb radiation.

Visible fluorescence occurs principally in two classes of substances: (*1*) a large variety of minerals and inorganic solid-state *phosphors* and (*2*) organic and organometallic compounds with extensive ultraviolet absorption.

In the first of these classes, we will mention only a method for the determination of uranium salts that has important application in nuclear technology (13). The sample is oxidized by evaporation with nitric acid and then fused with sodium fluoride to a melt containing the fluorides of sodium and uranium; this solidifies to a glass on cooling. The glass is examined directly in a specially designed fluorometer. The sensitivity is of the order of 5×10^{-9} g of uranium in a 1-g solid sample.

An example of a fluorometric analysis in inorganic chemistry is the determination of ruthenium in the presence of other platinum metals (14). The complex ion of ruthenium(II) with 5-methyl-1,10-phenanthroline fluoresces strongly at pH 6. Any other element of the platinum group can be present to the extent of at least 30 μg·ml^{-1} without interfering with the determination of ruthenium in the range 0.3 to 2.0 μg · ml^{-1}. The precision with the instrument reported was about ± 2 percent. Palladium forms a precipitate with the reagent, which can be removed by centrifuging. Iron must not be present, as it forms a complex that quenches the fluorescence. Figure 4-7 shows the excitation and fluorescence spectra of the ruthenium complex, together with the transmission spectra of a pair of Corning glass filters appropriate for a fluorometer with right-angle observation. Interference filters are available that permit in-line fluorometry.

Fig. 4-7 Tris-(5-methyl-1,10-phenanthroline)–ruthenium(II) complex ion. (*A*) Uncorrected excitation spectrum; (*B*) fluorescent emission spectrum corresponding to excitation at 450 nm; (*C*) Corning filter no. 5-59; (*D*) Corning filter no. 3-68. The vertical scale represents relative radiant power for *A* and *B* and percent transmittance for *C* and *D*. [*Redrawn from Analytical Chemistry (14) and literature of Corning Glass Works.*]

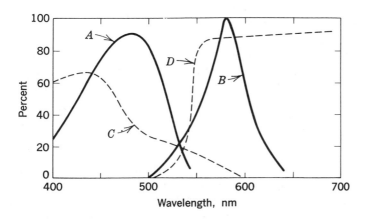

As another example, the determination of vanadium by reaction with benzoic acid in an acetate buffer at pH 2 may be mentioned. Fluorescence, excited at 300 nm, appears at 410 nm, and is linear over the range 0.5 to 400 ppb. Only iron and large amounts of titanium were found to interfere (15).

Most organic applications are in the determination of polycyclic molecules, including a great many substances of biochemical and pharmacological significance. The following paragraphs outline two such analytical methods.

Thiamine (vitamin B_1) is commonly assayed by the blue fluorescence of its oxidation product, thiochrome (16). The sample is treated with phosphatase, an enzyme that causes hydrolysis of the phosphate esters of thiamine, frequently present in foodstuffs. The phosphatase and other insoluble matter is filtered off, and the filtrate diluted to a known volume. Two aliquots are taken, one for analysis and one for a blank. To the first is added an oxidizing agent (ferricyanide), and to both equal quantities of NaOH and isobutyl alcohol. After shaking and removal of the aqueous layer, the alcoholic solution is examined in the fluorometer.

Riboflavin can also be determined fluorometrically (16). The fluorescent power is dependent to a large degree on the exact conditions and on the nature and amount of the impurities. In order to make certain that any impurities have the same effect on standard and unknown, the method of standard increment is adopted. The fluorescence of a portion of the standard is measured in the same solution with the unknown. The procedure also takes advantage of the fact that riboflavin can readily be oxidized to a nonfluorescing substance that is in turn easily reduced to regenerate the vitamin quantitatively.

A different sort of application of spectrofluorometry lies in the determination of ultraviolet absorption spectra in very small or very dilute samples. This takes advantage of the virtual identity of corrected fluorescence excitation spectra and conventional absorption spectra. The excitation is likely to show less fine structure but, on the other hand, is more selective than absorption spectra, because many contaminants do not fluoresce. If only uncorrected fluorescence spectra can be obtained, this method must be used with caution, to avoid drastic alteration of the spectra caused by the various types of quenching. Duplicate spectra should be obtained at several concentration levels; if these are identical, concentration quenching is probably absent. In favorable cases, absorption spectra can be obtained on nanogram to microgram quantities of fluorescent compounds.

PHOSPHORIMETRY

Phosphorescence is distinguished operationally from fluorescence, first by its relatively long decay time (milliseconds to seconds) and second by a shift toward longer wavelengths. An example showing the relation be-

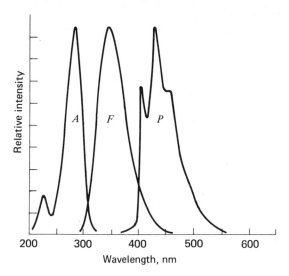

Fig. 4-8 Spectra of tryptophan. (*A*) Absorption; (*F*) fluorescence; (*P*) phosphorescence. [*Academic Press* (17).]

tween phosphorescence, fluorescence, and excitation spectra is given in Fig. 4-8 (17).

Few organic compounds show appreciable phosphorescence in solution at room temperature, but many more display the phenomenon if cooled with liquid nitrogen. The reduction in temperature appears to enhance the probability of transitions from excited singlet to metastable triplet states, the required condition for phosphorescence, and also to diminish competing mechanisms for nonradiative return to the ground state.

A report has recently appeared (18) indicating that valid phosphorescence spectra may be obtained at room temperature by evaporating a drop of solution on a strip of filter paper and examining the paper directly in a phosphorimeter. It will be interesting to see whether this development will result in a marked increase in the use of phosphorescence as an analytical tool.

Compounds that phosphoresce invariably fluoresce as well, and a phosphorimeter must be able to distinguish between the two. This can be done by means of a rotating shutter, called a *phosphoroscope* (Fig. 4-9). By this device a definite delay is introduced between the time intervals during which the sample is irradiated and observed. Several commercial instruments are available working on this principle. Theoretical aspects of optimum shutter design have been discussed in an interesting paper by O'Haver and Winefordner (19).

The rotating shutter can be eliminated by the use of pulse techniques (20). A flash tube is used as the source of primary radiation. An elec-

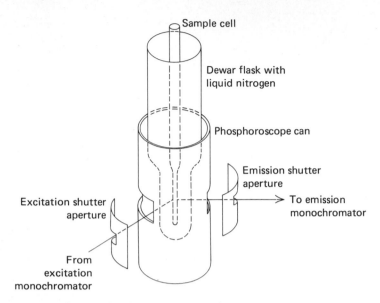

Fig. 4-9 Schematic diagram of a rotating-can phosphoroscope. [*Accounts of Chemical Research* (22).]

tronic timing circuit triggers the flash and at a variable time later, provides a high-voltage negative pulse to the photomultiplier, activating it for a predetermined interval. With this system, decay times as short as 10^{-5} s can be measured, a speed only attainable with difficulty with mechanical shutters.

A large number of organic compounds with conjugated ring systems phosphoresce intensely, and phosphorimetry provides excellent opportunities for their trace analysis. The subject has been discussed in some detail by Winefordner and Latz (21). They developed a simple, fast procedure for the determination of aspirin in blood serum and plasma, in which the presence of free salicylic acid does not interfere.

An additional technique possible with phosphorimetry is time resolution (22). This permits multicomponent analysis, provided the components have sufficiently different decay times. The time resolution can be achieved merely by changing the delay between activation and observation (i.e., changing the speed of the phosphoroscope), or more powerfully by logarithmic subtraction (23). For two components A and B the emitted power is

$$P_{\text{ph}} = P_{\text{A}} + P_{\text{B}} = P_{0,\text{A}} \exp \frac{-t}{\tau_{\text{A}}} + P_{0,\text{B}} \exp \frac{-t}{\tau_{\text{B}}} \qquad (4\text{-}4)$$

where τ, the decay time, is the time required for the phosphorescence of a substance to diminish to $1/e = 36.8$ percent of its initial value. If the values of P_{ph} are recorded as a function of time until both emissions have disappeared, it is possible to determine the amount of the species with the longer τ from the data taken after the radiation from the other component has died away. Then its contribution can be subtracted out, leaving a record of the emission from the shorter-lived species.

The best solvents to use in phosphorimetry are those which solidify to glasses at liquid-nitrogen temperatures, rather than freezing to a crystalline state. This eliminates the tendency toward segregation of the solute, giving solid solutions instead. A particularly useful solvent is EPA, a mixture of ethyl ether, isopentane, and ethanol in the volume ratio of 5:5:2.

PROBLEMS

4-1 A 2-g sample of pork is to be analyzed for its vitamin B_1 content by the thiachrome method. It is extracted with HCl, treated with phosphatase, and diluted to 100 ml. An aliquot of 15 ml is purified by adsorption and elution, during which process it is diluted to 25 ml. Of this, two 5-ml portions are taken, one of them is treated with $Fe(CN)_6^{-3}$, and both are made up to 10 ml for fluorometric examination. A standard solution of thiamine containing 0.2 $\mu g \cdot ml^{-1}$ is subjected to similar treatment, except that the portion introduced into the adsorption column is made up to its original volume after elution (i.e., not diluted). Two 5-ml aliquots are taken, one is oxidized, and both are made up to a final volume of 10 ml for measurement of fluorescence. The following observations are recorded:

Solution	Relative fluorescent power
A (standard, oxidized)	62.4
B (standard, blank)	7.0
C (sample, oxidized)	52.0
D (sample, blank)	8.0

Calculate the vitamin B_1 content of the pork in terms of micrograms per gram.

4-2 It has been shown (24) that the Hantzsch reaction can be used in a very sensitive analysis for either aldehydes or primary amines, including ammonia. In the presence of 2,4-pentanedione (acetylacetone) a cyclic compound is formed at pH 6 that is strongly fluorescent. The reaction is:

(*a*) If ammonia is in excess, the fluorescent power, after subtraction of a blank value, is nearly linear with concentration of formaldehyde from 0.005 to 1.0 $\mu g \cdot ml^{-1}$. What is the approximate linear range for the determination of traces of aniline with excess formaldehyde present? (*b*) The absorption spectrum of the fluorescent compound shows a fairly intense maximum at λ 410 nm (ϵ = 8000), too great to be accounted for by two unrelated 2,3-unsaturated ketone chromophores. Can you offer a possible explanation?

4-3 In the determination with a filter fluorometer of 7H-benz(*de*)anthracen-7-one (BO) as an air pollutant, considerable interference is encountered from fluorescent polynuclear hydrocarbons (25). The latter fluorescence is more readily quenched by *m*-dinitrobenzene (DNB), a fact that makes the analysis possible. (*a*) Addition of 15 percent DNB to the solution reduces the fluorescent power F of 5×10^{-6} *M* BO from 0.7 to 0.4 (relative units). If this amount of DNB diminishes the interfering materials fluorescence to 0.005 of its original value, how great an improvement in sensitivity would result from the use of the quenching agent? (*b*) Would you expect that the quencher could be dispensed with if the measurement were made with a spectrofluorometer?

4-4 The absorption spectrum of a fluorescent substance determined with a conventional spectrophotometer (e.g., those of Fig. 3-11 or 3-13) is likely to be in error because of the fluorescence. Suppose a compound has an absorption maximum (hence an excitation maximum) at 290 nm and a fluorescence maximum at 350 nm (e.g., tryptophan, Fig. 4-8). At what wavelength would you expect the greatest error? Would the observed absorbance be too large or too small at this point? Would your answers be different for a spectrophotometer in which the radiation from the lamp passes through the cuvet *before* dispersion in the monochromator? Explain.

REFERENCES

1. Conrad, A. L.: In I. M. Kolthoff and P. J. Elving (eds.), "Treatise on Analytical Chemistry," pt. I, vol. 5, chap. 59, Wiley-Interscience, New York, 1964.
2. West, W.: In W. West (ed.), "Chemical Applications of Spectroscopy," chap. 6, Wiley-Interscience, New York, 1956.
3. Hercules, D. M. (ed.): "Fluorescence and Phosphorescence Analysis," Wiley-Interscience, New York, 1966.
4. Smith, H. F.: *Res. Dev.*, July, 1968, p. 20.
5. Fletcher, M. H.: *Anal. Chem.*, **35**:278, 288 (1963); R. G. Milkey and M. H. Fletcher: *J. Am. Chem. Soc.*, **79**:5425 (1957).
6. Williams, R. T., and J. W. Bridges: *J. Clin. Pathol.*, **17**:371 (1964).
7. Sawicki, E.: *Talanta*, **10**:1231 (1964).
8. Chen, R. F.: *Anal. Biochem.*, **19**:374 (1967).
9. White, C. E., M. Ho, and E. Q. Weimer: *Anal. Chem.*, **32**:438 (1960).
10. Demas, J. N., and G. A. Crosby: *J. Phys. Chem.*, **75**:991 (1971).
11. Argauer, R. J., and C. E. White: *Anal. Chem.*, **36**:368 (correction, p. 1022) (1964).
12. Kordan, H. A.: *Science*, **149**:1382 (1965).
13. Byrne, J. T.: *Anal. Chem.*, **29**:1408 (1957).
14. Veening, H., and W. W. Brandt: *Anal. Chem.*, **32**: 1426 (1960).
15. Koh, K. J., and D. E. Ryan: *Anal. Chim. Acta*, **57**:295 (1971).
16. Association of Vitamin Chemists, Inc.: "Methods of Vitamin Assay," 3d ed., Wiley-Interscience, New York, 1966.
17. Udenfriend, S.: "Fluorescence Assay in Biology and Medicine," vol. 2, Academic, New York, 1969.

18. Paynter, R. A., S. L. Wellons, and J. D. Winefordner: *Anal. Chem.,* **46:**736 (1974).
19. O'Haver, T. C., and J. D. Winefordner: *Anal. Chem.,* **38:**602 (1966).
20. O'Haver, T. C., and J. D. Winefordner: *Anal. Chem.,* **38:**1258 (1966).
21. Winefordner, J. D., and H. W. Latz: *Anal. Chem.,* **35:**1517 (1963).
22. Winefordner, J. D.: *Acct. Chem. Res.,* **2:**361 (1969).
23. St. John, P. A., and J. D. Winefordner: *Anal Chem.,* **39:**500 (1967).
24. Belman, S.: *Anal. Chim. Acta,* **29:**120 (1963).
25. Sawicki, E., H. Johnson, and M. Morgan: *Mikrochim. Acta,* **1967:**297.

Chapter 5

The Absorption of Radiation: Infrared

Whereas the absorption of ultraviolet and visible radiation is conveniently considered a unit, the infrared region is better treated separately. There are two important reasons for this: first, the optical techniques are sufficiently divergent that no spectrophotometers are available to cover both the infrared and the visible-ultraviolet regions without modification; second, the absorption of infrared rests on a different physical mechanism than does that of the shorter wavelengths.

It was pointed out in Chap. 2 that the absorption of infrared radiation depends on increasing the energy of vibration or rotation associated with a covalent bond, provided that such an increase results in a change in the dipole moment of the molecule. This means that nearly all molecules containing covalent bonds will show some degree of selective absorption in the infrared. The only exceptions are diatomic elements such as H_2, N_2, and O_2, because only in these can no mode of vibration or rotation be found that will produce a dipole moment. Even these simple species show slight infrared absorption at high pressures, apparently as a result of distortions during collisions.

Infrared spectra of polyatomic covalent compounds are often exceedingly complex, consisting of numerous narrow absorption bands. This contrasts strongly with the usual visual and ultraviolet spectra. The difference arises in the nature of the interaction between the absorbing molecules and their environment. This interaction (in condensed phases) has a great effect on electronic transitions occurring within a chromophore, broadening the absorption lines so that they tend to coalesce into wide regions of absorption. In the infrared, on the other hand, the frequency and absorptivity due to a particular bond usually show only minor alterations with changes in its environment (which includes the rest of the molecule). The lines are not likely to be broadened so as to coalesce.

Exceptions to this generalization sometimes occur. For example, a long-chain molecule in the liquid phase is free to assume a limitless number of configurations, because of free rotation about the many C—C bonds. The spectra of these many forms will be nearly but not quite identical, so that a broadening of the bands will appear. Hence it is preferable to examine compounds of this type in the solid phase.

A typical infrared absorption spectrum, that of a film of polystyrene polymer, is reproduced in Fig. 5-1. Figure 5-2 shows the change in appearance of the spectrum of a long-chain compound, stearic acid, as it appears (a) in solution, (b) as a solid film at room temperature, and (c) at liquid-nitrogen temperature (1). The concentration of the solution and thickness of the film were adjusted to give convenient traces on the recorder. Notice that by comparison with the room-temperature solid, the solution shows considerable broadening of peaks and loss of detail through much of the range, while a drastic reduction in temperature has the opposite effect.

Infrared spectra are usually plotted as percent transmittance, as in Fig. 5-1, rather than as absorbance. This makes absorption bands appear as dips in the curve rather than as maxima, as is conventional in ultraviolet and visible spectra. This method of plotting is not universally employed, however, and Fig. 5-2 shows an inverted format. The independent variable in infrared spectrograms is sometimes the wavelength in micrometres (formerly called microns), and sometimes wave number in reciprocal centimetres. Many workers strongly favor the wave number treatment as being more easily correlated with the vibrations within the molecule. On the other hand, some prefer linear wavelength presentation because in the NaCl region the highly detailed fingerprint portion (about 5 to 15 μm) is spread out conveniently rather than bunched up at the short wavelength end. The two presentations are compared in Fig. 5-1.

INSTRUMENTATION

The majority of infrared spectrophotometers follow the same basic designs as those for use at shorter wavelengths, modified to allow for the differences in optical materials. There are in addition two other types without ultraviolet-visible counterparts, based on *multiplex* methods.

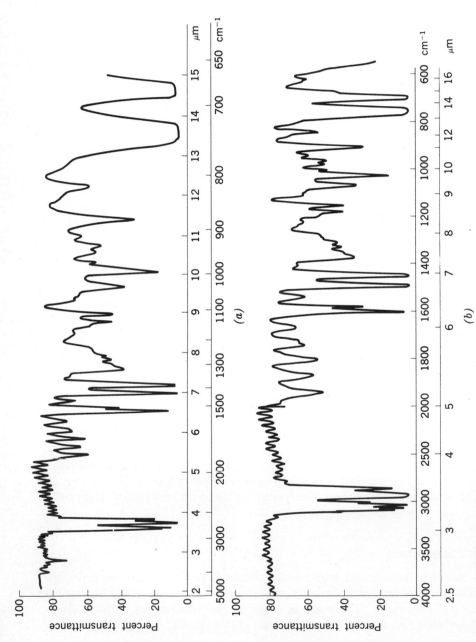

Fig. 5-1 The infrared spectrum of a polystyrene film. (a) On a linear wavelength scale, and (b) linear in wave number, with a scale change at 2000 cm^{-1}. (*Perkin-Elmer Corporation.*)

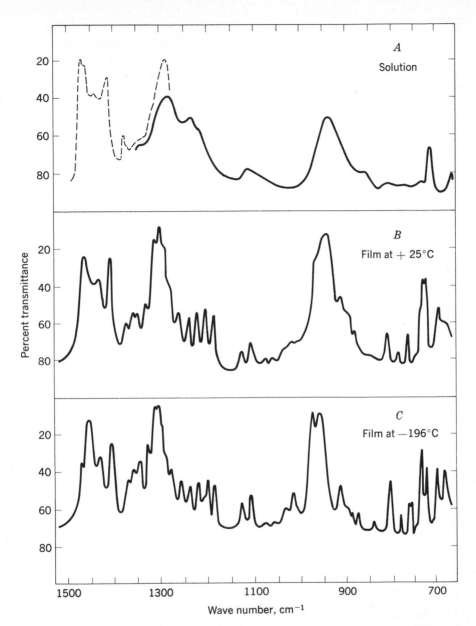

Fig. 5-2 Infrared spectra of stearic acid in different physical states. (*a*) In CCl₄ solution (dashed curve) and in CS₂ solution (solid curve); (*b*) film of the beta polymorph at room temperature; (*c*) at −196°C. [*Wiley* (1).]

OPTICAL MATERIALS

Lenses are almost never used in infrared spectrophotometers, but are replaced by front-surface concave mirrors. Mirrors have many advantages: They have no chromatic aberration, they can be made of sturdy materials such as metal or aluminized glass, without regard to optical transmission, and they are easier to mount. Radiation windows, such as the protective cover over a detector, sample cells,* and dispersive prisms, must be made of solid materials that transmit freely in the wavelength band of interest. Several materials with useful transmission are listed in Table 5-1.

It will be noticed that several of the entries in the table are water-soluble salts. This means that parts made of these materials must be carefully protected from moisture. If a prism is used for dispersion, it should be housed in an airtight compartment provided with a desiccant or other means for excluding moisture. Clearly, aqueous solutions cannot be handled in cells with salt windows, but this is no great restriction in practice, as water absorbs infrared too strongly to be of much use as a solvent. Salt cells should be stored in a desiccator.

* In the infrared region a cuvet is generally called an absorption cell.

Table 5-1   Infrared-transmitting solid materials

Material	Limiting wavelengths,* μm	Prisms	
		Beckman†	Perkin-Elmer‡
Optical glass	0.4–2.6		
Vitreous silica	0.16–4.0	. . . .	0.17–3.5
LiF	0.12–9.0	1.0–6.0	0.5–6.5
CaF₂	0.13–12.0	0.5–9.0	0.5–9.5
Si§	1.2–15		
Ge	1.8–23		
NaCl	0.2–25	1.0–16.0	1.0–15.5
KBr	0.25–40	10.0–25.0	11.0–25.0
CsBr	0.2–55	11.0–35.0	15.0–38.0
KRS-5¶	0.55–40		

* Taken from more extensive tabulations given by Barker (2). Figures indicate regions where transmission is greater than 10 percent for a 2-mm thickness.
† From literature of Beckman Instruments, Inc., referring to the interchanges for their IR-4 prism spectrophotometer.
‡ From literature of the Perkin-Elmer Corporation, referring to their Model 221, except for vitreous silica, which is for Model 450 with a specially selected prism.
§ Shows a transmission dip near 9 μm, due to Si—O bonds on the surface.
¶ Mixed crystal of thallium bromide and iodide.

SOURCES

In the part of the infrared region of greatest analytical importance, the *Nernst glower* is the most widely used spectrophotometric source. It consists of a rod or hollow tube about 2 cm long and 1 mm in diameter, made by sintering together oxides of such elements as cerium, zirconium, thorium, and yttrium. It is maintained at a high temperature by electrical heating, and can be operated in air, since it is not subject to oxidation. Another source, the *Globar*, is a rod of silicon carbide of somewhat larger dimensions than the Nernst glower. Although operated at lower temperatures (to avoid oxidation), it must be cooled by forced ventilation because of the tendency to overheat at the terminals. It has greater emissivity than the glower at wavelengths beyond about 30 μm. A simple coil of *Nichrome* wire can be used as an infrared source if the required wavelength range and intensity are not too great. For the near infrared a tungsten filament lamp is satisfactory.

DETECTORS

Infrared detectors (3) suitable for spectrophotometry can be classified in two broad groups: (*1*) those designated as thermal detectors, which depend on the integrated energy of a large number of incident photons to produce a measurable response via their heating effects; and (*2*) those in which an electron can be caused to move to a higher level by the energy of a single photon of infrared radiation.

In general photon detectors are faster and more sensitive, but severely restricted with respect to the wavelength ranges to which they can respond. Thermal detectors, on the other hand, are usable over a wide range of wavelengths, but have relatively low sensitivity and slow response.

The *thermocouple* is the most widely used infrared detector. It can be fabricated in a number of ways, usually with a tiny bit of blackened gold foil as the actual absorber of radiation. In one common type, the foil is welded to very fine wires of two metals of widely different thermoelectric power, which provide mechanical support as well as electrical connections. The thin wires connect to heavier supporting wires. The assembly is mounted in an evacuated housing (with a small infrared transmitting window) to increase sensitivity by minimizing conductive heat loss.

The *cold junction* of the couple actually consists of the heavy copper lead wires in contact with the thermocouple wires. Since the detector only needs to respond to chopped radiation, and to give an ac output, only *changes* in temperature are significant, hence the actual temperature of the cold junction is unimportant.

Thermocouple detectors are low-impedance devices and are usually coupled to a preamplifier through a high-turns ratio transformer (as high as 1000:1) with good low-frequency response.

A *bolometer* is a miniature resistance thermometer with a tiny platinum wire or a thermistor* as sensing element. The thermistor is more sensitive than platinum by about five times; it can be made to have a lower time constant, but is likely to be less reproducible.

The resistance of a bolometer can be measured by conventional circuits, one of which is shown in Fig. 5-3. One bolometer element is exposed to the radiation, while the other is shielded but subject to the same ambient conditions. The output voltage can be shown to be proportional to the ratio of the resistances of the two elements, the effects of changes in ambient temperature being essentially eliminated.

Another type of thermal detector, called a *Golay cell* (3), is based on the increase in pressure of a confined gas with temperature. The radiation to be measured is absorbed by a blackened film located in the center of a small gas chamber. The resulting pressure increase causes a thin, flexible mirror to bulge outward. The convex mirror forms part of an optical system by which light from a small incandescent bulb is focused on a photocell. The signal seen by the photocell is thus modulated in accordance with the power of the radiant beam incident on the gas cell.

The Golay detector is best suited to measurements of radiation chopped at a frequency of 10 to 15 Hz. It has about the same sensitivity as a thermocouple detector in the mid-infrared region, but has a wider range of utility. It is uniformly sensitive from the ultraviolet through the visible and infrared at least as far as a wavelength of 7.5 mm in microwaves, provided that appropriate window materials are selected. It is more expensive and bulky and somewhat less convenient to operate than many other infrared detectors.

Certain crystals, called *pyroelectrics*, have the property of developing a voltage across their opposed faces when subjected to heating. Examples are triglycine sulfate, barium titanate, and lithium niobate. Successful infrared detectors have been fabricated from such crystals. They have

---

*A thermistor is a resistor made by sintering together several metallic oxides. It has a large negative temperature coefficient.

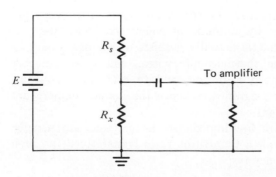

Fig. 5-3 Linear comparison circuit for a bolometer. R_x is the irradiated element, and R_s is an identical one, shielded from radiation but at the same ambient conditions. [*Journal of Chemical Education* (3).]

a smaller time constant than other thermal detectors, which means that radiation can be chopped at a higher rate and the spectrum scanned faster, a considerable advantage. As of this writing, they are more expensive than thermocouples, and are not yet widely used, but it is likely that they will be selected for many future designs.

Photon detectors for the infrared take the form of small wafers of semiconducting material such as lead sulfide, indium antimonide, or germanium doped with traces of copper or mercury. These are among the fastest and most sensitive detectors, but have limited wavelength ranges and, moreover, must be cooled with liquid nitrogen (some require liquid helium). Lead sulfide can be used in the near infrared at room temperature and is the detector of choice for this application.

SPECTROPHOTOMETERS

Automatic scanning spectrophotometers for the mid-infrared region are available in confusing abundance. They can be divided into two main classes: low-cost models for routine identification work, and research instruments equipped with various refinements, such as higher precision, higher resolution, greater wavelength span, and more versatility with respect to mode of presentation of the spectrum or with respect to type of sample holder. Most of these utilize the double-beam principle.

A surprising number of infrared spectrophotometers of various manufacturers and in all price classes have nearly identical optical layouts. The differences arise in the precision tolerances of the many optical and mechanical components, and are reflected primarily in the specifications of wavelength and photometric accuracy and of resolution, and secondarily in the variety of automated features for the convenience of the operator. [For a thorough discussion of resolution in the infrared, its significance, and how to measure it, see the treatise by Stewart (4).]

One of the most common optical systems is shown in Fig. 5-4. Energy from the source N, which may be a Nernst glower or a Nichrome wire, is reflected by two sets of mirrors to form two symmetrical beams, one of which passes through the sample cell, and the other through a blank or reference cell. The two beams are reflected by a further array of mirrors, so that they are both focused on the slit S_1. C is a motor-driven circular chopping disk, half-reflecting and half-cutaway (Fig. 5-4c), which permits the two beams to reach the slit during alternate time periods. Slit S_1 is the entrance slit to a Littrow monochromator equipped with either a 60° prism and mirror combination (Fig. 5-4a), or with a pair of back to back diffraction gratings (5-4b). The selected wavelength band passes out through the exit slit S_2 and is brought to focus on the thermocouple detector D. (Very few spectrophotometers are currently built with prisms in place of gratings, but many are in use.)

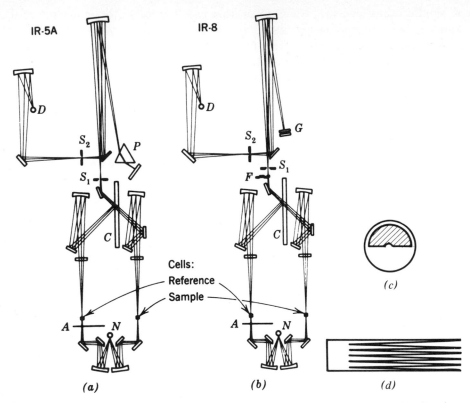

Fig. 5-4 Optical systems of two Beckman infrared spectrophotometers. (*a*) A prism instrument (IR-5A); (*b*) a grating model (IR-8); (*c*) detail of the beam-combining mirror (*C* in the diagrams); (*d*) detail of the optical attenuator (*A*). (*Beckman Instruments, Inc.*)

The detector sees a square wave corresponding in amplitude to the difference in powers of the two beams. This ac signal, after amplification, drives a servomotor to control the position of a comb attenuator *A* (Fig. 5-4*d*) which adjusts the power of the reference beam to equal that of the sample beam, an *optical null*. The motor simultaneously positions the pen on the built-in recorder.

In the grating instrument (Fig. 5-4*b*), two gratings are required to give the desired resolution over the whole range. The first grating has typically 300 lines per millimetre, blazed at 3 μm, and is in the operating position, slowly rotating, while the spectrophotometer scans from about 2 to 5 μm. When 5 μm is reached, the grating table abruptly turns to bring the second grating (100 lines per millimetre, blazed at 7.5 μm) into position, and the scan resumes, covering the range from 5 to 16 μm. Both gratings are operated in the first order. Four filters *F* are required to eliminate higher orders; they are positioned sequentially by the drive mechanism at the required wavelengths.

Either prism or grating spectra may be linear in either wavelength or wave number, according to the design of the cams and bars making up the mechanical linkage between the scanning motor and the prism or grating table. This is, of course, simplest to achieve for linear wavelength with grating dispersion. From the standpoint of the operator, however, one is as convenient to use as another.

RAPID-SCAN SPECTROPHOTOMETERS

One of the limitations of conventional infrared spectrophotometers is their inherently slow response. High-speed instruments are needed for kinetic studies and for on-line identification of constituents of a flowing stream. For example, it is highly desirable to be able to run an absorption spectrum on the effluent from a gas chromatograph as it is being eluted, rather than on separately collected fractions.

The most obvious way in which this objective of increased speed can be approached is simply to gear up a conventional spectrophotometer to run faster. This results in considerable loss of resolution in the spectra obtained, so that identification is much less certain. It is better to design an instrument specifically for a rapid scan. This requires the use of either a cooled semiconductor (5) or a pyroelectric detector (6).

One commercially available instrument* with a pyroelectric detector can record an entire spectrum (2.5 to 15 μm) in either 6 or 30 s, as selected. It is designed to be connected directly to a gas chromatograph, so that carrier gas fills the reference cell, while column effluent flows through the analytical cell (both cells are about 10 cm long). The resulting spectra possess resolution fully adequate for identification purposes. Quantitative measurements are taken from the chromatograph.

INTERFEROMETRIC SPECTROPHOTOMETERS

Since radiant energy consists of trains of electromagnetic waves, generally of many frequencies superimposed, the instantaneous electrical and magnetic fields at any point will be the resultant of those due to the individual frequencies. Therefore it should be possible in principle to retrieve *all* the information carried by a beam simply by letting it fall on a radiation detector and plotting the response as a function of time. The trouble with such a direct approach is that the alternating fields, measurable in units of 10^{14} Hz, are many orders of magnitude too fast to follow with any known detector (7).

* Norcon Instruments, Inc., South Norwalk, Conn.

The speed difficulty can be overcome through the use of an interferometer (Fig. 5-5). Light from the source is collimated by lens L and then divided into two equal parts by the beam splitter BS. The beams are reflected back by plane mirrors M_1 and M_2. Portions of both beams are finally incident on the detector. If the two mirrors M_1 and M_2 are equidistant from BS, the detector will see the same time-varying electromagnetic fields that it would have seen without the interferometer, but they will be only half as intense. If mirror M_2 is then moved to the right along the optical axis, the phases of the two beams reaching the detector will differ, and interference will result. The retardation of the phase for a given increase in path length depends on the wavelength of the radiation, and is observed by the detector as a series of successive maxima and minima of intensity.

To be used as a spectrophotometer, mirror M_2 is made to move at a constant speed for a distance long compared to the wavelength, perhaps a few millimetres. This causes the detector response to fluctuate at a rate dependent on the speed of mirror motion and the wavelength of radiation.

Ideally, monochromatic radiation should produce a cosine wave as the two optical paths deviate from equality (cosine rather than sine because the amplitude is maximal at zero deviation). Figure 5-6a and b shows this for two monochromatic frequencies. If both frequencies enter together into the interferometer, curve c will result. Carrying this synthetic approach further will produce great complications in the spectra (see Fig. 5-7). However, it follows that any combination of frequencies with corresponding amplitudes will produce a unique *interferogram* containing all the spectral information of the original radiation.

Fig. 5-5 Michelson interferometer, as used in a Fourier-transform spectrometer.

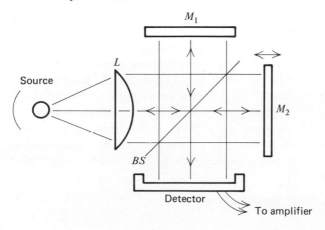

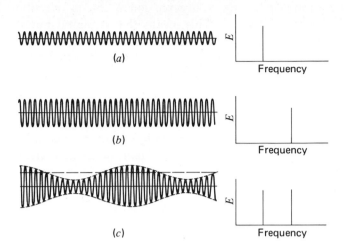

Fig. 5-6 Comparison of information in the time domain (left), and frequency domain (right). (a) First cosine wave, frequency ν_1; (b) second cosine wave, ν_2; (c) both frequencies present simultaneously.

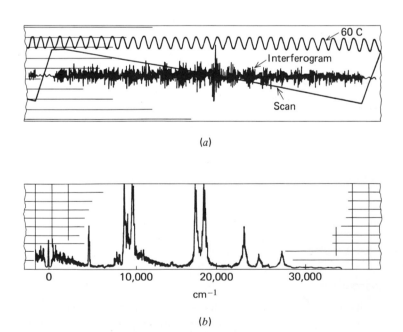

Fig. 5-7 (a) Interferogram and (b) spectrum of the same organic compound. The sine wave at the top of (a) is a reference curve for calibration; the sloping line marked "scan" tracks the movement of the mirror. [*Journal of Chemical Education* (7).]

The interferograms obtained by the above technique are not readily interpreted without electronic assistance. This means that a digital mini-computer (or time-shared access to a standard computer) is required. The interferogram, from a mathematical standpoint, is the *Fourier transform* of the spectrum, so the task of the computer is to apply the inverse Fourier transform.

Interferometric or Fourier-transform spectroscopy, as here described, has one great advantage over conventional techniques: it makes use of all frequencies from the source simultaneously, rather than sequentially as in a scanning instrument. (This is called the *Fellgett advantage,* after the scientist who first described it.) The Fellgett advantage can be shown to amount to an improvement in the signal-to-noise ratio of $\sqrt{M}$, where M is the number of resolution elements desired in a particular spectrum (7–9). In addition to the Fellgett advantage, the interferometric method has greatly increased sensitivity because of the large amount of radiation entering the slitless system.

There are several instruments of this type available. The optical diagram for one of them is shown in Fig. 5-8.

HADAMARD SPECTROSCOPY

Another method that permits a Fellgett advantage has been described by Decker (7, 10). All wavelengths of the spectrum, after dispersion in a spectrograph, are focused simultaneously in a plane and then reflected back upon themselves through the dispersing system. All the wavelengths are thereby brought together to impinge on the detector. A mask placed at the focal plane is pierced by perhaps 1000 parallel slits in a pseudorandom arrangement. The mask is in the form of a tape which is moved stepwise along the focal plane, each step equaling the width of the slits. At each of the 1000 steps a reading is taken of the detector response. The eventual result is a set of 1000 simultaneous equations in 1000 unknowns, easily solved with a computer. The 1000 unknowns correspond to successive points in the spectrum. Full details may be found in reference 10 and in prior articles listed therein.

ABRIDGED SPECTROPHOTOMETERS

LOW-RESOLUTION INSTRUMENTS

It is possible to make useful measurements with a manual, single-beam infrared photometer much like the photoelectric filter photometer in the visible range. A series of such instruments, manufactured under the trade-name Miran,* uses a Nichrome source and a pyroelectric detector. Wave-

* Wilks Scientific Corporation, South Norwalk, Conn.

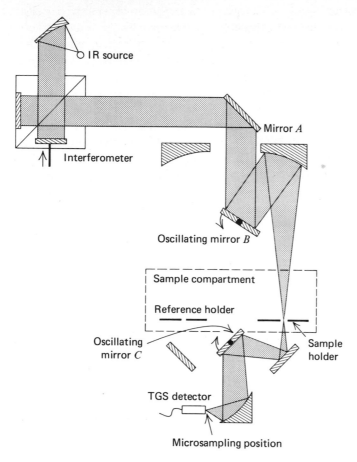

Fig. 5-8 Optical diagram of a Fourier-transform spectrometer, Model FTS-14, simplified. (*Digilab, Inc.*)

length selection is made with a circular interference-wedge filter, which provides considerably less resolution than a grating but is still adequate for many uses. The sensitivity is high, because a larger cone of radiation can be accepted than with a conventional spectrophotometer. The chief advantage, besides lower cost, is ruggedness and portability. Instruments of this type are appropriate for the estimation of a compound or a class of compounds when it is not necessary to distinguish between closely related species.

NONDISPERSIVE PHOTOMETERS

Instruments that depend on selective filtering for specificity are widely used to monitor gas streams and in air pollution studies. Consider as an example the determination of CO in the presence of other mixed gases. Since CO only absorbs certain characteristic frequencies, it is only these frequencies that are of interest in its measurement. Effective elimination of all other frequencies can be achieved in either of two ways, compared in Fig. 5-9. In Fig 5.9a beams of radiation from a pair of identical sources traverse sample and reference cells and impinge on a differential detector that contains CO (usually diluted with argon to reduce the heat capacity) (11). Any difference in power between the two beams results in a temperature differential between the two sides of the detector. It is highly selective, because the only radiation that will produce heating in the detector is identical in frequency to that absorbed by CO in the sample. An added filter cell is provided to desensitize the instrument to other possible components of the gas stream for which absorption bands overlap those of CO. The two chambers of the detector are separated

Fig. 5-9 Nondispersive infrared analyzers. (*a*) With differential detectors; (*b*) with an infrared-fluorescent source.

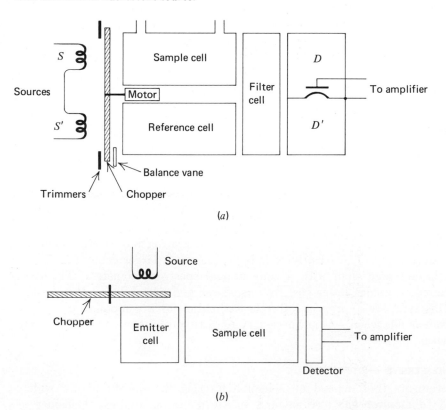

by a metallized flexible diaphragm which forms a variable capacitor with a fixed plate.

A motor-driven chopper is required, because the differential detector is a dynamic device; i.e., it responds to rapid changes more effectively than to gradual drifts. The chopper is designed to interrupt the two beams simultaneously. The sensing capacitor is incorporated in a high-frequency electronic circuit which energizes a small servomotor to drive a balancing vane across the reference beam to achieve an optical null. The amount of such compensation is indicated on a meter or recorded on a chart.

The single-beam instrument diagrammed in Fig. 5-9b makes use of a cell containing CO as the *source* of radiation rather than as the detector (12). The CO is heated by infrared from an incandescent source and then emits radiation at the characteristic CO frequencies, which passes through the sample to a nonselective detector. This instrument is less liable to interference from vibration and ambient temperature changes than the previous model. Either type can be sensitized to any polyatomic gas desired.

FAR-INFRARED SPECTROPHOTOMETERS

Dispersive instruments for the region beyond about 50 μm are much fewer in number than those for shorter wavelengths. Those that are available utilize a high-pressure mercury arc with a quartz envelope as source. The quartz filters out wavelengths below about 60 μm, but the useful range extends below this, the emission occurring from the hot quartz rather than the arc. Either a Golay or a pyroelectric detector is applicable. Atmospheric absorption is a problem in the far infrared, and spectrophotometers must be capable of being purged with dry nitrogen or evacuated.

Several gratings must be interchanged at successive wavelength intervals; for the longest wavelengths a grating may have as few as 2 or 3 lines per millimetre. To cut out background radiation and unwanted orders, filters must be present. There are two types of filters suited for far-infrared applications, both depending on selective reflection rather than transmission. One of these utilizes the phenomenon of *Reststrahlen*,* the narrow bands of high reflection in crystalline substances which correspond to the refractive index maxima associated with areas of high absorption (13, 14). The other type consists of a *scatter plate*, which may be a metallic film deposited on a rough-ground glass plate, or a grating with its rulings horizontal rather than parallel to the vertical slits. The scattering is very effective in reducing the amount of higher-frequency radiation passing through the optical system, but has practically no deleterious effect on the long waves of interest.

It is particularly important in the far infrared to chop the radiation prior to its passage through the sample and monochromator, because sponta-

* From the German, meaning "residual rays."

neous thermal radiation from the sample and optical elements may be far from negligible. Chopping, together with a tuned amplifier, enables the detector to respond only to radiation from the source itself.

Interferometric spectrometers are particularly valuable in the far infrared. The expense of their manufacture is somewhat less in this region, a trend opposite that observed with dispersive instruments.

CALIBRATION AND STANDARDIZATION

The calibration of a spectrophotometer with respect to both wavelength and transmittance may change gradually with continued use, as the result of mechanical wear, fogging of optical surfaces, or aging of components, so that periodic checks are highly advisable.

The wavelength or wave number scale can in theory be calibrated by the dispersion geometry of a grating of known spacing or a prism of known refractive index, but this is impracticable as a routine procedure. The most common check is to run the spectrum of a thin sheet of polystyrene as a secondary standard. Infrared manufacturers usually supply mounted samples for this purpose, along with a standard spectrum for comparison. The spectrum (see Fig. 5-1) shows easily recognized absorption bands distributed throughout the mid infrared.

Another convenient wavelength check is performed by switching to single-beam operation and scanning the spectrum with no sample present. Under this condition the absorption of atmospheric water vapor and carbon dioxide will be clearly seen (Fig. 5-10) and, as their wavelengths are known with precision, the scale can readily be checked at a number of points. The

Fig. 5-10 Single-beam trace without sample, showing the absorption bands due to atmospheric water and carbon dioxide. [*Plenum* (20).]

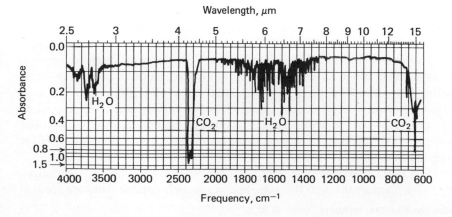

spectrum makes evident one of the major advantages of a dual-beam spectrophotometer—the cancellation of atmospheric absorption.

If the spectrophotometer is to be used for quantitative measurements, the linearity of its photometric scale must also be verified occasionally. This can be done roughly by measuring the polystyrene spectrum, and more precisely by measuring the transmittance of a series of thicknesses of a liquid such as benzene.

CHEMICAL APPLICATIONS

QUALITATIVE ANALYSIS

Since infrared absorption is related to covalent bonds, the spectra can provide much detailed information about the structure of molecular compounds.

It is possible, through a painstaking examination of a large number of spectra of known materials, to correlate specific vibrational absorption maxima with the atomic groupings responsible for the absorption. Such empirical correlations provide a powerful tool for the identification of a covalent compound.

Some sweeping generalizations can be made. It is useful to distinguish types of vibration as *stretching, distortion, bending,* etc. The shorter infrared wavelengths, from about 0.7 to 4.0 μm, include mostly stretching vibrations of bonds between hydrogen and heavier atoms; this includes the near-infrared region, and is especially useful in the identification of functional groups which contain hydrogen. The 4.0 to 6.5 μm region contains vibrations of double and triple bonds. Above this wavelength are found the "skeletal" distortion and bending modes, including C—H bending, etc.

The absorptions in the far-infrared region, beyond about 25 μm, correspond to vibrational modes involving heavy atoms and groups of atoms, including bonds of carbon to phosphorus, silicon, heavy metals, and also heavy metals to oxygen, and many others. Also found in this region are some low-lying distortional frequencies, such as the ring-puckering mode in four-membered rings, as well as *torsional* vibrations of methyl and other groups, and the majority of purely rotational bands.

Table 5-2 gives an indication of the infrared regions corresponding to frequently occurring bond types. Much more extensive and detailed correlations, taking account of the intramolecular environment of each bond, are available. These have been prepared in convenient chart form by Goddu (15) for the 1.0 to 3.1 μm, and by Colthup (16) for the 2.5 to 25 μm region. Bentley (17) has compiled a chart which extends out to 33 μm. All of these charts are reproduced in Meites' "Handbook of Analytical Chemistry." Most of the major manufacturers of infrared spec-

Table 5-2   Infrared positions of various bond vibrations*

Bond	Mode†,‡	Relative strength§	Wavelength, μm	Wave number, cm^{-1}
C—H	Stretch	s	3.0–3.7	2700–3300
C—H	Stretch (2ν)	m	1.6–1.8	5600–6300
C—H	Stretch (3ν)	w	1.1–1.2	8300–9000
C—H	Stretch (C)	m	2.0–2.4	4200–5000
C—H	Bend, in-plane	m–s	6.8–7.7	1300–1500
C—H	Bend, out-of-plane	w	12.0–12.5	800–830
C—H	Rocking	w	11.1–16.7	600–900
O—H	Stretch	s	2.7–3.3	3000–3700
O—H	Stretch (2ν)	s	1.4–1.5	6700–7100
O—H	Bending	m–w	6.9–8.3	1200–1500
N—H	Stretch	m	2.7–3.3	3000–3700
N—H	Stretch (2ν)	s	1.4–1.6	6300–7100
N—H	Stretch (3ν)	w	1.0–1.1	9000–10000
N—H	Stretch (C)	m	1.9–2.1	4800–5300
N—H	Bending	s–m	6.1–6.7	1500–1700
N—H	Rocking	s–m	11.1–14.3	700–900
C—C	Stretch	m–w	8.3–12.5	800–1200
C—O	Stretch	m–s	7.7–11.1	900–1300
C—N	Stretch	m–s	7.7–11.1	900–1300
C=C	Stretch	m	5.9–6.3	1600–1700
C=O	Stretch	s	5.4–6.1	1600–1900
C=O	Stretch (2ν)	m	2.8–3.0	3300–3600
C=O	Stretch (3ν)	w	1.9–2.0	5000–5300
C=N	Stretch	m–s	5.9–6.3	1600–1700
C≡C	Stretch	m–w	4.2–4.8	2100–2400
C≡N	Stretch	m	4.2–4.8	2100–2400
C—F		s	7.4–10	1000–1350
C—Cl		s	13–14	710–770
C—Br		s	15–20	500–670
C—I		s	17–21	480–600
Carbonates		s	6.9–7.1	1400–1450
Carbonates		m	11.4–11.6	860–880
Sulfates		s	8.9–9.3	1080–1120
Sulfates		m	14.7–16.4	610–680
Nitrates		s	7.2–7.4	1350–1390
Nitrates		m	11.9–12.3	820–840
Phosphates		w	9.0–10.0	1000–1100
Silicates		. . .	9.0–11.1	900–1100

* Approximate only; fundamentals unless noted; collected from various literature sources.

† (2ν) means second harmonic or first overtone, etc.

‡ (C) means combination frequency.

§ s = strong, m = medium, w = weak.

trophotometers publish charts corresponding to the ranges and dispersions of their own instruments. Detailed discussion of such correlations can be found in numerous texts and monographs intended to assist in the determination of the structure of organic compounds (15, 18–20).

It must be emphasized that there are many situations in which the infrared absorption of a compound is altered more or less extensively by the conditions under which it is observed. These variations in the location of absorption bands make it necessary to use empirical tables and charts with great caution when attempting to determine the structure of an unknown substance. The causes of such variations may be instrumental or chemical in origin. Figure 5-11 (20) illustrates the effects of changes in slit widths and scanning rates. Gross changes in either could interfere with an identification.

Another type of interaction is exemplified by the effect of hydrogen bonding on the absorption frequency of a carbonyl group. The frequency corresponding to the stretching mode of the $C{=}O$ bond in a compound dissolved in a nonpolar solvent is lowered considerably by the formation of hydrogen bonds with an added hydroxylic substance or by changing to hydroxylic solvent. The carbonyl absorption is also altered by its environment within its own molecule: The absorption of the $C{=}O$ bond is significantly different in a carboxylic acid in which an intermolecular hydrogen bond (formation of a dimer) can occur, as compared with an ester in which there is no possibility of such a bond. The anion is affected by resonance which renders the two oxygen atoms equivalent to each other, so that the double-bond character of the carbonyl is even more profoundly changed. The following tabulation shows representative values for the carbonyl stretching vibrations of a few aliphatic compounds (19).

Ketone	5.81–5.85 μm	1720–1710 cm^{-1}
Carboxylic acid monomer	5.65–5.71 μm	1770–1750 cm^{-1}
Carboxylic acid dimer	5.81–5.85 μm	1720–1710 cm^{-1}
Ester	5.76 μm	1735 cm^{-1}
Salt, asymmetric stretch	6.21–6.45 μm	1610–1550 cm^{-1}
Salt, symmetric stretch	~7.14 μm	~1400 cm^{-1}

This discussion of the carbonyl absorption is presented as an abbreviated example of the kind of structural considerations that can be of great significance to the organic chemist as well as to the analyst.

One result of the complexity of infrared spectra is that it is highly improbable that any two different compounds will have identical curves. Hence an infrared spectrum of a pure compound presents a sure method of identification, provided that the analyst has at hand an extensive compi-

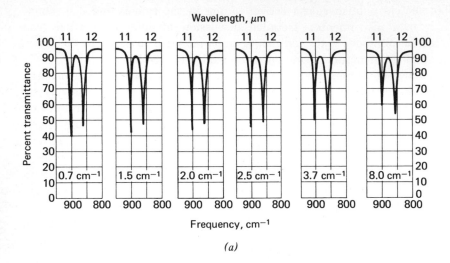

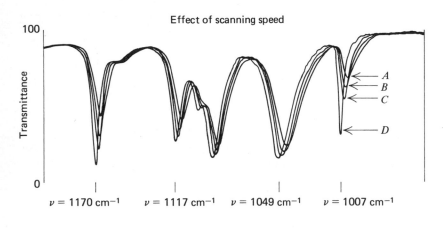

Fig. 5-11 The effects of instrumental variables on infrared spectra. (*a*) The 861 and 903 cm⁻¹ bands of cyclohexane at various spectral slit widths. (*b*) A spectrum recorded at several scanning speeds; effective times for full-scale scans: *A*, 8 min; *B*, 11 min; *C*, 16 min; and *D*, 3 h. [*Plenum* (20).]

lation or atlas of spectra of known compounds. Several such atlases are available (21, 22).*

---

* A series of reference spectra of particularly high reliability is being compiled by the Coblentz Society. Their requirements and necessary precautions make interesting reading (23, 24). The Coblentz spectra are commercially available (22).

QUANTITATIVE ANALYSIS

Beer's law, as presented in Chap. 3, applies equally to the infrared region:

$$A = \log \frac{P_0}{P} = abc \tag{5-1}$$

In the *near* infrared, 1- to 10-mm cuvets and dilute solutions are the rule, and no special difficulties arise in the use of this relation. Beyond this region, however, the path length b is usually much less, and the concentration greater, because of the lack of suitable solvents for the infrared, a matter which will be considered later. The greater concentration is apt to cause deviations from the law, as a result of molecular interactions. Furthermore, it is often difficult to make an accurate photometric measurement because of the overlap of absorption bands.

The path length can be measured by a number of methods. In some cell designs it can be done with a micrometer. In a cell which has plane parallel walls, interference fringes resulting from multiple internal reflections can easily be recorded by running the spectrophotometer so as to indicate the apparent transmission through the empty cell against air in the reference path (19). The resulting trace will resemble Fig. 5-12. The value of b can then be computed from the equation

$$b = \frac{n}{2(\tilde{\nu}_1 - \tilde{\nu}_2)} \tag{5-2}$$

where n is the number of fringes between wave numbers $\tilde{\nu}_1$ and $\tilde{\nu}_2$.

Some cells do not have walls of sufficient flatness to give sharp interference fringes. The effective or average path length can then be deter-

Fig. 5-12 Typical fringes used in the calculation of path length. [*Allyn and Bacon* (19).]

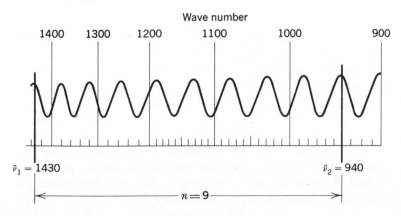

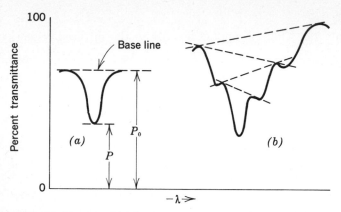

Fig. 5-13 Hypothetical infrared absorption spectra. (*a*) Illustrating the use of the base-line measurement technique; (*b*) showing uncertainty as to where to draw the base line.

mined by an absorption measurement at $\bar{\nu} = 850$ cm^{-1} with benzene in the cell. Experiment has shown that 0.1 mm of cell thickness corresponds to 0.22 in absorbance units (19).

Deviations from Beer's law due to high concentration can be handled only by careful calibration with a series of solutions of known concentration. The experimental determination of absorbance often presents difficulties on account of the background absorption and overlapping bands of other materials present. Suppose it is desired to find the absorbance corresponding to the first band shown in Figure 5-13. A *base line* is drawn across the shoulders of the band; the quantities P_0 and P can then be measured as shown, and the absorbance calculated. However, with an absorption such as the second in the same figure, the proper location of the base line is open to much doubt; a few possibilities are suggested. In this situation, which frequently occurs, the only procedure is to standardize on one particular way to draw the line. The certainty of one's results could be improved by analyzing several absorption bands in the same spectrum; consistent results would go far to prove the adequacy of the method.

The present and future importance of infrared spectrophotometry lies more and more with the qualitative identification of substances, pure or in mixtures, and as a tool in the establishment of structure. The quantitative aspects have diminished in importance, partly because of their inherent difficulty and largely because quicker and more convenient methods of quantitation have emerged, especially chromatography. A typical procedure for the examination of a sufficiently volatile mixture requires separation of the components on a gas chromatograph, which shows the number of components and the relative quantities of each, followed by qualitative identification of each by infrared spectrophotometry.

PREPARATION OF SAMPLES

Gaseous samples can be examined in an infrared spectrophotometer with no prior preparation except perhaps removal of water vapor. Most manufacturers provide accessory gas cells equipped with mirrors to allow the radiation to traverse the sample many times. Path lengths up to perhaps 10 m can be attained.

Liquids are frequently handled pure, without solvent, in thin layers. Solvents present difficulties because no liquids are available that are entirely free of absorption on their own account. Carbon tetrachloride is satisfactory over a considerable range, but dissolves only a limited number of substances. Chloroform, cyclohexane, and other liquids can be use in restricted wavelength ranges and in very thin layers. Figure 5–14 shows in chart form the spectral regions where various solvents can be used. The degree of absorption that can be tolerated in the solvent depends on the sensitivity of the spectrophotometer. It can be cancelled out, at least to a good approximation by the usual means of placing a blank cell in the reference beam.

It is often possible to select two solvents with complementary absorp-

Fig. 5-14 Chart showing the infrared transmission regions for a number of solvents. The rectangles designate the areas of transmission. (*Eastman Kodak Company.*)

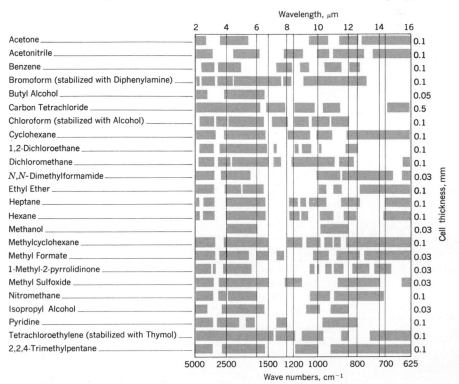

tion regions, so that two successive spectra will cover the entire wavelength span of interest. CCl_4 (4000 to 1335 cm^{-1}) and CS_2 (1350 to 400 cm^{-1}) form such a pair, and are highly recommended where solubilities permit (23). An example is seen in Fig. 5-2a.

A variety of absorption cells for liquids are available, ranging from inexpensive ones designed to be discarded as soon as they become at all fouled, to cells which can be disassembled for cleaning and cells provided with a screw thread to allow adjustment of thickness. The *cavity cell* is a convenient form, made by machining a parallel-sided hole in a salt block. The inner surfaces of a cavity cell cannot be polished, so the thickness cannot be measured by the interference fringe method.

A *demountable cell* is shown in Fig. 5-15. It consists of a pair of salt plates separated by a shim or gasket made of metal or sometimes of Teflon, the whole held together as a sandwich by metal clamps. Two holes are drilled through the metal frame and one salt plate, for filling and flushing. A cell can generally be used many times before it becomes necessary to disassemble it and clean or repolish the plates. The cells are usually filled, emptied, and rinsed with the aid of a hypodermic syringe (without needle), the nib of which fits the holes in the cell frame.

Liquids of high viscosity are often simply sandwiched as a layer between two salt plates, since it is not easy to introduce viscous liquids

Fig. 5-15 Cell for holding a liquid sample in an infrared spectrophotometer. (*Beckman Instruments, Inc.*)

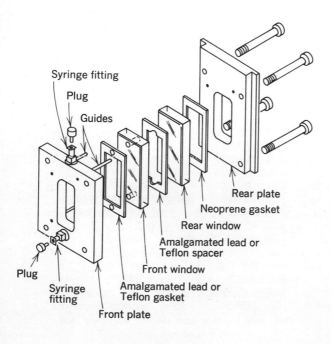

into preassembled cells.

Solid samples can be prepared for analysis by incorporating them into a pressed plate or pellet of potassium bromide or (less commonly) potassium iodide or cesium bromide. A weighed portion of the powdered sample is thoroughly mixed (as in a ball mill) with a weighed quantity of highly purified and desiccated salt powder. Then the mixture is submitted to a pressure of several tons in an evacuated die, to produce a highly transparent plate or disk which can be inserted into a special holder for the spectrophotometer. The disk typically is about 1 cm in diameter and perhaps 0.5 mm in thickness. The exact thickness is required for quantitative purposes; sometimes it can be determined from the dimensions of the die, or it can be measured with a micrometer caliper.

There are hazards in the use of the potassium bromide pellet technique. If the particle size is not sufficiently small, excessive scattering results. Some heat-sensitive materials (steroids, for example) may show signs of partial decomposition, presumably arising from the heat generated in crushing crystals of salt. The technique has been considered in critical detail in a paper by Hannah (25).

Solid samples can also be examined in the form of a thin layer deposited by sublimation or solvent evaporation on the surface of a salt plate. Another procedure with much to recommend it is called *mulling*. The powdered sample is mixed to form a paste with a little heavy paraffin oil (medicinal grade Nujol is often used). The oil has only a few isolated absorption bands, specifically at about 3.5, 6.9, and 7.2 μm. If these bands interfere, the mull may be made with Fluorolube, a fluorocarbon material which has no absorption at wavelengths shorter than about 7.7 μm. The mull is sandwiched between salt plates for measurement.

An advantage of either the mull or the pressed disk is that scattering of radiation is reduced to a minimum; this is a source of trouble when a powdered sample is run as such.

ATTENUATED TOTAL REFLECTION

It is well known that when a beam of radiation encounters an interface between two media, approaching it from the side of higher refractive index, total reflection occurs if the angle of incidence is greater than some critical angle, the value of which depends on the two refractive indices. Not so generally realized, though predicted by electromagnetic theory, is the fact that in total reflection some portion of the energy of the radiation actually crosses the boundary and returns. This is suggested by the oversimplified sketch of Fig. 5-16. If the less dense medium absorbs at the wavelength of the radiation, the reflected beam will contain less energy than the incident, and a wavelength scan will produce an absorption spectrum.

In principle, this will apply to any spectral region, but it has been

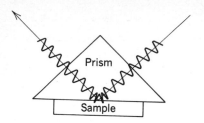

Fig. 5-16 Schematic representation of the total internal reflection of a beam of radiation, showing some degree of penetration into the substrate. (*Wilks Scientific Corporation.*)

found most useful in the infrared. The distance to which the radiation appears to penetrate in internal reflection depends on the wavelength, but is of the order of 5 μm or less in the sodium chloride region. The phenomenon goes under a variety of names, including *attenuated total reflectance* (ATR), *frustrated internal reflectance* (FIR), and when several reflectances are utilized, *multiple internal reflectance* (MIR) or *frustrated multiple internal reflectance* (FMIR); we will use the abbreviation ATR.

Special sample holders have been devised to make use of ATR for infrared analysis, mostly designed to fit the sample compartment of conventional spectrophotometers; a few are shown in Fig 5-17. In (*a*) is an arrangement for a single reflection at the sample surface, at a variable angle. By moving the lower prism (*b*), which has silvered surfaces and hence reflects at any angle, the angle of incidence at the sample can be varied, still maintaining the entering and exiting beams on the same straight line. The drawing (*c*) shows a linear multiple reflection device (26) especially convenient for solid samples, which are clamped in optical contact with the two surfaces of a slab of transparent solid of high index, several centimetres long. Figure 5-17*d* represents an elongated prism with plane parallel sides, so arranged that the radiation enters and leaves at the same end (27); this type can be surrounded by a liquid sample, as shown, or can be clamped between solids.

ATR spectra are not identical to those obtained conventionally, but are quite similar. The distortion becomes greater as the angle of incidence approaches the critical angle. The further from the critical angle, however, the less the absorption, and so a greater number of reflections must be introduced.

ATR has been found most useful with opaque materials which must be observed in the solid state. Applications include studies of rubber and other polymeric materials, adsorbed surface films, and coatings such as paints.

MICROWAVE ABSORPTION*

Absorption in the microwave region (28, 29) can be considered an extension of the far infrared, as the phenomena giving rise to the absorption are primarily rotational transitions in molecules possessing a permanent dipole

* Sometimes called *molecular rotational resonance spectroscopy* (MRR).

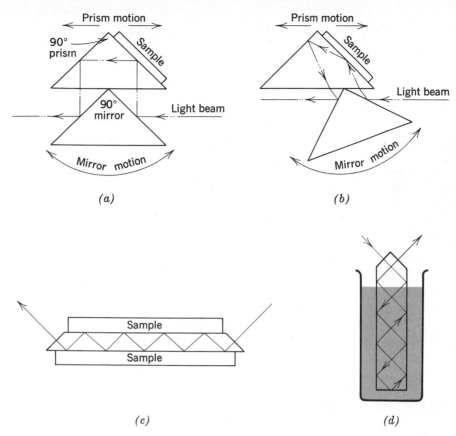

Fig. 5-17 ATR apparatus. (a) and (b) An arrangement whereby the entering and leaving beams are in the same straight line (the lower prism and the left face of the upper must be silvered; motion of the parts as indicated will permit the angle of reflection at the sample to be varied); (c) a device for multiple reflections; (d) an immersible plunger to permit multiple reflections with a bulk liquid sample. [(a) to (c), *Wilks Scientific Corporation*; (d) *Analytical Chemistry* (27).]

moment. Absorptions can also be observed under appropriate conditions in molecules possessing magnetic moments, such as O_2, NO, NO_2, ClO_2, and in free radicals. The microwave spectral region most fruitful for chemical spectroscopy lies between approximately 8 and 40 GHz. [1 GHz (gigahertz) $= 10^9$ Hz.] As we will see in a later chapter, this includes the frequencies useful in electron spin magnetic resonance (ESR), so that these two methods share some common features, though based on different molecular mechanisms.

There are no microwave absorptions which are characteristic of specific bond types or functional groups, such as we find in optical and infrared spectra. Qualitative analysis can be applied only by means of comparison with known spectra. The microwave region is excellent as a complement to the infrared in the identification of relatively small compounds by the fingerprint approach. This is partly due to the very great number of frequencies available. If absorption bands can be resolved at a separation of 200 kHz, a reasonable figure, this gives a total of 160,000 spaces or "channels" in the 8 to 40 GHz region. By comparison, if an average resolution of 1 cm^{-1} is attainable over the infrared region from the visible up to 200 cm^{-1} (50 μm), only 10,000 spaces are available. Of course there are many frequencies that do not appear in any known spectra, so these figures may be somewhat misleading: nevertheless, an extremely large number of compounds can in principle be examined in a microwave spectrometer with very little probability of overlap. Only gaseous samples can be studied, but the vapor pressure need not be high; 0.1 to 10 Pa (approximately 10^{-3} to 10^{-1} torr) is the usual pressure range.

The chief limitation arises from interactions between the rotational energy levels associated with various bonds within the molecule. If there are more than three or four rotors present, so many interactions will appear that the spectrum will become a mass of lines, few of which will be sufficiently intense to be useful. Hence large molecules cannot be studied unless cyclic structures prevent rotation around some of the bonds.

The quantitative absorption law (Lambert's law) can be expressed in the form

$$P = P_0 \times 10^{-\alpha b} \quad \text{or} \quad \log \frac{P_0}{P} = \alpha b \tag{5-3}$$

where P_0 and P represent, respectively, the radiant microwave power incident on and passing through the absorption cell, which has an effective length b. The absorption coefficient α corresponds to the absorbance A, as employed in the optical region, taken per unit length of the absorption cell. Ideally it is a function only of the number of absorbing molecules per unit path length, and hence should be related to the partial pressure of the substance in a mixture of gases.

The degree to which such a relation is valid depends largely on the way in which the absorption coefficient is measured and utilized. At low pressures and power levels, where saturation effects are not evident, the pressure of the absorbing species is proportional to $\alpha_{\max}$, the height of the absorption maximum. At higher pressures, the value of $\alpha_{\max}$ becomes constant, and an increase in pressure results in the broadening of the line as measured at half height. Above this point it can be shown that the partial pressure is very nearly proportional to the integrated absorption:

$$Kp = \int \alpha(\nu) \, d\nu \tag{5-4}$$

where ν = frequency, p = partial pressure, and K is a proportionality

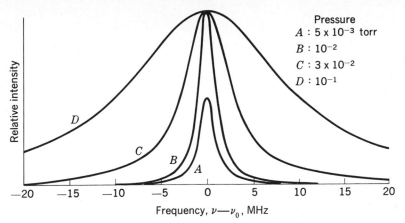

Fig. 5-18 Variation of microwave absorption with pressure change. Note that the peak intensity remains constant over a wide pressure range. [*Wiley* (29).]

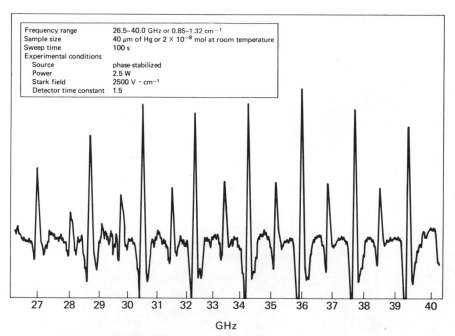

Fig. 5-19 Microwave absorption spectrum of *p*-chlorotoluene. Each peak appears as a doublet, corresponding to the two chlorine isotopes. (*Hewlett-Packard Company.*)

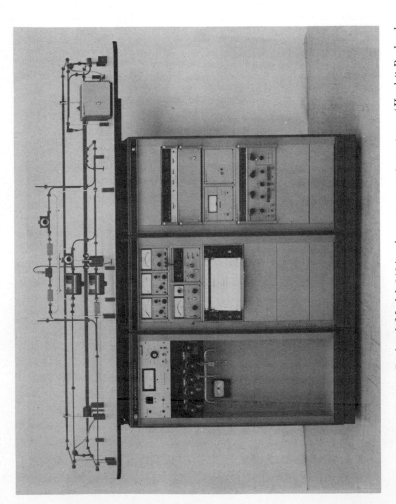

Fig. 5-20 The Hewlett-Packard Model 840A microwave spectrometer. (*Hewlett-Packard Company.*)

constant. Figure 5-18 shows the absorption curves of a typical band at a series of partial pressures.

In spite of pressure broadening, the absorption maxima are narrow enough that overlap with maxima of other compounds is exceedingly unlikely. This means that the partial pressure as determined from Eq. (5-4) can be taken directly as an analytical result, without the need of correction for impurities. Armstrong (30) has reported that a gas such as SO_2 produces a linear response from 100 percent concentration down to 20 ppb, a truly remarkable span.

Molecular conformations can be studied by microwave absorption; trans and gauche forms of a compound have different spectra and can be identified thereby. Barriers to intramolecular rotation can also be measured (30).

If isotopic substitution alters the dipole moment of a compound, this will show up in the spectra; Fig. 5-19 shows an example, which will also serve to illustrate the general appearance of a microwave spectrogram.

MICROWAVE INSTRUMENTATION

The details of theory and construction of apparatus cannot be discussed here for lack of space; they have been reviewed recently (31). Spectrophotometers have been constructed that are comparable to single- and double-beam instruments in the more familiar spectral ranges. The source is either a specialized vacuum tube (a klystron, for example) or a solid-state oscillator based on a tunnel diode or a Gunn diode. These oscillators can be swept through a considerable frequency range, generating essentially monochromatic radiation at each point. The beam chopper is replaced by an electronic system of modulation based on the Stark effect. A crystal rectifier serves as detector of the radiation.

Figure 5-20 shows the external appearance of a microwave spectrometer, Model 840A, from Hewlett-Packard. It can operate from 8 to 40 GHz with four range changes.

PROBLEMS

5-1 A sample of ethyl bromide suspected of containing trace amounts of water, ethanol, and benzene was examined in an infrared spectrophotometer. The following relative absorbances were obtained by a base-line method: at 2.65 μm, $A = 0.110$; at 2.75 μm, $A = 0.220$; at 14.7 μm, $A = 0.008$. The instrument employed, cells, etc., exactly duplicated those for which reference data have been reported (32, 33). Calculate the amounts of the three impurities present.

5-2 It has been reported (34) that ferro- and ferricyanides can be determined in the presence of each other in aqueous solutions by their infrared absorption. The following observations were made:

	Ferrocyanide	Ferricyanide
Molar absorptivity ϵ, $l \cdot mol^{-1} \cdot cm^{-1}$	4.23×10^3	1.18×10^3
Linear range, $mol \cdot l^{-1}$	0.00031–0.04	0.0012–0.1
Wave number, max, cm^{-1}	2040	2115

Cells of 50-μm path length between windows of Irtran-1 (a proprietary preparation of MgF_2) were utilized. In a particular photographic bleach solution, the following data were read from the spectrogram: base line, flat at $A = 0.043$; $A = 0.258$ at 2040 cm^{-1} and 0.515 at 2115 cm^{-1}. What are the two concentrations in moles per liter?

5-3 The following observations have been reported (19) for the apparent absorbance of cyclohexanone in cyclohexane solution at 5.83 μm. (The cell thickness was 0.096 mm.)

Concentration, $g \cdot l^{-1}$	Absorbance
5	0.190
10	0.244
15	0.293
20	0.345
25	0.390
30	0.444
35	0.487
40	0.532
45	0.562
50	0.585

Plot these data. (*a*) Over what range is Beer's law followed, after the necessary background correction is made? (*b*) Compute the value of ϵ, the molar absorptivity.

5-4 Determine (in millimetres) the path length of the cell which produced the interference fringes shown in Fig. 5-12. At what wave number is this cell thickness equal to just 10 wavelengths?

5-5 A high-temperature gas stream in a plant manufacturing "water gas" contains as major constituents H_2, CO, CO_2, and H_2O, with a possible small quantity (less than 0.5 percent by volume) of CH_4. It is desired to sensitize a nondispersive infrared analyzer to monitor the CH_4 content. The gas is cooled prior to introduction into the analyzer, so that excess H_2O condenses out and is removed. Look up the spectra of these substances in an atlas, and describe in detail what should be placed in each compartment of the analyzer of Fig. 5-9a.

5-6 Calculate the approximate number of photons ($\lambda = 10$ μm) needed to just be detectable with a thermistor having the following specifications: Receptor: gold, $1.0 \times 0.1 \times 0.005$ mm; thermistor: 10 kΩ (at 25°C), coefficient of resistance, -500 $\Omega \cdot K^{-1}$. The least measurable resistance change is $\Delta R = 1$ Ω. The heat capacity of gold is 0.0308 $cal \cdot g^{-1} \cdot K^{-1}$, and its density is 19.32 $g \cdot cm^{-3}$. (Note that, because of heat loss to the surroundings, the actual number of photons may have to be considerably greater than this calculation will predict.)

REFERENCES

1. Jones, R. N., and C. Sandorfy: In W. West (ed.), "Chemical Applications of Spectroscopy," p. 308, Wiley-Interscience, New York, 1956.
2. Barker, J. D.: *Electro-Opt. Syst. Design,* October, 1970, p. 32.
3. Ewing, G. W.: *J. Chem. Educ.,* 48:A521 (1971).
4. Stewart, J. E.: "Infrared Spectroscopy, Experimental Methods and Techniques," p. 276 ff., Dekker, New York, 1970.
5. Brown, R. A., J. M. Kelliher, J. J. Heigl, and C. W. Warren: *Anal. Chem.,* 43:353 (1971).
6. Penzias, G. J.: *Anal. Chem.,* 45:890 (1973).
7. Ewing, G. W.: *J. Chem. Educ.,* 49:A377 (1972).
8. Low, M. J. D.: *J. Chem. Educ.,* 47:A163, A255, A349, A415 (1970).
9. Bell, R. J.: "Introductory Fourier Transform Spectroscopy," Academic, New York, 1972.
10. Decker, J. A., Jr.: *Anal. Chem.,* 44(2):127A (1972).
11. Walters, S. H.: In D. M. Considine (ed.), "Process and Controls Handbook," p. 0-73, McGraw-Hill, New York, 1957.
12. Link, W. T., E. A. McClatchie, D. A. Watson, and A. B. Compher: AIAA Paper No. 71-1047, Joint Conference on Sensing of Environmental Pollutants, American Institute of Aeronautics and Astronautics, New York, 1971.
13. Kimmitt, M. F.: "Far-Infrared Techniques," Pion, London, 1970.
14. Ref. 4, p. 165.
15. Goddu, R. F., and D. A. Delker: *Anal. Chem.,* 32:140 (1960).
16. Colthup, N. B.: *J. Opt. Soc. Am.,* 40:397 (1950).
17. Bentley, F. F., and E. E. Wolfarth: *Spectrochim. Acta,* 15:165 (1959).
18. Bellamy, L. J.: "The Infrared Spectra of Complex Molecules," 2d ed., Wiley, New York, 1958.
19. Conley, R. T.: "Infrared Spectroscopy," 2d ed., Allyn and Bacon, Boston, 1972.
20. Alpert, N. L., W. E. Keiser, and H. A. Szymanski: "IR: Theory and Practice of Infrared Spectroscopy," 2d ed., Plenum, New York, 1970.
21. American Petroleum Institute: "Catalog of Infrared Spectrograms," API Research Project 44, Texas A & M University, College Station, Texas.
22. "Sadtler Standard Spectra," a continually updated subscription service, Sadtler Research Laboratories, Inc., Philadelphia.
23. Coblentz Society: *Anal. Chem.,* 38(9):27A (1966).
24. Smith, A. L., and W. J. Potts: *Appl. Spectrosc.,* 26:262 (1972).
25. Hannah, R. W.: *Inst. News,* 14(3–4):7 (1963). (Perkin-Elmer Corp.)
26. Hansen, W. N., and J. A. Horton: *Anal. Chem.,* 36:783 (1964).
27. Harrick, N. J.: *Anal. Chem.,* 36:188 (1964).
28. Gordy, W.: In W. West (ed.), "Chemical Applications of Spectroscopy," p. 169, Wiley-Interscience, New York, 1956.
29. Gordy, W., W. V. Smith, and R. F. Trambarulo: "Microwave Spectroscopy," Wiley, New York, 1953; reprinted by Dover, New York, 1966.
30. Armstrong, S.: *Appl. Spectrosc.,* 23:575 (1969).
31. Ewing, G. W.: In "Topics in Chemical Instrumentation," p. 120, Chemical Education, Easton, Pa., 1971.
32. McCrory, G. A., and R. T. Scheddel: *Anal. Chem.,* 30:1162 (1958).
33. Williams, V. Z.: *Anal. Chem.,* 29:1551 (1957).
34. Drew, D. M.: *Anal. Chem.,* 45:2423 (1973).

Chapter 6

The Scattering of Radiation

The term *scattering*, as applied to the interaction of radiant energy with matter, covers a variety of phenomena. The word always implies a more or less random change in the direction of propagation. The mechanism involved depends on the wavelength of the radiation, the size and shape of the particles responsible for the scattering, and sometimes their spatial arrangement.

The detailed electromagnetic theory of scattering has been worked out by Mie (1–4), but is too complex to utilize directly. Simplifications can be achieved in restricted areas of application, so it is convenient to distinguish scattering without change in frequency, including *Rayleigh* scattering (in which the particles are small compared to the wavelength), *Tyndall* scattering (for larger particles), and scattering in which the frequency is shifted (the *Raman* effect).

RAYLEIGH SCATTERING

It was shown by Lord Rayleigh in 1871 that radiation falling on a small transparent particle induces in the particle an electric dipole oscillating at the frequency of the radiation. The oscillating dipole then acts as a source, radiating energy in all directions at the same frequency (but not with uniform power in all directions).

It can be shown from the Rayleigh-Mie theory that scattering from small particles is proportional to the inverse fourth power of the wavelength; this neatly accounts for the blue of the sky and the redness of the setting sun, where scattering is predominantly due to particles of molecular dimensions. In chemical systems the exponent of the wavelength may vary from −4 to −2, principally because of larger particles. This variation indicates a gradual transition from Rayleigh to Tyndall scattering.

Measurements are nearly always carried out with visible light. The sample is illuminated by an intense beam, of power P_0 (Fig. 6-1). The transmitted power P_t can then be measured just as in absorptiometry, or the power scattered at a specific angle (such as P_{90} at 90°) may be determined. The ratio P_t/P_0 decreases with increasing number of particles in suspension, while ratios such as P_{90}/P_0 will increase, at least up to moderate concentrations. For very dilute suspensions, measurement at an angle is much more sensitive than in-line measurement, as it involves observation of the faint scattered light against a black background, rather than the comparison of two large quantities of nearly equal values.

In-line measurement is called *turbidimetry*, and can be performed with any standard absorptiometer. Measurement at an angle is usually restricted to 90°, for which standard square cuvets can be used; this method is called *nephelometry*.

The rigorous mathematical treatment of these techniques is not easy. For straight-line measurements a quantity called *turbidance*, corresponding to absorbance, can be obtained, that follows the relation:

$$S = \log \frac{P_0}{P_t} = kbc \qquad\qquad (6\text{-}1)$$

where S represents turbidance, k is a proportionality constant called the *turbidity coefficient*, b is the path length, and c the concentration in grams per liter. (The turbidity function τ, sometimes reported, is equal to $2.303k$.) This expression is valid only for very dilute suspensions because, as c in-

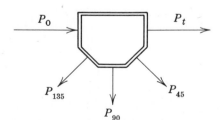

Fig. 6-1 Power relations in light scattering. P_0 is the power of the incident beam, and P_t that of the transmitted beam; P_{45}, P_{90}, and P_{135} are the powers scattered at the corresponding angles.

creases, more and more of the scattered light finds its way into the measuring photocell through multiple scattering.

It is not practicable to derive a theoretical equation for use in nephelometry, relating the scattered light at a given angle of observation to the concentration and other variables. The best we can do is to write as a working relation

$$P_\alpha = K_\alpha c P_0 \tag{6-2}$$

where K is an empirical constant for the system and the subscript α defines the angle of measurement.

The 90° angle of observation is not necessarily the optimum. In some instances a large increase in sensitivity has been found at angles as small as physically possible, just a few degrees from "straight through." However, there are instrumental problems involved which make precision measurements difficult to achieve. If a cylindrical cuvet is employed, the curved surfaces will act as a lens to decollimate the primary beam and to gather into the photocell rays originating from a considerable region in the cuvet, so that the angle of scattering is uncertain and dependent on the refractive indices of both solvent and dispersed particles. On the other hand, prismatic cuvets, either square or multifaced (like that in Fig. 6-1), permit only specified angles to be observed.

Any filter photometer or spectrophotometer can be utilized for turbidimetric measurements, but with limited accuracy and sensitivity. If the solvent and dispersed particles are both colorless, then a wavelength in the blue or near ultraviolet should be selected for maximum sensitivity. If colored, the optimum wavelength had best be determined by trial.

A turbidimeter with unusual properties is the Du Pont Model 430 (Fig. 6-2). This is a double-beam instrument that depends for its opera-

Fig. 6-2 Schematic diagram of Du Pont Model 430 turbidimeter. (*E. I. Du Pont de Nemours & Company.*)

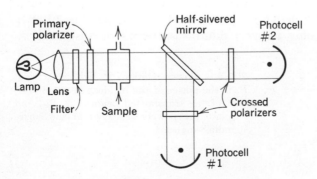

tion on the relative degree of polarization of transmitted and scattered light. If a suspension is illuminated with plane-polarized radiation, the transmitted beam will be found to be depolarized in proportion to the concentration of particles in suspension. In the Du Pont instrument the beam first passes through a polarizer and then the sample. The transmitted light is split by a half-silvered mirror into two beams. One of these passes through another polarizer with its axis parallel to that of the primary polarizer and the other through a similar polarizer with its axis crossed with the primary. The response of photocell no. 1 is *decreased* by the scattering, while that of no. 2 is *increased*. The electronics of the instrument automatically determines the ratio between the two signals, which is directly proportional to the concentration of suspended matter in the sample. The instrument is insensitive to color of solvent or particles or to lamp fluctuations, but cannot be used with solutions that contain optically active constituents.

Kaye (5) has described a low-angle light-scattering photometer using a helium-neon laser as a light source. The laser radiation forms a narrow beam with very little divergence, and the detector can measure scattering as close as 2° to the primary beam. Another advantage of the laser technique is a great reduction in sample size; 2×10^{-5} cm^3 is enough, compared to approximately 0.5 cm^3 as the smallest amount that can be handled in a conventional instrument.

Many manufacturers offer turbidimeters and nephelometers, some in simplified versions intended for particular applications.

ANALYTICAL APPLICATIONS

Light scattering is so highly dependent on the size of the scattering particles that conditions must be rigidly standardized to permit valid comparisons between unknowns and standards. If such precautions are taken, a considerable variety of analytical procedures of great sensitivity and adequate precision become available. Phosphate, for example, can be detected at a concentration of 1 part in more than 300 million parts of water as a precipitate with a strychnine-molybdate reagent. One part of ammonia in 160 million parts of water can be detected by a mercuric chloride complex (Nessler's reagent). Sulfur can be determined by conversion to sulfate and precipitation as the barium salt under conditions which lead to a colloidal suspension; the method is valid to a few parts per million.

MOLECULAR WEIGHTS AND PARTICLE SIZES

Debye has shown that scattering can be used to advantage to determine the weight-average molecular weight of a polymer in solution. The Debye equation calls for knowledge of the turbidity, concentration, index of refraction, wavelength, derivative of the index with respect to concentration, and the second virial coefficient, which is a measure of the nonideality of the solution. It is subject to the restriction that the molecules must be small compared to the wavelength; the details are beyond our present scope.

The diameter of scattering particles can be determined with good precision, provided they are less than about one-twentieth of the wavelength, by means of the equations worked out by Gans from the basic Rayleigh theory. The distribution of scattered light with angle is required, as well as most of the factors previously mentioned.

Several companies manufacture light-scattering photometers for work with high-molecular-weight polymers. These provide a high degree of versatility, with a choice of cuvets (including the semioctagonal type of Fig. 6-1), several angles of observation, polarization-measuring devices which give information about the shape of the particles, etc.

SCATTERING IN GASES

Smoke and fog are visible largely because of light-scattering effects, so it is not surprising that instruments for measuring these effects are useful tools in monitoring atmospheric pollution. Studies have been reported on the diminution of the power in a laser beam over a long path, as from one building to another; turbidity was found to correlate with automobile traffic in the area.

A portable instrument produced by MRI* is diagrammed in Fig. 6-3. A beam of light from the lamp is passed through a distance of about 60 cm to a light trap. Air is pulled through this region by a suction pump. A photocell is situated to one side and baffled so as to receive scattered radiation from the shaded area only. The least detectable turbidity is of the order of 5 μg · m^{-3} for a typical particle size. This instrument is portable enough to be operated in an automobile or aircraft.

* Meteorology Research, Inc., Altadena, Calif.

Fig. 6-3 Nephelometric analyzer for atmospheric particulates. (*Meteorology Research, Inc.*)

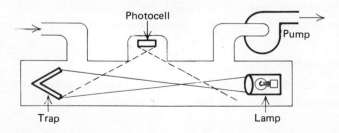

TURBIDIMETRIC TITRATIONS (6, 7)

Ever since the work of Gay-Lussac on silver chloride precipitation (1832), titration to a tubidimetric end point has been extensively employed. Such titrations can be carried out visually or in an absorptiometer.

Titrations of the form $A + B \rightarrow C$, where C is insoluble, might be expected to give a curve of either turbidance or scattered intensity which would consist of two intersecting straight lines (curve 1 of Fig. 6-4), as the amount of precipitate must increase to a maximum and then stay constant. Meehan and Chiu (8), however, have pointed out that this can be true only if the *number* of particles increases linearly to the equivalence point, while all remain the same size. This is not likely to be the case; more probably, added reagent will simultaneously form some new particles and add to those nuclei previously formed. In this situation, titration curves cannot be predicted in detail. If the particles become too large, a poorly defined end point or none at all (curves 2 and 3 of Fig. 6-4) will result. A complicating factor is that, unless the suspension is very dilute, the particles will continue to grow between added increments of titrant, so that excessively long times (several minutes) must be allowed between additions. The method is usable in the 10^{-5} to 10^{-6} M range, with an average relative error of ± 5 percent or more.

Bobtelsky and his coworkers (9, 10) have been successful in carrying out many hundreds of turbidimetric titrations at somewhat higher concentrations, usually 10^{-3} to 10^{-4} M. They made no attempt to control the particle size, and ran the titrations at normal speed. The curves give reproducible and useful results, but do not indicate true turbidance. In view of this, Bobtelsky is careful not to use the term turbidimetry but calls his method *heterometry* and his simplified photometer a *heterometer*.

RAMAN SPECTROSCOPY (11–13)

A sample that is completely free of suspended matter nevertheless scatters some light, the major part following Rayleigh theory. At the same time

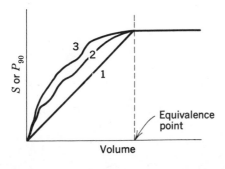

Fig. 6-4 Turbidimetric titration curves. Curve 1 is idealized; curves 2 and 3 might result from precipitates with mixed particle sizes, poor stirring, etc.

there may be Raman scattering, in which a change in frequency occurs. This phenomenon results from an interaction between the incident photons and the vibrational energy levels (or in some cases rotational levels) of the molecules. The interaction is not simply absorption, but rather a transfer of a portion of the energy of the photon to the molecule, or vice versa. In terms of wave numbers (proportional to energy) we can write:

$$\tilde{\nu}_R = \tilde{\nu}_i \pm \Delta\tilde{\nu} \tag{6-3}$$

where the subscript R designates the Raman (scattered) radiation, and i the incident radiation. $\Delta\tilde{\nu}$ is designated the *Raman shift*, and corresponds to the energy difference between vibrational levels, a quantity characteristic of the molecules.

Figure 6-5 shows the transitions possible between the vibrational levels in a polyatomic molecule. The dashed lines represent unstable "virtual" energy levels. The most intense Raman lines result from *Stokes* transitions, those in which energy is taken from the incident radiation by the sample [i.e., the negative sign of Eq. (6-3) applies]. If a significant number of molecules is already in the $\tilde{\nu}_1$ state, *anti-Stokes* lines can be observed. The Raman spectrum of CCl_4 presented in Fig. 6-6 shows both Stokes and anti-Stokes lines, together with the line of unchanged frequency due to Rayleigh scattering. Stokes lines are always more intense than anti-Stokes, and Rayleigh scattering is much more powerful than either.

Interactions of this kind take place if the *polarizability* of the molecule is altered by the vibration, i.e., the shape of the molecule is changed, but

Fig. 6-5 Energy levels in Raman scattering. The dashed lines indicate nonexistent states at levels dictated only by the energy of the incoming photons.

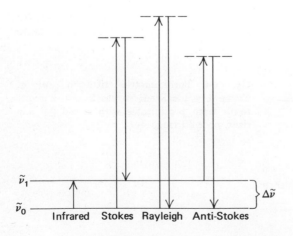

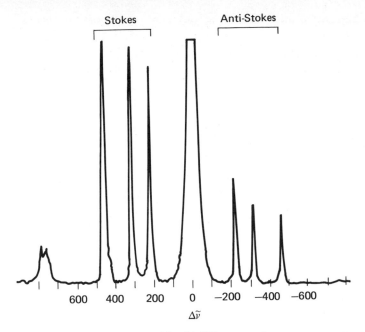

Fig. 6-6   Raman spectrum of liquid CCl_4 scanned at 500 cm^{-1} · min^{-1} using 3-μl sample, helium-neon 632.8-nm excitation. [*Journal of Chemical Education* (13).]

without generating a dipole moment.   Since production of a dipole moment is the requirement for vibrational absorption of infrared, it follows that frequencies observed in the infrared are likely not to be active in the Raman spectrum, and vice versa.   There may be some vibrations that alter both the polarizability and the dipole moment, and hence will be evidenced in both techniques.   Comparison spectra of 1,4-dioxane in the two modes (Fig. 6-7) illustrate these features.

Note that the value of $\Delta\tilde{\nu}$ in Eq. (6-3) does not depend on the actual magnitude of $\tilde{\nu}_i$.   Any incident radiation will give Raman shifts of exactly the same value.   The exciting radiation must be monochromatic if the Raman lines are to be sharp.   Also, if very small Raman shifts are to be observed, close to the intense Rayleigh line, a highly monochromatic source is essential.   Raman spectroscopy is always carried out in the visible region, where the required high dispersion is easily available with less costly optics, and where fluorescence is less of a problem than in the ultraviolet.

Raman work prior to the introduction of lasers used a high-intensity mercury lamp with a bandpass filter to isolate the line at 435.8 nm.   Now, however, lasers are universally employed, as they produce greater power with a much greater degree of monochromaticity.   The helium-neon laser

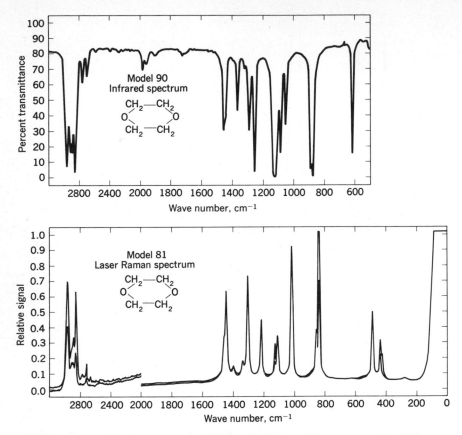

Fig. 6-7 Comparison of Raman and infrared spectra. The spectra of 1,4-dioxane, printed with wave number scales in register, show many points of similarity. Of interest is the band at 1220 cm^{-1}, which is strong in the Raman and almost invisible in the infrared. Conversely, the band at 620 cm^{-1} is active in the infrared but not in the Raman. (*Cary Instruments Division of Varian Associates.*)

line at 632.8 nm is often used; this laser is inexpensive and convenient to use, and the red light seldom excites fluorescence in the sample. However, there are two disadvantages to the helium-neon laser. One is the fourth-power dependence of scattered power on the frequency, which makes a blue line more sensitive than a red line. Also, most standard photomultiplier tubes fall off markedly in sensitivity beyond about 650 nm. A laser with ionized argon as its working material gives an output at 488.0 nm, and this makes a good Raman source unless it causes fluorescence in the sample. Fluorescence, if present, is likely to mask the faint Raman radiation.

Figure 6-8 is a schematic diagram of the Cary 83 Laser Raman Spectrophotometer.

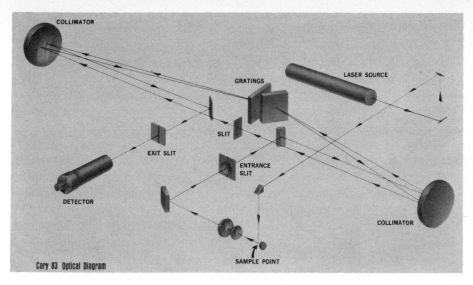

Fig. 6-8. Optical system of the Cary 83 laser Raman spectrophotometer. (*Cary Instruments Division of Varian Associates.*)

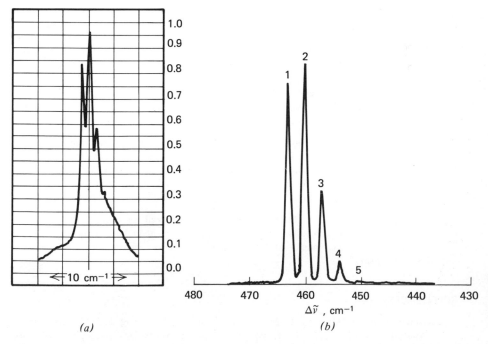

(a)   (b)

Fig. 6-9 Raman spectra of CCl$_4$ symmetric stretch vibration, taken at 0.5 cm^{-1} resolution. (*a*) At room temperature and (*b*) at 77 K, deposited on a cooled copper block. Complex structure arises from the five possible combinations of chlorine isotopes, the intensities reflecting their natural abundances. The spectrum is clearly resolved at low temperature, as bands are narrower and emission originating from hot bands is eliminated. [*Applied Spectroscopy* (15).]

APPLICATIONS OF RAMAN SPECTROSCOPY

As evident from Fig. 6-7, a Raman spectrum can be just as useful in fingerprint identification as the infrared spectrum. Atlases of Raman spectra are available (14). In this application, and in structure elucidation, Raman and infrared supplement rather than duplicate each other (15). The Raman method is not, at the present state of the art, easily applicable to quantitative analyses.

Sample preparation, for liquids, requires careful filtration or equivalent, to remove all traces of suspended matter that would increase Rayleigh or Tyndall scattering. Fluorescent impurities must be rigidly excluded, particularly if a blue light source is used. Filtered gases can be examined in multipass cells. Solids, or thin films deposited on solid supports, can be studied by reflecting the laser beam from the surface. Temperature control is not usually required, but the spectra of cooled samples may give interesting fine structure (Fig. 6-9).

PROBLEMS

6-1 The sulfur content of organic sulfonates and sulfonamides can be determined turbidimetrically by the $BaSO_4$ precipitate following digestion to destroy the organic matter (16). A 25.0-mg sample of a particular preparation of p-toluenesulfonic acid monohydrate, $C_7H_7SO_3H \cdot H_2O$ (formula weight 190.2), was subjected to the digestion, and then exactly one-tenth of it made up to a volume of 10.00 ml with a conditioning solution. Then 5.00 ml of 1.34 F $BaCl_2$ was added with controlled shaking. After 25 min the turbidance at 355 nm in a Spectronic-20 was found to be 0.295. A standard $(NH_4)_2SO_4$ containing 200 μg of sulfur, similarly treated, gave a turbidance of 0.322 and showed a linear relation on dilution prior to treatment. Calculate the percent purity of the original preparation.

6-2 A Raman spectrum of a pure compound is observed, the exciting radiation being the 435.8-nm line of mercury. Raman lines are observed at wavelengths 442.0, 443.5, 449.8, and 462.0 nm. (*a*) Compute the value of the Raman shift for each line in terms of wave number. (*b*) What is the wavelength of the anti-Stokes line corresponding to the Stokes line at 462.0 nm?

6-3 It is reported (15) that Raman lines can be observed as close as 2 cm^{-1} from the exciting radiation in favorable cases, when the helium-neon laser is used as illuminant. What are the wavelength and frequency for the corresponding transitions observed in absorption?

REFERENCES

1. Billmeyer, F. W., Jr.: In I. M. Kolthoff and P. J. Elving (eds.), "Treatise on Analytical Chemistry," pt. I, vol. 5, chap. 56, Wiley-Interscience, New York, 1964.
2. Mie, G.: *Ann. Physik*, **25**:377 (1908).
3. Van de Hulst, H. C.: "Light Scattering by Small Particles," Wiley, New York, 1957.
4. Gravatt, C. C., Jr.: *Appl. Spectrosc.*, **25**:509 (1971).

5. Kaye, W.: *Anal. Chem.*, **45**:221A (1973).

6. Coetzee, J. F.: In I. M. Kolthoff and P. J. Elving (eds.), "Treatise on Analytical Chemistry," pt. I, vol. 1, chap. 19, Wiley-Interscience, New York, 1959.

7. Underwood, A. L.: In "Advances in Analytical Chemistry and Instrumentation," vol. 3, p. 31, Wiley-Interscience, New York, 1964.

8. Meehan, E. J., and G. Chiu: *Anal. Chem.*, **36**:536 (1964).

9. Bobtelsky, M.: *Anal. Chim. Acta*, **13**:172 (1955).

10. Bobtelsky, M.: "Heterometry," American Elsevier, New York, 1960.

11. Rosenbaum, E. J.: In I. M. Kolthoff and P. J. Elving (eds.), "Treatise on Analytical Chemistry," pt. I, vol. 6, chap. 67, Wiley-Interscience, New York, 1965.

12. Szymanski, H. A. (ed.): "Raman Spectroscopy: Theory and Practice," Plenum, New York, 1967.

13. Bulkin, B. J.: *J. Chem. Educ.*, **46**:A781, A859 (1969).

14. Sadtler Research Laboratory, Inc., Philadelphia.

15. Sloane, H. J.: *Appl. Spectrosc.*, **25**:430 (1971).

16. Zdybek, G., D. S. McCann, and A. J. Boyle: *Anal. Chem.*, **32**:558 (1960).

Chapter 7

Atomic Absorption

In this chapter we shall consider absorption by atomic rather than molecular species. Only in the gaseous state can the optical properties of free atoms be observed, so sample preparation in the majority of cases requires volatilization followed by dissociation of molecules to atoms.

Atomic absorption (AA) obeys the same general laws as absorption by molecules, described in previous chapters. Hence an AA spectrophotometer must have the same sequence of components, altered where necessary to meet different requirements. The most striking departure is in the sample itself, so we will start there in our discussion.

ATOMIZATION

The conversion of the metallic elements of a sample from solution to the dissociated vapor can usually be accomplished by heat energy, either in a flame or by an electric furnace. Careful control of the temperature is needed for optimum conversion to the atomic vapor. Too high a temperature can be just as unfavorable as too low, as it will cause a fraction of the atoms to become ionized, and ions do not absorb at the same wavelengths as neutral atoms.

FLAME ATOMIZERS

A typical burner for AA is shown in Fig. 7-1. The fuel and oxidant gases are fed into a mixing chamber, whence they proceed through a series of baffles, which ensure complete mixing, to the burner head. The flame orifice is in the form of a long, narrow slot, so that a ribbon flame is produced. The sample, in solution, is aspirated into the mixing chamber by a small air jet. A fault of an aspirator of this kind is that it produces droplets of widely differing sizes. In this burner, the baffles have the added function of intercepting the larger drops, so that those reaching the flame are more nearly uniform, leading to improved reproducibility.

Such a burner, using premixed gases, obviously presents a hazard, for if the flame should strike back into the mixing chamber, a violent explosion would ensue. The likelihood of strike-back is minimized by making the burner slot as narrow as possible so that gases will blow through at a high velocity and by making the metal parts around the slot rather massive so that heat will be conducted away readily. Even then, explosion can occur if the gas flow is not adjusted correctly. The burner should be so made that it will come apart harmlessly and can be reassembled. A heavy safety shield should always be in place when the burner is operating.

Compressed air and acetylene are most commonly chosen as oxidant and fuel for AA. The maximum temperature obtainable is about 2200°C. When higher temperatures are needed, nitrous oxide can be substituted for the air. This gas decomposes to give a 2:1 mixture of nitrogen and oxygen, compared to the 4:1 ratio in air; the highest temperature it will give when burning acetylene is almost 3000°C. Pure oxygen cannot be used with acetylene in a premix burner, because the flame propagates so rapidly that strike-back cannot be avoided.

Although the flame is a convenient and reproducible source of heat, it is less than ideal as a sampling device for AA, in that the two sequential endothermic processes, solvent evaporation and atomization, must take

Fig. 7-1 Laminar-flow, premix burner for AA.

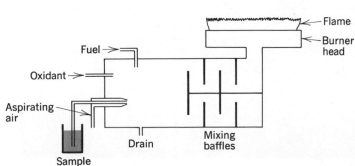

place within the very short time interval that it takes for a particle to shoot through the flame. In addition, the flame introduces significant random fluctuations in the effective optical path, because of turbulence, and this causes excessive electrical noise, reducing sensitivity.

An improvement can be achieved by using the flame only as a heat source. The dissolved sample is placed in a metal boat or cup, called a *Delves cup* (1), and dried by holding it for a few moments close to, but not in, the flame. Alternatively, drying can be done independently on a hot plate. The cup is then held in the hottest region of the flame to vaporize the sample (Fig. 7-2). The vapor enters a horizontal quartz tube heated by the ribbon flame and is measured there. The quartz tube serves as a holding tank and appreciably lengthens the time that atoms can be observed, hence greatly increasing sensitivity. This technique, using nickel cups, has been found particularly useful for determinations of trace lead. Nickel cups cannot be used with acidic materials, so such samples must be neutralized or the analysis performed in a cup made of tantalum or other inert material.

Fig. 7-2 Delves microsampling cup, showing quartz tube for delaying the vapor. (*Perkin-Elmer Corporation.*)

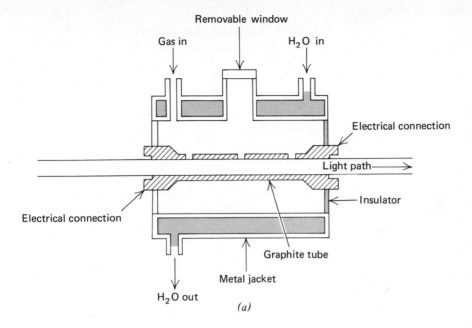

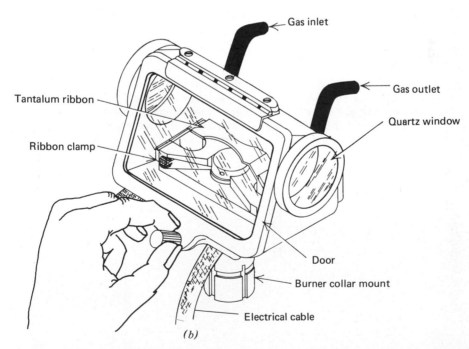

Fig. 7-3 (*a*) A graphite furnace for atomization of samples. The sample is inserted from above, through the removable window. The furnace is enclosed in a water-cooled metal jacket. (*Perkin-Elmer Corporation.*)   (*b*) A tantalum ribbon heater for AA sampling. (*Instrumentation Laboratories, Inc.*)

NONFLAME ATOMIZERS

A relatively recent development has been the introduction of electrical heating devices to replace the flame (2). In one form, the sample (1 to 2 μl) is placed by micropipet in a depression on a horizontal rod of graphite (3) or ribbon of tantalum (4). In another a graphite rod is drilled axially to provide a path for the light beam, and the sample is introduced through a hole in one side (5). Figure 7-3 shows these modifications. In each, the temperature can be controlled electrically with much greater precision than is possible in a flame. It is usually desirable to raise the temperature in steps, first just warm enough to evaporate the solvent, then hot enough to decompose organic matter, if present, followed by a suitably high temperature to atomize the desired metals.

Nonflame methods have come to be about equally as prevalent as the use of flames, and it seems likely that their use will increase in the future.

SOURCES OF RADIATION

In principle, a continuous source should be usable in AA, just as in molecular absorption. Work has been reported along these lines (6, 7), but so far it has not been found practicable. The difficulty is caused by the extreme narrowness of the absorption line, of the order of 0.001 nm, which removes only a minute fraction of the energy from the narrowest spectral region passed by a conventional monochromator. Figure 7-4 shows this

Fig. 7-4 AA of radiation from a continuous source with a monochromator. Curve *a* represents the band of wavelengths passed by the monochromator, and curve *b* the absorption by an atomic species in the flame. Curve *b* should be many times narrower than shown, in comparison with curve *a*. (*Unicam Instruments, Ltd.*)

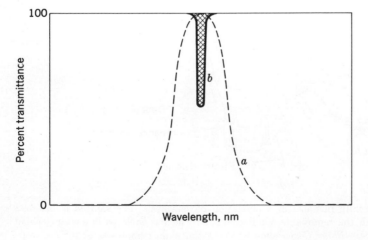

relation. The area beneath curve *a* is only slightly diminished by the absorption of the narrow line at *b*. In order to obtain results with adequate precision, a spectrophotometer with high dispersion is necessary, together with an expanded-scale recorder, so that integrated intensities can be read with precision.

More practical, at the present state of development, is the use of a series of sources that give sharp emission lines for specific elements. The most successful of such sources is the hollow-cathode, glow-discharge lamp (8). This consists of two electrodes, one of which is cup-shaped and made of the specified element (or an alloy of that element) (Fig. 7-5); the material of the anode is not critical. The lamp is filled with a low pressure of a noble gas. Application of 100 to 200 V will produce, after a short warm-up period, a glow discharge with most of the emission coming from within the hollow cathode. Positive ions provided by the inert gas bombard the cathode, removing metal atoms by a process known as sputtering. These atoms then accept energy of excitation and emit radiation. The emissions consist of discrete lines of the metal, plus those of the fill gas. The gas is selected by the manufacturer to give the least spectral interference with the metal concerned. The spectrum of a typical hollow-cathode lamp is shown in Fig. 7-6. Note that the sensitive 232-nm line is surrounded by numerous other nickel lines that are not absorbed by nickel vapor. The desired line can be isolated by a narrow bandpass monochromator.

Figure 7-7 shows schematically the relation between the emission spectrum of a hollow-cathode lamp (*a*), the same, as seen through a monochromator (*b*), and the effect of absorption in the metal vapor (*c*). Compare this series with Fig. 7-4. In the present case, only the peak heights need be measured, since the absorption and emission lines have nearly equal half widths.

It is possible to fabricate hollow-cathode lamps with a mixture (or alloy) of several metals lining the cathode cup, so that a number of elements can be determined without the necessity of changing lamps. Examples are: Ca, Mg, and Al; Fe, Cu, and Mn; Cu, Zn, Pb, and Sn; and Cr, Co, Cu, Fe, Mn, and Ni.

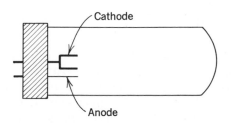

Fig. 7-5 Hollow-cathode lamp.

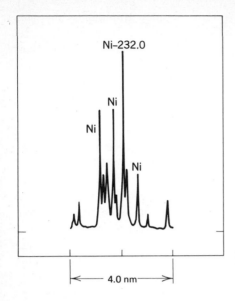

Fig. 7-6 Emission from a lamp with a hollow nickel cathode. Only the line at 232.0 nm is absorbed by Ni atoms in a flame. (*Westinghouse Electric Corporation.*)

PHOTOMETRIC SYSTEMS

The double-beam arrangement as used with conventional spectrophotometers is not immediately applicable in AA. This is particularly true with flame atomizers, because there is no way in which a blank flame can be made to duplicate exactly the analytical flame. Double-beam photometry still has the merit of correcting for fluctuations in the lamp output, and since hollow-cathode lamps require several minutes to stabilize when turned on cold, considerable time can be saved when changing lamps.

An improved method that corrects for background effects involves the use of a continuous source simultaneously with the line source (9, 10) (Fig. 7-8). Radiation from a hydrogen or deuterium lamp passes through the sample along with the resonance radiation from the hollow-cathode lamp. The electronic system sorts out the signals from the two sources and takes their ratio. One manufacturer (Perkin-Elmer) alternates the two by means of a chopper; another (Instrumentation Laboratories) sends both beams through the optical system at the same time but chops them at different rates. The two beams are attenuated equally by background absorption or scattering, but only the resonance radiation is appreciably absorbed in the sample. Figure 7-9 shows an example of the application of this principle.

A significant fraction of metal atoms in the flame or vapor will be raised to excited levels, either thermally or by absorption of energy from the incident light beam. These atoms will emit resonance radiation in all directions, at exactly the same wavelength that the monochromator is set to transmit. If the radiation from the hollow-cathode lamp were

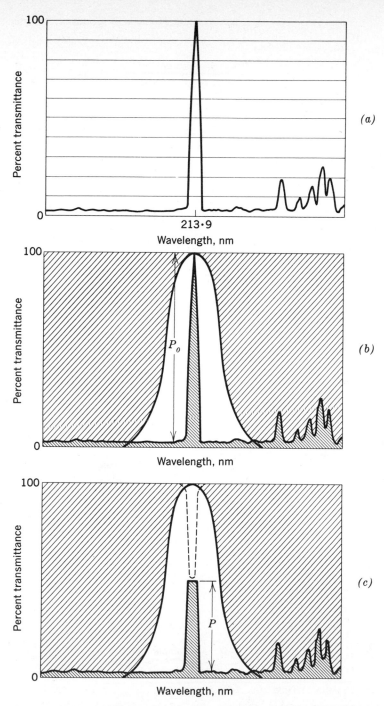

Fig. 7-7 AA. (*a*) The spectrum of a zinc cathode lamp (the width of the 213.9-nm line has been exaggerated for clarity). (*b*) Extraneous zinc lines are eliminated by a monochromator centered at 213.9 nm. (*c*) The power of the 213.9-nm line is sharply reduced by absorbing zinc atoms. (*Unicam Instruments, Ltd.*)

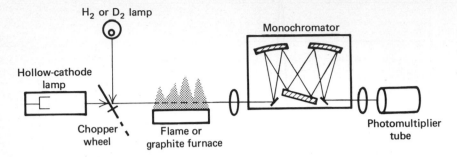

Fig. 7-8   AA spectrometer with H_2 (or D_2) lamp for background correction.

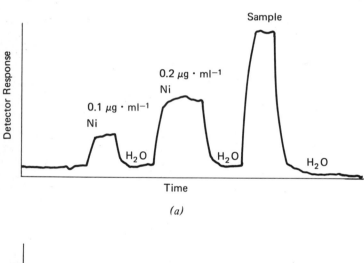

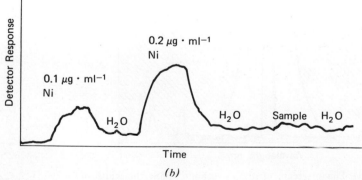

Fig. 7-9   AA traces taken sequentially on two standard Ni solutions and a sample of KCl.   (*a*) Without background corrector, 0.4 ppm Ni is indicated; (*b*) with D_2 background corrector, the sample is shown to be free of Ni.   (*Perkin-Elmer Corporation.*)

continuous, there would be no way to distinguish between this spurious radiation and that transmitted from the lamp. Hence it is essential that the lamp's radiation be interrupted, either by a chopping wheel or by a pulsed potential applied to the lamp itself. The detector electronics can be tuned to respond only to the corresponding frequency, thus discriminating against thermally excited emission. The component due to resonance fluorescence will vary at the chopping frequency, and so cannot be distinguished from the transmitted radiation; it is concentration-dependent, and its only effect is to diminish somewhat the AA sensitivity. Atomic fluorescence, as an analytical tool itself, will be considered in the next section.

ATOMIC FLUORESCENCE

Atomic vapors, whether produced in a flame or graphite furnace or otherwise, will fluoresce under suitable excitation. Most important for analytical purposes is resonance fluorescence, in which the exciting and emitted radiations are at the same wavelength (11, 12).

The equipment necessary to measure atomic fluorescence (AF) is similar to that for AA, except that observation of light emitted from the cell is at an angle, usually 90°. Since the faint fluorescent light is observed against a dark background, the arguments against a continuous source for AA no longer apply. However, most continuous sources, such as the high-power xenon arc, do not have sufficient brightness in the ultraviolet below about 250 nm, where the resonance lines of many elements occur. Hollow-cathode lamps are generally too weak for this application. The most used source is an *electrodeless discharge lamp*. This consists of a sealed quartz tube containing a small quantity of the pure metal and a low pressure of argon. It is excited by an intense microwave field in a waveguide cavity and emits approximately the same spectrum as a hollow-cathode lamp of the same metal, but with greater intensity. Such lamps can be made with two or more metals in one tube, and more than one tube can be operated at the same time (13), so that several analyses can be made without changing lamps.

As an example of analysis by AF, consider the simultaneous determination of four elements (13). The sample was aspirated into an air-acetylene flame and illuminated at slightly different angles by radiation from two dual-element electrodeless discharge lamps. The wavelength range including the resonances of the four elements was scanned automatically. With Zn–Cd and Ni–Co lamps, the span of 210 to 245 nm was suitable. Detection limits of 0.02 ppm for Ni and Co, 0.003 ppm for Zn, and 0.002 ppm for Cd were reported.

DETECTION LIMITS IN AA AND AF

The sensitivities of these methods depend in a complicated way on the optical properties of the atomic vapor, the temperature, the relative line

Table 7-1   Detection limits (μg · ml^{-1}) for metals by AA, AF, and AE*

Limits	AA	AF	AE
<10^{-4}	....	Ag, Cd, Zn	Ca, Li, Na
<10^{-3}	Ag, Ca, Cd, Mg	Cu, Hg, Mg	Sr
<10^{-2}	Au, Be, Co, Cr, Cu, Fe, K, Li, Mn, Na, Ni, Pb, Rb, Sr, Zn	Au, Be, Bi, Co, Fe, Ga, Mn, Ni, Pb, Tl	Al, Ba, Cr, Cs, Cu, Fe, Ga, In, K, Mg, Mn, Rb, V
<10^{-1}	Al, As, Ba, Bi, Cs, Ga, Ge, In, Mo, Pt, Sb, Se, Si, Sn, Ti, Tl, V	Al, As, Ca, Cr, Ge, In, Sb, Se, Sn, Sr, V	Ag, Be, Co, Mo, Tl, V
<1	Hg	Mo, Si	Ge, Ni, Pb, Sn

* AE = atomic emission; see Chap. 8.

widths of lamp and absorber, and the geometry of the optical system. In AF, in addition, the quantum efficiency of fluorescence is involved. The governing equations for these cases have been worked out (14), but are not needed for practical analyses.

Table 7-1 gives detection limits for common elements for which limits less than 1 μg · ml^{-1} have been reported (11, 14) (see the references for a more complete list). It will be seen that AA is more sensitive for some elements (Ca, Cr, Mo, Si, Sr) and AF is more sensitive for others (Ag, Bi, Cd, Cu, Ga, Hg, Tl, Zn), while for the others listed the two methods are comparable.

INTERFERENCES

The chemical reactions taking place in flames can give rise to interferences in both AA and AF using a flame as sample medium. The chief difficulty is due to incomplete dissociation or to the formation of refractory compounds. Some elements, like Ti, Al, and V, are oxidized in the flame, forming compounds that withstand the temperature of the air-acetylene flame; most such compounds yield to the N_2O-acetylene flame. In other cases, the element being sought forms stable compounds with some other constituent of the sample. For example (8), consider the determination of Sr in the presence of Al or Si. Figure 7-10 shows how greatly the analysis is interfered with. This interference can be almost eliminated by addition to the sample of a small quantity of a La salt. La preferentially binds the Al or Si, leaving the Sr free.

Another sort of interference is found in the examination of Ca salts. Ca compounds are completely dissociated in the N_2O-acetylene flame, but the temperature is so high that an appreciable fraction of the Ca atoms are ionized and hence lost to the analysis. This effect can be controlled

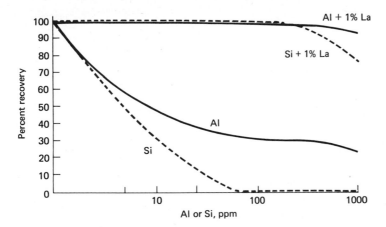

Fig. 7-10 Removal of chemical interferences; Al and Si both severely depress the absorption of Sr. Addition of La to the solution removes the interference by binding the Al and Si. (*Perkin-Elmer Corporation.*)

by the addition of a more easily ionized element such as Na, to keep the Ca in its reduced state.

Another source of interference arises from differences in viscosity or other bulk property of the solution, altering the ease with which it is aspirated and transported through the flame. Thus two solutions with the same concentration of metal but varying amounts of other extraneous materials may give divergent instrument readings.

With nonflame atomizers, the problems are similar. In graphite furnaces, it is carbon rather than oxygen that forms refractory compounds. The carbides of aluminum and silicon can be dissociated thermally, but not those of tungsten and boron. It is reported (15) that by coating the graphite surface with pyrolytic graphite or vitreous carbon (highly impervious forms of the element), carbide formation is prevented; this coating also decreases penetration of the sample into the porous graphite, hence making for easier cleaning and prolonged life.

Interference in AA due to excessive background can be compensated by use of a hydrogen or deuterium lamp, as described earlier.

A form of interference of particular importance to AF is scattering of the primary beam of radiation by particulates and residual droplets of solution. The latter source can be removed by proper adjustment of the burner, but there is no easy way to correct for scattering from solid particles. This phenomenon can usually be recognized by running a series of solutions containing increasing aliquots of the desired metal. If the supposed fluorescent light is due to scattering, its magnitude will change but little with increased concentration.

APPLICATIONS

AF is a relative newcomer that is just establishing its place as an analytical tool. As more instruments become commercially available, it can be expected to supplement AA in determinations in which it provides added sensitivity.

AA, on the other hand, is a well-established analytical method. It is the method of choice for the determination of a large number of metals, especially but not exclusively at trace levels. It is widely used in such fields as water analysis (16), pharmaceutical analysis (17), where it is applied to the determination of essential metallic salts as components of various preparations, and metallurgy (18, 19), where it is found useful for analysis of major constituents of alloys and ores, as well as for minor and trace elements.

An interesting application of AA is the isotopic analysis of Li (20). The Li line at 670.8 nm actually consists of a doublet with a spacing of 0.015 nm. The isotope shift is also 0.015 nm, with the ^{6}Li lines at longer wavelengths than the ^{7}Li. Examination of a Li sample first with a ^{6}Li hollow-cathode lamp and then with a ^{7}Li lamp gives sufficient data to determine the isotope ratio to three significant figures.

Each of the major manufacturers of AA equipment publishes an extensive series of analytical methods, giving all necessary details of the chemistry involved, as well as the optimum settings for their instruments.

PROBLEMS

7-1 Sodium can be measured in the 0.5 to 2 percent range by AA using the 330.259- and 330.294-nm nonresonant radiation from a zinc hollow-cathode lamp. This is about one-fiftieth as sensitive as the sodium secondary resonances at 330.232 and 330.299 nm from a sodium lamp, and the brightness at this wavelength is about half as great for the zinc lamp as for the sodium lamp. The presence of zinc in the absorption flame gives no interference in the determination of sodium. (*a*) Account for the seeming contradiction that, although the lamp is half as bright, the sensitivity is only one-fiftieth in the zinc lamp as compared to the sodium lamp. (*b*) Why does not zinc interfere?

7-2 To compare quantitatively the effectiveness of continuous and line (hollow-cathode) sources in AA, the following data were taken. A xenon lamp with a monochromator gave a band of which the full width at half-maximum height (FWHM) was 5 nm, whereas the radiation from a hollow-cathode lamp showed FWHM = 0.01 nm. The absorption curve of the corresponding element in a flame showed an absorbance FWHM = 0.02 nm (see Fig. 7-11). In a particular pair of experiments, one with the hollow-cathode and another with the xenon lamp, the concentration was such that the peak height of radiation within the absorption band was diminished

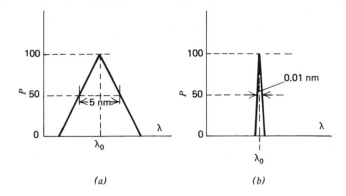

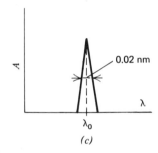

Fig. 7-11 (*a*) Emission band from a continuous source with a mono-chromator; (*b*) emission from a hollow-cathode lamp; (*c*) corresponding absorption line.

by absorption to one-half its value at zero concentration. How does the integrated area beneath the power-wavelength curves compare for radiation from the two sources after passage through the flame? Assume triangular line shapes and neglect stray light effects.

7-3 To monitor the presence of mercury vapor in the atmosphere in a metallurgical laboratory, a mercury-vapor lamp with its principal output at 253.7 nm, the resonance line, was set up as shown in Fig. 7-12, with a suitable photodetector at a total distance of 4 m. For calibration, the same lamp and photocell were placed close to each other, so that the radiation passed through a silica cuvet with a 2-cm path. With the calibration cuvet at 100°C, a drop of mercury produces a vapor pressure of 36.4 Pa (= 0.273 torr). If the observed absorbance in the long-path experiment was just 1.0 percent of that with the cuvet, what is the indicated partial pressure of mercury vapor in the room?

7-4 Aldehydes can be determined by an indirect AA procedure (21). The aldehyde is oxidized by Ag^+ ion (Tollen's reagent), resulting in the formation of 2 moles

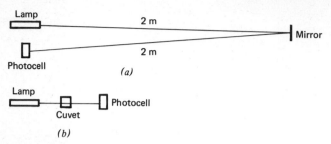

Fig. 7-12 Mercury lamp and photocell. (*a*) At a distance; (*b*) oriented for calibration.

of elementary silver for each mole of aldehyde. The silver is subsequently dissolved in HNO_3 and determined by AA at 328.1 nm. The linear range for the silver determination is stated to be 2 to 20 $\mu g \cdot ml^{-1}$. What range of quantities of aldehydes would this correspond to (in micromoles) if the samples were diluted to 10 ml?

REFERENCES

1. Delves, H. T.: *Analyst* (London), **95**:431 (1970).
2. Robinson, J. W., and P. J. Slevin: *Am. Lab.*, **4**(8):10 (1972).
3. Anderson, R. G., H. N. Johnson, and T. S. West: *Anal. Chim. Acta*, **57**:281 (1971).
4. Hwang, J. Y., C. J. Mokeler, and P. A. Ullucci: *Anal. Chem.*, **44**:2018 (1972).
5. Welz, B., and E. Wiedeking: *Z. anal. Chem.*, **252**:111 (1970).
6. Fassel, V. A., V. G. Mossotti, W. E. L. Grossman, and R. N. Kniseley: *Spectrochim. Acta*, **22**:347 (1966).
7. Veillon, C., J. M. Mansfield, M. L. Parsons, and J. D. Winefordner: *Anal. Chem.*, **38**:204 (1966).
8. Kahn, H. L.: *J. Chem. Educ.*, **43**:A7, A103, (1966).
9. L'vov, B. V.: *Spectrochim. Acta*, **24B**:53 (1969).
10. Hwang, J. Y., P. A. Ullucci, and C. J. Mokeler: *Anal. Chem.*, **45**:795 (1973).
11. Winefordner, J. D., and R. C. Elser: *Anal. Chem.*, **43**(4):24A (1971).
12. Smith, R.: In J. D. Winefordner (ed.), "Spectrochemical Methods of Analysis," chap. IV, Wiley-Interscience, New York, 1971.
13. Norris, J. D., and T. S. West: *Anal. Chem.*, **45**:226 (1973).
14. Winefordner, J. D., V. Svoboda, and L. J. Cline: *CRC Crit. Rev. Anal. Chem.*, **1**:233 (1970).
15. Amos, M. D.: *Am. Lab.*, **4**(8):57 (1972).
16. Parker, C. R.: "Water Analysis by Atomic Absorption," Varian Techtron, Palo Alto, Calif., 1972.
17. Smith, R. V.: *Am. Lab.*, **5**(3):27 (1973).
18. Taylor, R. W.: *Am. Lab.*, **2**(11):33 (1970).
19. Smith, D. C., J. R. Johnson, and G. C. Soth: *Appl. Spectrosc.*, **24**:576 (1970).
20. Wheat, J. A.: *Appl. Spectrosc.*, **25**:328 (1971).
21. Oles, P. J., and S. Siggia: *Anal. Chem.*, **46**:911 (1974).

Chapter 8

Emission Spectroscopy:
Electrical and
Flame Excitation

Ever since the work of Bunsen and Kirchhoff (1860), it has been known that many elements under suitable excitation emit radiations of characteristic wavelengths. This fact is utilized in the familiar qualitative flame tests for the alkali and alkaline earth elements. By employing more powerful electrical excitation in place of the flame, the method can be extended to all metallic and many nonmetallic elements. With some, such as Na and K, the spectra are simple, consisting of only a few wavelengths, while in others, including Fe and U, thousands of distinct reproducible wavelengths are present.

Quantitative analysis with the spectrograph is based on an empirical relation between the power of the emitted radiation of some particular wavelength and the quantity of the corresponding element in the sample. The radiant power is influenced in a complicated way by many variables, including the temperature of the exciting arc and the size, shape, and material of the electrodes. Hence procedures must be rigidly standardized, and the spectra of unknowns must always be compared with those of standard samples prepared with the same apparatus under identical conditions.

Spectrography utilizing photographic detection saw extensive development largely during the period between the two world wars. The present trend is to replace large emission spectrographs with other types of instruments that will achieve the same (or better) results more quickly, with a saving in laboratory space, and frequently without requiring as skilled an operator. These newer techniques include notably x-ray fluorescent emission and AA. Emission spectroscopy retains a place in direct-reading photoelectric instrumentation and in flame photometry. Photographic instruments now manufactured are primarily small ones intended for the quick qualitative identification that must often precede quantitative analysis. For these reasons, then, only a brief account of classical spectrographic methodology will be given here. There are many excellent books available with more complete treatments (1–3).

EXCITATION OF SAMPLES

There are several ways in which emission spectra can be produced. Gaseous substances are easily excited by passing a high-voltage electric discharge through the sample contained in a glass tube.

More elaborate methods are necessary for solid samples. In the usual procedure, a powerful discharge is passed between two portions of the sample or between the sample and a *counterelectrode* which does not contain the elements being determined. If the sample is not obtainable in the form of a rod, but rather as a powder or in solution, it may be placed in a recess drilled in the end of a graphite electrode, which is then connected as the lower electrode (usually positive). The upper electrode may be a pointed rod of graphite. Graphite has the advantage for this purpose of being highly refractory (that is, it does not melt or sublime at the temperature of the arc); it is a sufficiently good electric conductor; and it introduces no spectral lines of its own. A drawback, however, is that the hot carbon reacts slightly with the nitrogen of the air to form cyanogen gas, which becomes excited to give bands of luminosity in the region 360 to 420 nm. This can be avoided, if necessary, by enclosing the arc in a mantle containing steam or an inert gas.

The *dc arc* is the most sensitive method of excitation and is widely used for qualitative analysis of metals. A current of 5 to 15 A at 220 V is caused to flow through the arc in series with a variable resistor (10 to 40 Ω). The chief disadvantage of the dc arc is that it is less reproducible than might be desired. The discharge tends to become localized in "hot spots" on the surface of the electrode, which results in uneven sampling. This defect is overcome in the *ac arc*, wherein the discharge is automatically interrupted 120 times every second. A source of 2000 to 5000 V at 1 to 5 A is required. The flow of current is controlled by a variable inductor or resistor in the circuit. The ac arc is particularly suitable for the analysis of the residues from solutions which have been evaporated onto the surface of an electrode.

The *spark* source is energized from the ac power lines by means of the *Feussner* circuit (Fig. 8-1). High voltage (15,000 to 40,000) is obtained through a step-up transformer or Tesla coil. The secondary winding is connected across a capacitor and in series with the spark gap, an inductance coil, and an auxiliary spark gap operated by a motor rotating synchronously with the alternations of the line current. The purpose of the synchronous gap is to make sure that the spark only passes at the moment of highest voltage, thereby insuring reproducibility. The spark excitation is better suited than the arc to precise timing of exposures, and in addition does not destroy the sample, as only minute amounts are vaporized.

Many modifications of arc and spark circuits, including intermediate forms, have been described, some of which are quite ingenious and complex. An excellent summary is given by Jarrell in the "Encyclopedia of Spectroscopy" (4).

A recent development of significance is the application of the laser as a high-energy source (5). The laser beam can be focused on a minute area of the sample, causing localized vaporization of even the most refractory materials. The vapor may have received sufficient excitation thermally, or an electric discharge may be superimposed. The localization of the effect can be an advantage, permitting examination of areas as small as 50 μm in diameter, or it can be a disadvantage, rendering more difficult the representative analysis of a macro sample. An advantage of laser excitation is that the sample need not be electrically conducting.

PREPARATION OF ELECTRODES AND SAMPLES

The electrodes between which the arc or spark is to be struck can be of several kinds. If the sample is a metal and an adequate amount is available, it can be used directly as a pair of *self-electrodes*, with the arc passing between two pieces of the sample, usually in rod form. If only a single electrode can be fashioned from the sample, a *counterelectrode* of some non-

Fig. 8-1 Feussner circuit for a spark source.

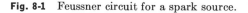
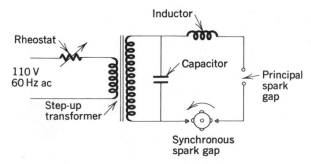

interfering material may be employed. The counterelectrode may be graphite or it may be one of the major constituents of the sample, in pure form. For example, in analyzing a steel for alloyed elements, the counterelectrode could be a simple steel, shown spectrographically to be free of the elements sought.

Samples other than metals are deposited in or on a noninterfering electrode. By one method, the sample is brought into solution, with acid or other reagent if necessary; a few drops are placed in a hole drilled in the end of a carbon electrode (Fig. 8-2e), and the solvent is evaporated by preheating the electrode in a small oven. This electrode is then positioned in the lower holder with a plain carbon counterelectrode (Fig. 8-2c) above it, for arcing. In another procedure, the dissolved sample is placed in a deep hole drilled nearly all the way through a porous carbon electrode (Fig. 8-2g), which is made the upper electrode; the sample is allowed to seep out through the pores directly into the arc. In still another method, the

Fig. 8-2 Representative shapes of graphite electrodes. Forms *a* and *c* are intended as counterelectrodes. Forms *b*, *d*, *e*, and *f* are provided with cups into which the sample may be inserted; *b* and *d* have a central post to which the arc is struck. The narrow neck is to reduce thermal conductivity. Form *f* has a particularly large heat capacity. Form *g* is drilled nearly to the bottom and is porous. (*Ultra Carbon Corporation.*)

sample in powder form is mixed with pure graphite powder and poured dry into the cup of an electrode (Fig. 8-2e). The added graphite, being an electric conductor, tends to stabilize the arc. Many modified procedures have been published which show advantages in particular kinds of work.

For laser excitation, compression of a powdered sample together with an inert matrix to form a pellet has been reported (5). A press intended for KBr pellet formation was used.

The spectroscopist must always keep in mind possible differences in volatility of components of the sample. In some cases one or more components may volatilize and be completely burned within a half-minute or so after the arc is started, before other substances present have been heated sufficiently to appear in the arc. This may be a disadvantage, especially in the case of trace amounts, where the sample may have lost some constituents before sufficient exposure has been given to the photographic plate. It is sometimes possible to take advantage of differential volatility, in that spectra of volatile constituents may be recorded without interference from the less volatile. An example is the determination of lithium, aluminum, and other oxides as impurities in uranium oxide (6); uranium is especially rich in spectral lines, and hence determination of trace impurities is difficult. In this method, the uranium is first converted to the nonvolatile U_3O_8, and 2 percent of its weight of Ga_2O_3, a comparatively volatile oxide, is added. The Ga_2O_3 acts as a carrier to sweep out even minute quantities of the impurities into the arc. This results in high sensitivity and accuracy, even for impurities present only to the extent of a few parts per million.

IDENTIFICATION OF LINES

For qualitative spectrographic analysis it is necessary only to identify the element responsible for the emission of each wavelength present in the spectrum of the unknown. This is done by comparison with spectra obtained from authentic samples of pure elements. All the known wavelengths for all the elements are listed in reference tables, but to take advantage of the tables it must be possible to determine accurately the wavelengths of the lines produced by the unknown. Alternatively, spectra of various possible elements may be photographed with the same spectrograph and compared line by line with the unknown.

The most convenient method of determining wavelengths of unknown lines is by comparison with the known lines of a standard photographed on the same plate. An example is given in Fig. 8-3. An arc struck between iron electrodes provides the most common standard. It is chosen because it produces thousands of lines fairly evenly spaced throughout the visible and ultraviolet. All these iron lines have been measured repeatedly with great care on both prism and grating spectrographs of high dispersion, and

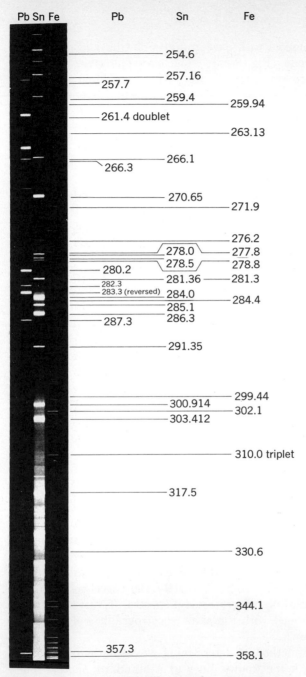

Fig. 8-3 Spectra of lead, tin, and iron photographed on a large Littrow prism spectrograph. Some of the brighter lead lines are visible in the tin spectrum on the original plate, showing that the sample contained some lead. The 283.3-nm line of lead is so intense that it shows photographic reversal (dark center) on the original plate. (*Bausch & Lomb, Inc.*)

thus can serve as a measuring scale for other spectra. In practice the spectrum of the unknown is sandwiched between duplicate spectra of iron recorded immediately next to it on the same plate.

QUANTITATIVE ANALYSIS

In a photographic instrument, a quantitative analysis can be carried out only by comparison of the optical density of the silver deposit caused by an emission line with the similar deposit derived from a standard. Because of the many variables introduced by the photographic process, this can be done to best advantage by the method of *internal standards*. This method depends on measurement of the ratio of the radiant power of a given line of the unknown to that of some line of another constituent of the sample which is present in known (or at least constant) amount. This standard may be an element already present in the specimen, such as the iron in a steel sample, or it may be an extraneous element added in known quantity to all samples. This procedure completely eliminates errors due to inequalities of plate characteristics and development. The line to be used as a standard should be as close as possible to the unknown in wavelength and also in power, so that nonlinearity of the photographic emulsion with respect to these factors will not be a serious source of error. Two lines selected as particularly appropriate for this purpose are called a *homologous pair*.

A typical analytical procedure will be presented in some detail, as an illustration of this technique. The example chosen is the analysis of traces of magnesium in solution, with molybdenum as internal standard. It is taken from a series of similar procedures for many elements described by Nachtrieb (2), to whose book the student is referred for further details.

A condensed spark discharge between Cu electrodes has been found convenient for this work. The electrodes are rods 3 to 4 cm long and 5 mm in diameter. The tips must be carefully machined smooth and flat in a lathe to ensure uniformity and to remove surface contamination. After machining, they must be protected from dust and handled only with forceps. In preparation for use, two electrodes are placed upright in a small heating coil, and a 50-μl portion of test or standard solution is placed on the tip of each by means of a pipet. The solution is then carefully evaporated to dryness, and the electrodes are ready for excitation.

A series of standard solutions is prepared from pure $MgCl_2$ at concentrations of the order of 0.1 ng to 10 μg of Mg per ml. To each solution is added ammonium molybdate to the extent of 20 μg of Mo per ml. The distilled water must be shown spectroscopically to be free of Mg.

The electrodes are clamped in position with an accurately measured spacing of 2 mm. A 25-kV spark is struck across them through a Feussner circuit.

The spectrum contains lines from both Mg and Mo, as well as Cu and any other metals that may be present. Several homologous pairs are available and are about equally good. The lines at 279.81 nm (Mg) and 281.62 (Mo) represent one such pair. The difference between the optical densities of these two lines, i.e., the ratio of intensities of the emissions at the two wavelengths, is plotted against the amount of Mg in the sample on log-log paper (Fig. 8-4).

This method is fast and convenient and, at the same time, highly sensitive. It is possible to identify 1 ng of Mg in a volume of 1 ml. The precision of the determination is of the order of 5 to 10 percent, which is quite satisfactory in view of the small quantities involved.

Another example of a quantitative spectrographic analysis (7) is the determination of many transition metals in the parts-per-billion range in KCl intended for electrochemical purposes. The procedure includes a preconcentration by precipitation with 8-quinolinol in the presence of $InCl_3$ as carrier and $PdCl_2$ as internal standard.

FLAME PHOTOMETRY

Atomic emission (AE) from atoms thermally excited in a flame has developed along a separate track from that of other emission techniques. It has been found particularly convenient for the determination of alkali and alkaline earth metals and a few others, as listed in the final column of Table 7-1.

Fig. 8-4   Working curve for the analysis of magnesium by the copper spark method, with molybdenum as internal standard. [*McGraw-Hill* (2).]

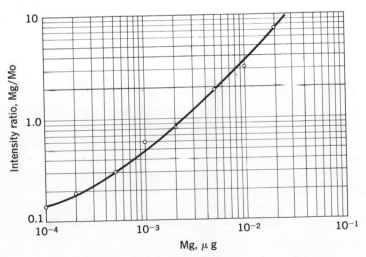

The flame must be hotter in AE than for the same element in AA, because as large a fraction as possible of the vaporized atoms must be energetically excited, rather than simply dissociated. Too high a temperature, however, will cause loss of atoms by ionization.

The burner used for AE is usually of the type known as *total consumption* (Fig. 8-5). The sample is aspirated into the flame by either the fuel or oxidant, but the gases are not premixed. The resulting flame concentrates the source of emitted radiation into a small space. An oxygen-acetylene flame can be used, in addition to the several combinations mentioned in connection with AA.

The power of the radiation from the flame at a wavelength characteristic of a particular element is found to be very closely proportional to the concentration of the metal, if a background correction is first made. The background luminosity is caused largely by the presence of other metals, since in general each excitable cation will give some radiation over a wide spectral region, even at a considerable distance from its discrete lines. Any scattering in the monochromator or photometer will also contribute. The effect of background can best be eliminated by the application of a base-line technique of measurement analogous to that discussed previously in connection with absorption spectra. This is easy to do if the spectra are observed with a spectrophotometer, but is not so convenient with a filter flame photometer. In the latter case, corrections are usually made on an empirical basis.

In addition to the background effect, a specific interaction may sometimes be observed between metals, either enhancing or repressing the normal luminescence. This difficulty can be overcome, though with some loss in sensitivity, by intentionally adding a large excess of any likely contaminating cations before analysis. In the analysis of natural waters (10) containing sodium, potassium, calcium, and magnesium, interference with the analysis of each element by the other three can be avoided by the use of

Fig. 8-5 A "total-consumption" burner, as used in AE.

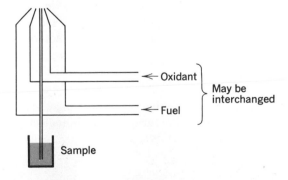

radiation buffers. For example, in the determination of sodium, to a 25-ml sample is added 1 ml of a solution which is saturated with respect to the chlorides of potassium, calcium, and magnesium. Any slight variation in the amounts of these elements in the sample will be negligible compared with the amount added. The mixed solution is then examined with the flame photometer, and the response at the wavelength of sodium radiation, corrected for background luminosity, is compared with a calibration curve prepared from standards. Concentration differences of 1 or 2 ppm can easily be detected for sodium or potassium, and 3 or 4 ppm for calcium. In the case of magnesium, the method is less sensitive.

Another procedure to eliminate the results of interferences is an application of the method of standard addition. After measurement of the emission from the sample, a known amount of the desired element is added to a second aliquot of the unknown and the measurement repeated. The added standard will be subject to the same interferences as the sample constituent; hence analysis by direct comparison is possible.

A significant replacement for a flame is a plasma produced electrically in argon gas. The excitation can be by radio-frequency induction (8) or by a dc discharge (9). The plasma that results is very energetic, capable of dissociating and activating a large number of elements including several, such as chlorine, not accessible to gas flames. The dissolved sample is aspirated into the plasma just as into a chemical flame. The argon plasma, besides its higher temperature, which can reach 10 kK,* has the advantage of requiring a single nonflammable gas rather than two hazardous gases, and it also shows relative freedom from chemical interferences.

COMPARISON OF AE WITH AA

As shown in Table 7-1, only a small number of elements can be observed at lower concentration limits by AE than by AA. It should be noted that the response of an emission photometer is *linear* within limits, with respect to concentration, whereas the results from an atomic absorption photometer, following Beer's law, bear a logarithmic relation to concentration. Hence the absorption technique provides a much longer concentration range over which measurements can be made.

Atomic absorption is at present less convenient with respect to qualitative scanning of samples, because of the need to change lamps. On the other hand, quantitative analysis is usually more satisfactory, as AA does not suffer so greatly from chemical interferences.

INSTRUMENTATION

Instruments for observing and measuring emission spectra range from hand-held direct-vision spectroscopes to giant photographic or photoelectric spectrographs of high resolution and great sensitivity.

* Kilokelvin.

In the area of flame photometers (for AE), many instruments are designed especially for clinical applications, where the major interest is in Na, K, and Ca in body fluids, etc. Sufficient precision can be obtained with interference filters in place of a monochromator, as the lines are easily separated. Some instruments use the simplest of photometric circuits, a photovoltaic cell directly connected to a galvanometer. Others gain sensitivity by means of a photomultiplier and stability by the application of added lithium as an internal standard. Still other instruments include a monochromator, or use an existing spectrophotometer by means of an adaptor.

Representative of the smaller spectrometers for qualitative and semiquantitative analysis of solid samples in the visible region is the Vreeland spectroscope.* This is a bench-top instrument provided with ac arc excitation and a concave grating of 42.5-cm focal length, 15,000 lines per inch. The spectrum is focused on a translucent plastic screen fitted to the Rowland circle (Fig. 8-6). Adjacent to the spectrum on each side can be placed reference spectra supplied as positive films for direct line-by-line comparison.

A somewhat larger bench-top instrument is the SpectraSpan,† which makes use of a unique design of argon plasma "flame" (9), and an echelle grating together with a quartz prism to separate orders. The plasma source is so arranged that it can be used in the absorption mode with a hollow-cathode lamp, as well as for emission.

Figure 8-7 shows a large, modern, photoelectric, direct-reading spectrometer. This instrument uses a concave grating with a ¾-m focal length

* Spectrex Company, Palo Alto, Calif.
† SpectraMetrics, Inc., Burlington, Mass.

Fig. 8-6 Optical diagram of the Vreeland spectroscope. The sample is held on a small hearth beneath the horizontal electrodes. The grating and slit are mounted on a Rowland circle in a vertical plane. (*Spectrex Company.*)

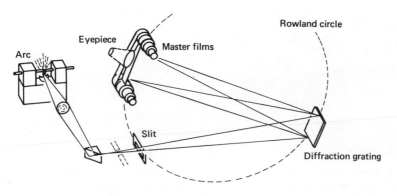

Fig. 8-7 The Model 750 Atomcounter, a direct-reading spectrometer controlled by the Model PDP-8 computer visible in the left-hand panel. The computer controls need be manipulated only for initial setup and subsequent testing; during analysis, the operator only needs access to the Teletype unit. (*Jarrell-Ash Division, Fisher Scientific Company.*)

Fig. 8-8 The spectrometer bed, heart of the 750 Atomcounter. The adjustable spark source is at lower center, and the concave grating at upper left. The focal curve is visible along the right, with a number of photomultipliers in place. (*Jarrell-Ash Division, Fisher Scientific Company.*)

in a Rowland-circle mounting. Along the focal curve is located a series of photomultipliers (Fig. 8-8), each behind its own exit slit. These slits are positioned so as to accept suitable wavelengths for a series of elements. All the selected elements are measured simultaneously without scanning.

The operation of the spectrometer is entirely under the control of a *minicomputer,* a type of instrument discussed in Chap. 27. After inserting the sample and turning on the spark, the operator can instruct the instrument by typing commands on the teletypewriter; then the analytical results, corrected for background if necessary and calculated in reference to an internal standard, are printed out automatically on the same typewriter.

PROBLEMS

8-1 A shipment of "chemically pure" aluminum metal is to be analyzed for its magnesium content. A 1.000-g sample is dissolved in acid and enough ammonium molybdate added to contain 2.000 mg of molybdenum. The solution is diluted to 100.0 ml, and 50 μl is evaporated onto the tips of two copper electrodes and sparked before the spectrograph slit. The resulting plate, examined with a densitometer, shows a density of 1.83 for the 279.81-nm line and 0.732 for the 281.62-nm line. With the aid of Fig. 8-4, compute the percent of magnesium in the sample.

8-2 The sketch in Fig. 8-9 shows a portion of a spectrograph plate as seen through a magnifier with a built-in scale of millimetres. The field of view includes portions of the spectra produced by arcing, first, iron electrodes, and then electrodes fashioned from an unknown thought to be an aluminum alloy. The iron lines, marked λ_1 and λ_2, have been identified by comparison with a standard iron spectrum. Line x of the unknown is to be identified. The scale readings are as follows: $\lambda_1 = 9.90$ mm, $\lambda_2 = 8.37$ mm, and $\lambda_x = 10.25$ mm. (*a*) Calculate the dispersion of the spectrograph in this region in terms of nanometres per millimetre. (*b*) Determine the wave-

Fig. 8-9 Magnified region of a spectrum plate, with linear scale for interpolation. (*Bausch & Lomb, Inc.*)

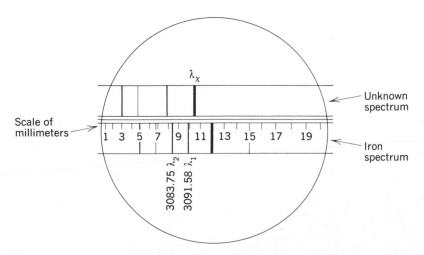

length of the unknown line. (c) Refer to a table of wavelengths and make a tentative identification of this line, with due regard for the elements likely to be found in this type of alloy. In what region of the spectrum would you search for strong lines to confirm your identification?

8-3 Plot the dispersion curve for the spectrograph on which the spectra of Fig. 8-3 were taken.

REFERENCES

1. Harrison, G. R., R. C. Lord, and J. R. Loofbourow, "Practical Spectroscopy," Prentice-Hall, Englewood Cliffs, N.J., 1948.
2. Nachtrieb, N. H.: "Principles and Practice of Spectrochemical Analysis," McGraw-Hill, New York, 1950.
3. Sawyer, R. A.: "Experimental Spectroscopy," 3d ed., Dover, New York, 1963.
4. Jarrell, R. F.: In G. L. Clark (ed.), "Encyclopedia of Spectroscopy," p. 158, Reinhold, New York, 1960.
5. Ishizuka, T.: *Anal. Chem.*, **45**:538 (1973).
6. Scribner, B. F., and H. R. Mullin: *J. Res. Natl. Bur. Stand.*, **37**:379 (1946).
7. Farquhar, M. C., J. A. Hill, and M. M. English: *Anal. Chem.*, **38**:208 (1966).
8. Dickinson G. W., and V. A. Fassel: *Anal. Chem.*, **41**:1021 (1969).
9. Elliott, W. G.: *Am. Lab.*, **3**(8):45 (1971).
10. West, P. W., P. Folse, and D. Montgomery: *Anal. Chem.*, **22**:667 (1950).

Chapter 9

Polarimetry, Optical Rotatory Dispersion, and Circular Dichroism

In preceding chapters we have been concerned with radiant energy principally from the standpoint of its absorption, its emission, and its wavelength distribution. Its wave nature has only interested us in connection with diffraction effects. We will now discuss phenomena concerned with polarized radiation.

Many transparent substances characterized by a lack of symmetry in their molecular or crystalline structure have the property of rotating the plane of polarized radiation. Such materials are said to be *optically active*. Probably the most familiar examples are quartz and the sugars, but many other organic and inorganic compounds also possess this property. The extent to which the plane is rotated varies widely from one active compound to another. The rotation is said to be *dextro* (+) if it is clockwise to an observer looking toward the light source, and *levo* (−) if counterclockwise. For any given compound, the extent of rotation depends on the number of molecules in the path of the radiation or, in the case of solutions, on the concentration and the length of the containing vessel. It is also dependent on the wavelength of the radiation and the temperature. The *specific rotation*, represented by the symbol $[\alpha]^t$, is defined by the formula

$$[\alpha]^t = \frac{\alpha}{dc} \tag{9-1}$$

where α is the angle (measured in degrees) through which the plane of the polarized light is rotated by a solution of concentration c grams of solute per milliliter of solution contained in a cell d decimetres in length; t designates the temperature. The wavelength is commonly specified as 589.3 nm, the D line of a sodium-vapor lamp. Some representative values for specific rotation are given in Table 9-1.

POLARIMETERS (1)

The most general instrument in this field is the *polarimeter* (Fig. 9-1). A typical manual instrument is shown schematically in Fig. 9-2. Monochromatic radiation from a sodium lamp is made parallel by a collimator and polarized by a calcite prism. Following the polarizer, there is placed a small auxiliary calcite arranged to intercept one-half of the beam. (The function of this feature will be explained below.) The radiation then passes through the sample, which is contained in a glass tube of known length, closed at both ends by glass plates and then through the analyzer to the eyepiece for visual observation.

In principle, the polarimeter could function without the small auxiliary prism. The polarizers would initially be crossed without any sample in the beam, and then again with the sample present. The angle through which the analyzer is turned would then be the quantity sought. However, this simple arrangement is unsatisfactory because it requires the observer to identify the position where the transmitted radiation is zero, which cannot be done with precision. The added prism in half the beam makes it possible to avoid this difficulty. It is permanently oriented with its polarizing axis at an angle of a few degrees from that of the polarizer. There is then a particular position of the analyzer at which the radiations passed in the

Table 9-1 Specific rotations of solution (at 20°C)

Active substance	*Solvent*	$[\alpha]_D^{20}$
Camphor	Alcohol	$+\ 43.8°$
Calciferol (vitamin D₂)	Chloroform	$+\ 52.0$
Calciferol (vitamin D₂)	Acetone	$+\ 82.6$
Cholesterol	Chloroform	$-\ 39.5$
Quinine sulfate	$0.5\ F$ HCl	$-220.$
l-Tartaric acid	Water	$+\ 14.1$
Sodium potassium tartrate (Rochelle salt)	Water	$+\ 29.8$
Sucrose	Water	$+\ 66.5$
β-d-Glucose	Water	$+\ 52.7$
β-d-Fructose	Water	$-\ 92.4$
β-Lactose	Water	$+\ 55.4$
β-Maltose	Water	$+130.4$

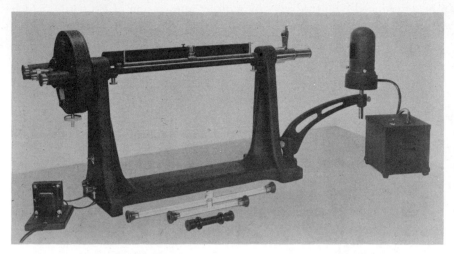

Fig. 9-1 Precision polarimeter, with sodium lamp.   (*O. C. Rudolph and Sons.*)

two halves of the beam are just equal in power.   This provides a more satisfactory reference point than does the position of complete extinction, since the visual observation consists of matching exactly the powers of two half-beams at some intermediate level, for which application the eye is well suited.

It is possible to design photoelectric polarimeters with or without automatic recording features, and there are several such instruments on the market.   They are single-beam devices, differing primarily in the servomechanism for rotating the plane of polarization to compensate for the rotation of the sample.

APPLICATIONS

The most extensive application of analysis by optical rotation is in the sugar industry (2).   In the absence of other optically active material, sucrose

Fig. 9-2 Diagram of a conventional polarimeter.   The Nicols are compound prisms of calcite.

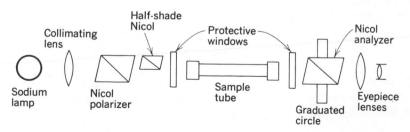

can be determined quantitatively by a direct application of Eq. (9-1), which
may be written in the form

$$c = \frac{\alpha}{d[\alpha]_{\mathrm{D}}^{20}} = \frac{\alpha}{(2)(66.5)} = \frac{\alpha}{133.0} \qquad\qquad (9\text{-}2)$$

for sucrose in the customary 2-dm tube, at a temperature of 20°C.

If, however, other active substances are present, a more elaborate
treatment is required. Sucrose, alone among common sugars, can be made
to undergo a hydrolysis reaction in the presence of acid, according to the
equation

$$C_{12}H_{22}O_{11} \quad + H_2O \xrightarrow{\text{acid}} C_6H_{12}O_6 + C_6H_{12}O_6$$

Sucrose	Glucose	Fructose
$[\alpha]_{\mathrm{D}}^{20} = +66.5°$	$+52.7°$	$-92.4°$

The resulting mixture of glucose and fructose is called *invert sugar*, and the
reaction *inversion*. During the inversion process, the specific rotation
changes from $+66.5$ to $-19.8°$, corresponding to an equimolar mixture of
the products. By measuring the rotation before and after inversion, it is
possible to calculate the amount of sucrose present. The usual procedure
is to start with a sample of 100-ml volume, measure its rotation, and then
add 10 ml of concentrated hydrochloric acid. The acidified solution must
stand at least 24 h at 20°C, 10 h at 25°C, or 10 min at 70°C, to ensure
completion of the reaction. The rotation is then redetermined. It can
be shown that under these conditions the weight of sucrose in the original
sample is

$$w_s = -1.17\,\Delta\alpha - 0.00105[\alpha]_{\mathrm{D}(X)}w_x \qquad\qquad (9\text{-}3)$$

where $\Delta\alpha$ is the observed change in angle of rotation, $[\alpha]_{\mathrm{D}(X)}$ is the specific
rotation of any other active material which may be present, and w_x is the
weight of this active impurity. If nearly all the active material present
is sucrose, the second term becomes negligible. On the other hand, if one is
determining the sucrose present as a minor constituent in a large portion of
another active substance, the second term can be evaluated and the weight
of sucrose calculated from the observed $\Delta\alpha$.

Another example of an analytical procedure based on optical rotation
is the simultaneous determination of penicillin and the enzyme penicillinase
(3). Penicillin is destroyed quantitatively by the enzyme at a rate which
is directly dependent on the amount of enzyme present but independent
of the penicillin concentration. A graph of rotation against time gives
a straight line which terminates when the penicillin is all used up. The

slope of the line is a measure of the concentration of the enzyme. Figure 9-3 shows such a graph for five values of enzyme concentration. It will be seen that the curves tail off, an effect due to secondary reactions. The true time of disappearance of penicillin is found by the intersection of the extrapolated straight portions. The concentration can be determined with a precision of about ± 1 percent for the penicillin, and about ± 10 percent for the enzyme.

OPTICAL ROTATORY DISPERSION AND CIRCULAR DICHROISM (4–6)

The wavelength dependence of optical activity, known as *optical rotatory dispersion* (ORD), is a more fruitful source of structural information about asymmetric compounds than is the specific rotation at a single wavelength. It is closely related to the phenomenon called *circular dichroism* (CD) (7, 8).

In Chap. 2 it was shown that an ordinary beam of light can be resolved into two plane-polarized beams. We will now carry that line of thought forward another step. Plane-polarized radiation can be further resolved into two beams said to be circularly polarized in opposite senses (Fig. 9-4a). The indices of refraction of a medium for the left- and right-hand circular components may not be the same, and will be designated n_L and n_R, respectively. The corresponding absorptivities a_L and a_R may also differ. For an isotropic medium, such as glass or water, $n_L = n_R$, and $a_L = a_R$, and we say that the index and the absorptivity are independent of the state of polarization. If $n_L \neq n_R$, a phase difference appears between the two components, which is tantamount to saying that the plane of polarization has been rotated, i.e., that the angle $\alpha \neq 0$ (Fig. 9-4b). If $a_L \neq a_R$,

Fig. 9-3   The enzymatic destruction of penicillin with varying concentrations of penicillinase in a phosphate buffer at pH 7. The figures indicate relative concentrations of the enzyme. [*Analytical Chemistry* (3).]

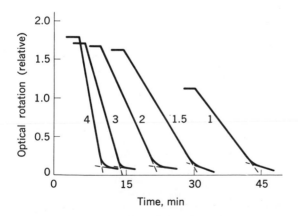

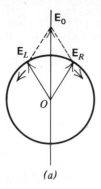

(a)

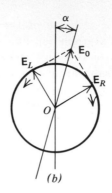

(b)

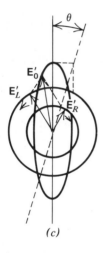

(c)

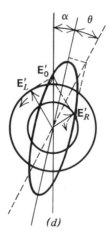

(d)

Fig. 9-4 (*a*) Resolution of plane-polarized radiation into left- and right-hand circularly polarized components. The two radius vectors $\mathbf{E}_L$ and $\mathbf{E}_R$, rotating in opposite directions, are the equivalent of vector $\mathbf{E}_0$ vibrating in a vertical plane. The beam of radiation propagates perpendicularly into the plane of the paper at the center of the circle. (The vectors referred to are the *electric* vectors of the respective beams of radiation.) (*b*) Corresponding vector diagram of beam of radiation in (*a*) after passing through an optically active material. A phase difference has appeared, corresponding to a rotation of the plane of polarization by angle α. (*c*) Vector diagram for radiation in (*a*) after traversing a sample for which the right-circularly polarized component is absorbed more than the left. The resultant vector $\mathbf{E}_0'$ now describes an ellipse with its major radius equal to $\mathbf{E}_L' + \mathbf{E}_R'$ and its minor radius equal to $\mathbf{E}_L' - \mathbf{E}_R'$. This is a hypothetical case of CD not accompanied by optical rotation. The eccentricity of the ellipse is angle θ. (*d*) Corresponding vector diagram for the radiation of (*a*) after passing through a sample that shows both optical activity and CD.

182

then one component is absorbed more strongly than the other; the vector diagram representing the polarization then turns out to be elliptical, with the eccentricity θ a measure of $a_L - a_R$ (Fig. 9-4c and d). The electric vector **E** of an electromagnetic wave is proportional to the square root of the intensity, hence the vectors in Fig. 9-4c and d are, respectively, $\mathbf{E}'_L = \mathbf{E}_L \sqrt{T_L}$ and $\mathbf{E}'_R = \mathbf{E}_R \sqrt{T_R}$, where T_L and T_R are the transmittances (9).

It can be shown that the angle of rotation, as used in Eq. (9-1), expressed in radians, is given by

$$\alpha = \frac{\pi}{\lambda}(n_L - n_R)d \tag{9-4}$$

and that the *ellipticity* θ is

$$\theta = \tfrac{1}{4}(a_L - a_R)d \tag{9-5}$$

The quantity $n_L - n_R$ is called the *circular birefringence*, and $a_L - a_R$ the *circular dichroism* of the medium. The value of α can be measured directly with a polarimeter, but θ can only be determined indirectly through the CD.

ORD and CD are closely related, as can be seen in the curves of Fig. 9-5. The combined phenomenon is known as the *Cotton effect*. It appears

Fig. 9-5 The ultraviolet absorption, CD, and ORD spectra of a compound for which the CD is positive.

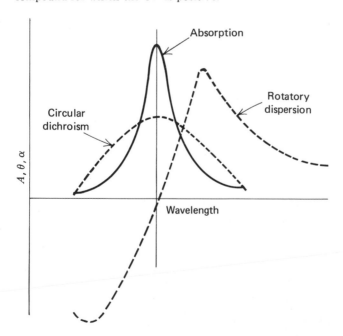

in connection with chromophoric absorptions in optically active compounds. An example is shown in Fig. 9-6. The maxima of CD and absorption and the inflection of the ORD curve theoretically coincide in wavelength, but the superimposed effects of other chromophores may cause shifts in the observed maximum in absorbance (as is the case in Fig. 9-6). CD curves will often pinpoint "hidden" absorption maxima, and with less ambiguity than will the corresponding ORD curves. On the other hand, ORD curves are affected to a considerable degree by more distant chromophoric bands, and hence are more characteristic of specific compounds than CD. Both ORD and CD can give essential information about stereochemical features of optically active materials. In practice, the relation between the several curves for the same substance may be more complex than the example in Fig. 9-6; as the theory is not within the scope of this book, the student is referred to the literature for further details (4, 8).

ORD PHOTOMETERS

There are several recording spectropolarimeters in present manufacture. They can be considered modified single-beam spectrophotometers, as shown schematically in Fig. 9-7. In this instrument (as in most others) the beam

Fig. 9-6 ORD, CD, and ultraviolet absorption spectra of the weak transition due to the episulfide group in 3β-hydroxycholestan-5α,6α-episulfide. [*Tetrahedron* (10).]

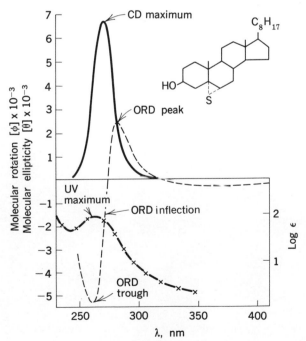

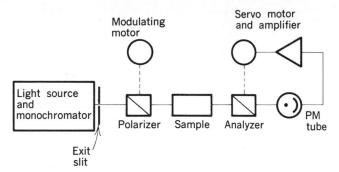

Fig. 9-7 Schematic of Durrum-Jasco spectropolarimeter. (*Durrum Instrument Corporation.*)

of light from a conventional monochromator passes sequentially through a polarizer, the sample, and an analyzer to a photomultiplier tube. The beam is modulated with respect to its state of polarization at 12 Hz by a motor-driven device which causes the polarizer to rock back and forth through an angle of $\pm 1°$. The servoamplifier responds only to the 12-Hz frequency, and causes the servomotor to adjust the analyzer continuously to the point where the 12-Hz signal will be symmetrically disposed about the null point. The servomotor also positions the recording pen.

The Cary Model 60 spectropolarimeter is similar in principle, except in the method of modulating the beam, for which a *Faraday cell* is employed. This is a vitreous silica rod surrounded by a coil carrying 60-Hz current, which causes a rotation of the plane of polarization through a few degrees at the 60-Hz frequency. In this instrument, the analyzing prism is immovable, and the servo system operates the polarizer to find the null point.

CD APPARATUS

Since conventional spectrophotometers are designed to determine the difference between the absorbances of a sample and a standard, they can easily be adapted to measure CD which is likewise the difference between two absorbances.

Circular polarization is accomplished in two steps. First, the beam of radiation must be plane-polarized and, second, the polarized beam must be passed through a device which will resolve it into right- and left-circularly polarized components, and retard one component relative to the other by exactly one-quarter wavelength. The most important circular resolvers are of two types: (*1*) those which depend on total internal reflections, such as the *Fresnel rhomb* (Fig. 9-8), and (*2*) the *Pockels electrooptical modulator*. In the latter, a high potential (in the kilovolt range) is applied across a plate of potassium dihydrogen phosphate or similar piezoelectric

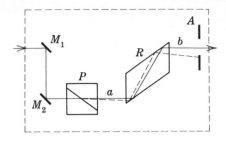

Fig. 9-8 Optical assembly for measurement of CD. Radiation enters at the left; is displaced downward by mirrors M_1 and M_2; is plane-polarized by the compound prism P; and is passed through the Fresnel rhomb R, where it undergoes two internal reflections, introducing a phase retardation of one-quarter wavelength and thus producing circular polarization. The mask A eliminates the extraordinary ray, while permitting the ordinary ray to pass. The entire unit fits into the sample chamber of a standard spectrophotometer; a second unit, oppositely oriented, is required for the reference path. The sample is placed at b for CD measurements or at a to study plane-polarized transmission. (*Cary Instruments Division of Varian Associates.*)

crystal cut perpendicularly to its optical axis. The first type has a somewhat limited wavelength range over which the retardation is near enough to one-quarter wavelength; in the Pockels type, the retardation can be altered by choice of potential; hence it can be programmed to give continuously correct output as the wavelength is changed. The Pockels modulator with its power supply is much more elaborate and expensive than the Fresnel rhomb.

The Cary CD accessory for the Model 14 spectrophotometer utilizes two matched assemblies, each consisting of a polarizer, rhomb, and sample space, with the geometries so arranged that the radiation in the reference path is right-circularly polarized, while that in the sample path is left-circularly polarized. The CD adaptor for the Cary Model 60 spectropolarimeter makes use of a Pockels cell.

PROBLEMS

9-1 A 10.00-g sample of impure Rochelle salt is dissolved in enough water to make 100.0 ml of solution. A 20-cm polarimeter tube is first filled with distilled water (to establish a true zero of the scale), and then with a portion of the solution. The following observations were made at 20°C:

Scale setting for water   +2.020°
Scale setting for solution   +5.750°

What is the percent by weight of Rochelle salt in the sample?

9-2 Compute the percent error introduced into the determination of sucrose by Eq. (9-3), with the second term omitted, in each of the following samples: (*a*) 30 g sucrose plus 2 g lactose, (*b*) 16 g sucrose plus 16 g lactose, and (*c*) 2 g sucrose plus 30 g lactose. The variation of the specific rotation of lactose between 15 and 20°C may be neglected.

REFERENCES

1. Gibb, T. R. P., Jr.: "Optical Methods of Chemical Analysis," McGraw-Hill, New York, 1942.
2. Bates, F. J., et al.: Polarimetry, Saccharimetry, and the Sugars, *Natl. Bur. Stand. Circ.* C440, 1942.
3. Levy, G. B.: *Anal. Chem.,* **23**:1089 (1951).
4. Beychok, S.: *Science,* **154**:1288 (1966).
5. Djerassi, C.: "Optical Rotatory Dispersion: Applications to Organic Chemistry," McGraw-Hill, New York, 1960.
6. Struck, W. A., and E. C. Olsen: Optical Rotation: Polarimetry, in I. M. Kolthoff and P. J. Elving (eds.), "Treatise on Analytical Chemistry," pt. I, vol. 6, chap. 71, Wiley-Interscience, New York, 1965.
7. Abu-Shumays, A., and J. J. Duffield: *Anal. Chem.,* **38**(7):29A (1966).
8. Velluz, L., M. Legrand, and M. Grosjean, "Optical Circular Dichroism," Academic, New York, 1965.
9. Duffield, J. J., A. Abu-Shumays, and A. Robinson, *Chem. Instrum.,* **1**:59 (1968).
10. Djerassi, C., H. Wolf, D. A. Lightner, E. Bunnenberg, K. Takeda, T. Komeno, and K. Kuriyama, *Tetrahedron,* **19**:1547 (1963).

Chapter 10

X-Ray Methods

When a beam of electrons impinges on a target material, the electrons in general are slowed down by multiple interactions with the electrons of the target. The energy lost is converted into a continuum of x-radiation, called *Bremsstrahlung*.* This continuum shows a sharp minimum wavelength λ_{min} (maximum frequency) corresponding to the maximum energy of the electrons, which cannot be exceeded. This cutoff wavelength is given (in nanometres) by

$$\lambda_{min} = \frac{hc}{Ve} = \frac{1240}{V} \tag{10-1}$$

where h = Planck's constant, c = the velocity of electromagnetic radiation in vacuo, e = the electronic charge, and V = the accelerating potential across the x-ray tube, in volts.

As the potential is increased, a point is reached where the energy is sufficient to knock a planetary electron completely out of the target

* From the German, meaning "braking radiation."

188

atom. Then as another electron falls back into the vacancy, a photon of x-radiation is emitted with a wavelength dependent on the energy levels, and hence characteristic of the element. Since high energies are involved, the electrons closest to the nucleus are the principal ones affected. Thus a K electron may be ejected and its place taken by an electron from the L shell. Since these inner electrons are not concerned with the state of chemical combination of the atoms (except for the lighter elements), it follows that the x-ray properties of the elements are independent of chemical combination or physical state, to a close degree of approximation. (Chapter 11 discusses some exceptions to this generalization.) The wavelengths corresponding to such high energies are small, of the order of 1 to 1000 pm.* The range of 70 to 200 pm includes the wavelengths most useful for analytical purposes.

The x-ray emission spectrum of a given target material will resemble that shown in Fig. 10-1, a continuum with superimposed discrete lines. The sharp lines are designated $K\alpha$, $K\beta$, etc., for transitions to the K level.† Heavy elements show other groups of lines corresponding to the L, M, and higher levels. If excitation is brought about by fluorescence, i.e., irradiation with x-rays of shorter wavelength, the continuum does not appear; only the characteristic lines are present. This is a favorable situation for x-ray emission analysis, greatly increasing the signal-to-noise ratio.

For analytical purposes, x-radiation can be utilized in several distinct ways. (*1*) The absorption of x-rays will give information about the absorbing material, just as in other spectral regions. (*2*) The diffraction of x-rays permits analysis of crystalline materials with a high degree of specificity and accuracy. (*3*) Wavelength measurements will identify elements in the excited sample. (*4*) Measurement of radiant power at a given wavelength can be a quantitative indicator of the composition of the sample.

THE ABSORPTION OF X-RAYS

In common with other regions of the electromagnetic spectrum, x-rays can be absorbed by matter, and the degree of absorption is controlled by the nature and amount of the absorbing material.

* X-ray wavelengths were formerly given in terms of *kX units,* where 1 kX unit = 0.100202 nm. The usual wavelength unit in x-ray studies is the angstrom (Å). In furtherance of the SI unitary system, this book uses nanometres (nm) or picometres (pm).

† X-rays due to transitions from the L to the K shell are called $K\alpha$ x-rays, $K\alpha_1$ and $K\alpha_2$ corresponding to electrons originating in different sublevels of the L shell; x-rays due to transitions from the M to the K shell are called $K\beta$, and so on.

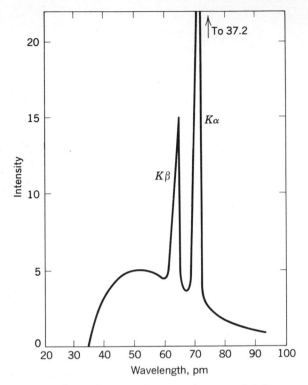

Fig. 10-1 Intensity curve for x-rays from a molybdenum
target operated at 35 kV.   The continuous background
is due to *Bremsstrahlung* radiation; the peaks are $K\alpha$
and $K\beta$ transitions.   [*Wiley* (2).]

The absorption follows Beer's law, which may be written in the form

$$P_x = P_0 e^{-\mu x} \tag{10-2}$$

or

$$\ln \frac{P_0}{P_x} = \mu x \tag{10-3}$$

for monochromatic x-rays, where P_0 is the initial power of the radiation
and P_x is the power after passage through an absorbing sample x centimetres
in length.   If the beam of x-rays has a cross section of 1 cm², then μ,
which is called the *linear absorption coefficient,* represents the fraction of
energy absorbed per centimetre.   Frequently more convenient is a *mass
absorption coefficient,* defined as

$$\mu_m = \frac{\mu}{\rho} \tag{10-4}$$

where ρ is the density of the absorbing material. The coefficient μ_m is found empirically to be related to the wavelength and atomic properties of the absorbing substance by the formula

$$\mu_m = \frac{CNZ^4\lambda^n}{A} \tag{10-5}$$

where N is Avogadro's number, Z is the atomic number, A is the atomic weight of the absorbing element, λ is the wavelength, n is an exponent between 2.5 and 3.0, and C is a constant which is approximately the same for all elements, within limited regions as will appear.

The variation of μ_m with wavelength follows an exponential law, so that if the logarithms are plotted, a straight line should result with slope equal to the exponent of λ. Figure 10-2 shows such a plot for the absorption coefficient of argon. The most striking feature of this graph is the discontinuity at $\lambda = 387.1$ pm. This is known as the *K critical absorption wavelength* of argon. Radiation of greater wavelength has insufficient energy to eject the K electrons of argon; hence, it is not absorbed so greatly as is radiation of slightly shorter wavelength. Larger atoms than argon show similar discontinuities at longer wavelengths corresponding to the photoelectric ejection of L and M electrons.

It is instructive to plot μ_m against the quantity Z^4/A for x-rays of a particular wavelength, as in Fig. 10-3, which shows the values for Cu $K\alpha$ radiation for elements from sodium to osmium. The critical absorption edges also appear in this plot. The "constant" C of Eq. (10-5) changes value abruptly at the edge discontinuities, and is not perfectly constant between them, as shown by the slight curvature most easily visible between the K and L_I edges. This curvature can alternatively be ascribed to variations in the exponent n. Equation (10-5) will need refinement as our

Fig. 10-2 The x-ray absorption of argon in the region 50 to 1000 pm. [*From data of Compton and Allison* (1).]

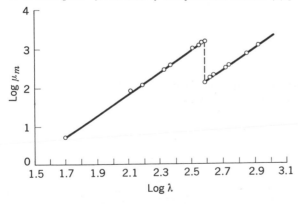

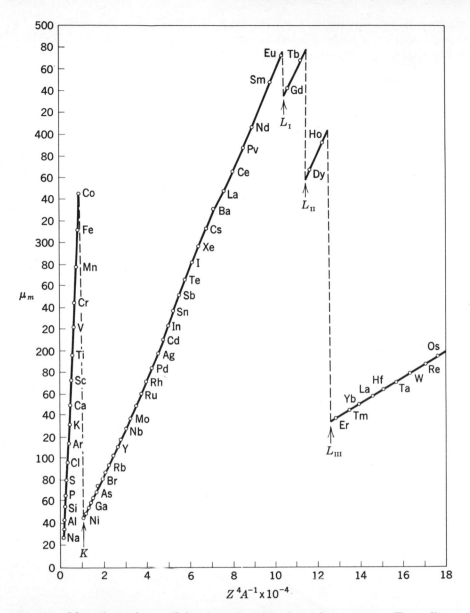

Fig. 10-3   Mass absorption coefficients as a function of atomic constants.   The radiation is Cu $K\alpha$, 154.18 pm.   [*From data of Compton and Allison* (1).]

knowledge of the underlying phenomena improves. Empirical absorption data for most elements are to be found in the literature (1, 2).

MONOCHROMATIC SOURCES

It is a simple matter to obtain narrow bands of wavelengths at various discrete points in the x-ray region, but not so simple to construct a mono- chromator which can be varied at will. Narrow bands can be obtained by three methods: (*1*) by choosing characteristic emission lines which are much stronger than the background, and isolating them with the aid of filters; (*2*) by means of a monochromator in which a crystal of known spacing acts as a diffraction grating; and (*3*) by using radioactive sources.

A monochromatic beam can often be obtained by means of a filter consisting of an element (or its compound) which has a critical absorption edge at just the right wavelength to isolate a characteristic line from a source target. For example, in Fig. 10-1, if a material could be found

Fig. 10-4 The zirconium absorption curve superimposed on the 35-kV molybdenum emission spectrum results in the nearly monochromatic radiation of Mo Kα. [*Wiley* (2).]

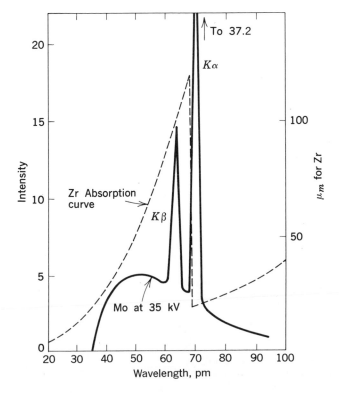

with a critical edge between the wavelengths of Mo $K\alpha$ (70.9 pm) and Mo $K\beta$ (63.2 pm), it would absorb the β while passing the α line. Such a filter can be made of zirconium, which has its K absorption edge at 68.9 pm (see Fig. 10-4). The ratio of the absorption coefficients of zirconium at 63 and 71 pm is only about 4.75, but the intensity of the Mo $K\alpha$ is about 5.4 times that of Mo $K\beta$, so that a fair degree of isolation can be achieved, the exact relation depending on the thickness of the filter. Table 10-1 lists filters suitable for $K\alpha$ isolation from several targets.

CRYSTAL MONOCHROMATORS

If a greater degree of monochromaticity is required than can be obtained with filters, a grating monochromator must be employed. Design of these instruments will be detailed in connection with x-ray diffraction.

Such a monochromator could produce a narrow band of wavelengths at any point in the spectrum if a continuous source of x-rays were available. However, the continuous radiation produced at low tube voltages is not sufficiently powerful to be of much use, so for practical purposes we are still restricted to the various characteristic emission wavelengths. The advantage of the monochromator over the filter lies in its great reduction of background, which results in an improved signal-to-noise ratio in spite of a considerably attenuated signal.

RADIOACTIVE SOURCES

Radiations in the x-ray region are emitted from some radioactive elements by either of two mechanisms (4). The first of these, correctly called

Table 10-1 Characteristic wavelengths and filters for x-ray tubes with commonly available targets*

Target	λ, $K\alpha_1$, pm	Element for a filter	K-edge wavelength, pm	Thickness of foil, μm	Percent $K\beta_1$ absorbed
Cr	229.0	V	226.9	15.3	99.0
Fe	193.6	Mn	189.6	15.1	98.7
Co	178.9	Fe	174.3	14.7	98.9
Ni	165.8	Co	160.8	14.3	98.4
Cu	154.1	Ni	148.8	15.8	97.9
Mo	70.9	Zr	68.9	63.0	96.3
Ag	55.9	Pd	50.9	41.3	94.6
W	20.9	†			

* The data are from tabulations in references 2 and 3.
† No suitable filter material is available for tungsten radiation.

gamma radiation, involves intranuclear energy levels, and so is not truly x-radiation. The other is known as *electron capture* (EC) or *K capture.* Since an electron in a *K* orbital has a finite probability of spending some of its time within or very close to the nucleus, there is in many atoms a finite probability of the capture of a *K* electron by the nucleus. This process lowers the atomic number by one unit and leaves a vacancy in the *K* shell. Hence true x-rays of the next lower element result, unaccompanied by any significant continuous radiation. Table 10-2 lists a few isotopes which have proven useful as x- or gamma-ray sources. Filters may still be needed to separate individual lines.

Radioactive sources have the advantage of not requiring an elaborate high-voltage power supply and an evacuated tube with limited lifetime. On the other hand, they have the disadvantage that they cannot be turned off; radiation hazard is present at all times, whether an experiment is in progress or not.

X-RAY DETECTORS

X-rays were first detected by their latent image production in photographic materials. Photography is no longer used for quantitative purposes, but it is convenient for certain diffraction techniques in which a two-dimensional pattern is produced and its geometry must be precisely established. Film for this purpose is coated on both sides with layers of emulsion, to increase absorption of the penetrating radiation. Other detectors are based on the ability of x-rays to produce flashes of light (scintillations) in certain materials, or to cause ionization in others.

SCINTILLATION DETECTORS

Certain materials have the property of emitting a tiny flash of light when an x-ray photon is absorbed, a case of fluorescence. A prominent example is crystalline NaI into which a small amount (1 or 2 percent) of TlI has

Table 10-2   Isotopes useful as sources of x-rays or gamma rays*

Isotope	*Half-life*	*Radiations*
^{55}Fe	2.60 years	EC: Mn $K\alpha$, 210.3 pm
^{57}Co	270 days	EC: Fe $K\alpha$, 193.7 pm; Fe $K\beta$, 175.7 pm
		Gamma: 86.1, 10.17 pm
^{60}Co	5.26 years	Gamma: 1.06, 0.931 pm
^{241}Am	458 years	EC: Np $L\alpha$, 88.9 pm; Np $L\beta$, 69.8 pm; Np $L\gamma$, 59.7 pm
		Gamma: 20.8, 47.0 pm

* The data are taken from references 3, 4, and 5, recalculated to consistent units as necessary. This table is not intended to be exhaustive.

been incorporated. Observation of these flashes with a photomultiplier gives a reliable measure of the number of photons incident on the crystal. The duration of each flash is very short (of the order of 10^{-8} s), but the resulting pulse of anode current in the photomultiplier may take a longer time to die away (approximately 10^{-6} s). Hence separate scintillations can be distinguished and counted up to at least 10^6 per second. The scintillation detector is inherently linear in response, meaning that the radiant energy in each flash is proportional to the energy of the photon causing the flash. The basic circuitry for use with the photomultiplier has been discussed in Chap. 3; additional electronics are described in Chap. 26.

GAS-IONIZATION DETECTORS

Since x-rays ionize gases through which they pass, their presence can be detected by the conductivity of the gas. This can be done with an *ionization chamber,* a simple metallic container filled with dry gas. A potential of 100 V or more is impressed across the container and an insulated central electrode. The resulting current is measured with an electrometer. The response of this system is slow; hence individual photons cannot be discerned, and the output signal represents an averaged, or steady-state, condition.

Much improved response can be obtained by increasing the voltage to the point where electrons liberated in the initial ionizing event are accelerated sufficiently to ionize additional molecules by collision. The number of ions so produced depends on the energy of the incident x-ray photon, so the observed current is proportional to the photon energy. When used in this mode, the device is called a *gas proportional counter.* Its action is fast enough to permit electronic counting of photons. If the voltage is raised still higher, a saturation effect becomes evident and all pulses are of equal magnitude, regardless of the energy of the original photon. In this mode, the detector is called a *Geiger-Müller* (or *Geiger*) *counter.* See Chap. 23 for a more detailed treatment of gas-ionization detectors.

SOLID-STATE IONIZATION DETECTORS

Germanium and silicon, as free elements, can be sensitized to ionizing radiation, such as x-radiation, by the addition of lithium. The lithium is allowed to diffuse into the crystalline material (*drifting* is the technical term), scavenging impurities as it penetrates. An x-ray photon entering the crystal dislodges electrons from the lattice, leaving vacancies commonly called *holes*, equivalent in effect to mobile positive charges. The number of such charge-separation events is directly related to the energy content of the photon; hence the signal obtained due to the presence of the mobile charges (the conductivity) is also proportional. Detectors of this type must be cooled with liquid nitrogen (some with liquid helium) even when not in use to prevent further diffusion of the lithium, which would greatly reduce the sensitivity and eventually destroy the detector.

The electrical output from one of these solid-state detectors is much smaller than the corresponding signal from a gaseous detector or from a scintillation detector, so that high-gain electronic amplification is required.

Figure 10-5 shows the relative spectral response of typical detectors of the three most important kinds. Note that the abscissa scale is laid off in energy units, the reciprocal of wavelength. Each detector gives a somewhat broadened peak for a nominally monochromatic wavelength, as shown in Fig. 10-6. The Fe $K\alpha$ and Fe $K\beta$ lines at 193.7 and 175.7 pm (6.40 and 7.05 keV) are reproduced by a Si(Li) detector, but not by either gas proportional or scintillation detectors.

ANALYSIS BY X-RAY ABSORPTION

As an analytical tool, x-ray absorption is of most value where the element to be determined is the sole heavy component in a material of low atomic weight. A number of important analyses fall in this category and make the method a significant one for industrial control purposes. Lead in gasoline can be determined this way (6), as can chlorine in organic compounds (7) and uranium in solutions of its salts (8).

Analysis by direct x-ray absorption has largely been displaced by the x-ray fluorescent methods to be described subsequently. The latter

Fig. 10-5 Relative spectral response of proportional, scintillation, and Si(Li) detectors. (*EDAX.*)

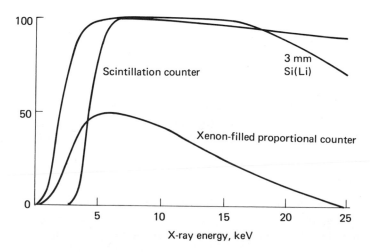

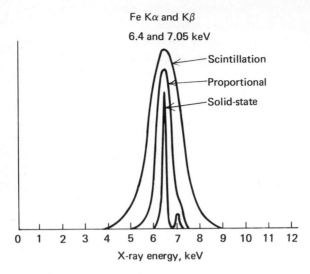

Fig. 10-6 Comparison of resolution between scintillation, proportional, and Si(Li) detectors. (*EDAX.*)

gives qualitative as well as quantitative information, with comparable apparatus.

ABSORPTION EDGE ANALYSIS

Another method of applying x-ray absorption makes use of critical absorption edges as means of identification and quantitative analysis. Since the absorption of an element in a sample is markedly greater at a wavelength just below one of its absorption edges than just above it, and since the location of such edges on the wavelength scale is characteristic of the absorbing element, a pair of absorption measurements bracketing the wavelength of the edge will serve to determine both the presence and the amount of the element sought.

If the power of an x-ray beam is plotted against wavelength in the vicinity of the K edge of an element, a curve like that of Fig. 10-7 will typically be obtained. The jump at λ_E, the wavelength of the absorption edge, is somewhat curved as shown, rather than vertical, because of the necessity of using slits of finite width to gain sensitivity. The quantity desired is the vertical distance between the intersections X and Y obtained by extrapolating, respectively, from lower and higher wavelengths. It is possible to estimate this height with good accuracy from measurements taken at two equally spaced wavelengths, λ_1 and λ_2. Mathematical details and justification for this short-cut procedure are given by Dunn (9), who

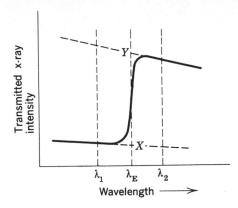

Fig. 10-7 Absorption at a critical edge. [*Analytical Chemistry* (9).]

found that a relative error as low as 1 percent can be obtained at concentrations down to 0.1 percent for many elements.

ABSORPTION APPARATUS

There appears to be no general-purpose x-ray absorptiometer on the market, though General Electric formerly produced a double-beam, nondispersive photometer with ionization-chamber detection. Laboratory analyses by x-ray absorption are carried out by the use of general-purpose equipment which will be described in a later section. For industrial control purposes, apparatus is usually designed specifically for each installation. A portable x-ray absorption analyzer for S and Pb in petroleum products, made by Columbia,* uses a radioactive source.

X-RAY DIFFRACTION (10, 11)

Since x-rays are electromagnetic waves of the same nature as light, they can be diffracted in a similar manner (Fig. 10-8). The equation given in Chap. 2 for diffraction by a grating

$$n\lambda = d \sin \theta \tag{10-6}$$

can also be applied to x-rays. In this case, however, the wavelength is smaller by a factor of 1000 or more, so that to obtain reasonable values of θ, the grating space d must also be made smaller by about the same factor. It is impracticable to rule a grating finely enough to meet this requirement, but it fortunately happens that the spacing between adjacent planes of atoms in crystals is of just the required order of magnitude. There is a variety of crystals suited for x-ray gratings; the most widely

* Columbia Scientific Industries, Inc., Austin, Tex.

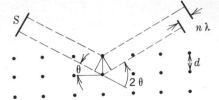

Fig. 10-8 Diffraction of x-rays by successive layers of atoms in a crystal. S represents an incoming plane wave front.

used include lithium fluoride, sodium chloride, calcite, gypsum, topaz, ethyl-enediamine d-tartrate (EDDT or EDT), and ammonium dihydrogen phosphate (ADP). In the simplest arrangement, the x-ray beam is reflected from a plane crystal, and the wavelength selected by varying the angle. Since the waves reflected at successive crystal planes must pass twice across the space between planes, Eq. (10-6) becomes the Bragg equation:

$$n\lambda = 2d \sin \theta \tag{10-7}$$

where d is now the distance between adjacent planes in the crystal. This equation can easily be derived with the aid of Fig. 10-8.

It is also possible to employ concave gratings with Rowland-circle focusing (10). For this purpose a crystal must be bent to conform to the geometric requirements. For improved focus, a crystal may first be bent so that its diffracting planes are curved with a radius equal to twice that of the Rowland circle, and then ground so that the surface is given a curvature with radius equal to that of the circle (Fig. 10-9). The focus-

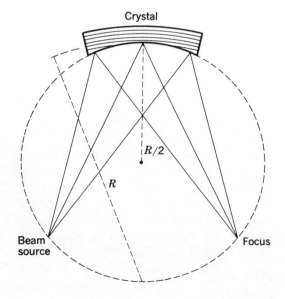

Fig. 10-9 An x-ray monochromator based on the Rowland circle. The crystal is bent and ground for accurate focus. [*Wiley* (2).]

ing monochromator is more expensive, largely because of the difficult handwork required in shaping the crystal, but it can give monochromatic intensity perhaps 10 times that of a plane crystal.

The diffraction of x-rays is of greatest analytical interest as it is applied to the study of the crystalline material producing the diffraction. No two chemical substances would be expected to form crystals in which the spacing of planes is identical in all analogous directions; so a complete study, in which the sample assumes all possible angular positions in the path of the x-rays, should give a unique result for each substance.

DIFFRACTION APPARATUS

Equipment for x-ray diffraction is essentially comparable to an optical grating spectrometer. Lenses and mirrors cannot be used with x-rays, so the instrument is quite different in appearance from its optical analog. A collimated beam can be obtained from an x-ray tube with an extended target by passage through a bundle of metal tubes or, if collimation in one plane only is required, through the spaces between a stack of parallel metal sheets. In some designs the emitting surface of the target is observed at a glancing angle (Fig. 10-10), which gives a close approximation to a line source with maximum intensity. Such a tube can be mounted vertically, anode up, and beams can be taken through each of several ports in different horizontal directions, permitting two, three, or even four independent experiments to be carried on simultaneously.

The diffracted beam of x-rays may be detected photographically or by one of the detectors previously described. The photographic method is typified by an apparatus known as a *Debye-Scherrer powder camera* (Figs. 10-11 and 10-12). The sample is prepared in the form of a fine homogeneous powder, and a thin layer of it is inserted in the path of the x-rays. The powder can be mounted on any noncrystalline supporting material, such as paper, with an organic mucilage or glue as adhesive. The powdered sample contains so many particles that some are oriented in every

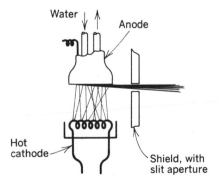

Fig. 10-10 The electrodes of a typical x-ray tube, showing the beam taken at a glancing angle.

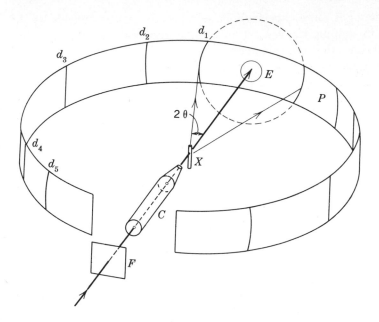

Fig. 10-11 Geometrical features of the Debye-Scherrer powder camera.   Radiation enters through filter F and collimator C to strike the sample X.   The undeviated central beam exits through a hole E cut in the film strip P.   Diffracted beams impinge on the film at points d_1, d_2, etc. [*Wiley* (2).].

Fig. 10-12 Debye-Scherrer x-ray pow-der camera. (*General Electric Com-pany.*)

possible direction relative to the beam of x-rays. There will therefore be diffracted rays corresponding to all sets of planes in the crystals. A strip of x-ray film is held in a circular position around the sample, as shown. Upon development, this film will show a series of lines symmetrically arranged on both sides of the central spot produced by the undeviated beam. The distance on the film from the central spot to any line will be a measure of the diffraction angle θ. Then, if the wavelength and order n are known, the spacing d can be calculated from the Bragg equation.

Several examples of films from a powder camera are reproduced in Fig. 10-13. They are described specifically in the caption (12). X-ray powder diffraction has been found a convenient means of identification for any compounds that can be obtained essentially pure in crystalline form. It can provide a powerful tool in inorganic qualitative analysis, where both the compounds themselves and their chemical derivatives can often be identified.

Ionization or scintillation detectors are most conveniently mounted in a housing that can be moved in a circular path around the sample. This apparatus is called a *goniometer* (Figs. 10-14 and 10-15). In the model illustrated, the linear target of the x-ray tube provides a line source 0.06 by 10 mm in size with high intensity. The angular aperture of the beam is indicated by the divergent lines in Fig. 10-15. It is defined by a single divergence slit that also limits the primary beam to the specimen area. An aperture of 1° is usually employed. Flat specimens up to 10 by 20 mm can be accommodated, or a cylindrical sample can be rotated by a small

Fig. 10-13 X-ray diffraction patterns from films exposed in a powder camera. The radiation was Cu K; a nickel filter 15 μm in thickness was employed for films a to d and the upper portion of e. The specimen in each case was mounted on a 0.075-mm glass fiber, except in e, where it was packed in a 0.3-mm glass capillary. (a) Pb(NO$_3$)$_2$; (b) W metal; (c) NaCl; (d) and (e) quartz. [*Science* (12).]

(a)

(b)

(c)

(d)

(e)

Fig. 10-14 Norelco goniometer. The sample is mounted on the needle projecting to the left and is rotated by the motor visible to the right. The x-ray beam enters from the slit beyond the sample; the x-ray tube is not shown. The detector is located in the tubular housing at the top and can be turned in an arc around the sample by a motor not visible. (*Philips Electronics, Inc.*)

motor. The receiving slit defines the width of the reflected beam detected by the counter. Equally spaced thin metal foils (parallel-slit assembly) limit the divergence of the beam in any plane parallel to the line source; two sets are used, as shown, which permit high resolution to be achieved. The scatter slit serves to reduce the background response caused by radiation other than that of the desired beam. The output from the detector

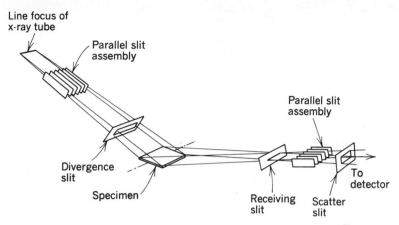

Fig. 10-15   Optical system of the Norelco goniometer. (*Philips Electronics, Inc.*)

is amplified and fed into a pen recorder, so that an automatic record is produced.   As both the recording paper and the arm bearing the detector are turned by synchronous motors, the recorded graph may be interpreted as intensity of the diffracted beam plotted as a function of the angle of diffraction, usually denoted by 2θ.

X-ray diffraction leads primarily to the identification of *crystalline* compounds.   Elements as such will be observed only if in the free crystalline state.   This contrasts sharply with x-ray absorption and with emission spectroscopy, in which the response is to the elements present, without much concern about their states of chemical combination.   For example, each of the oxides of iron gives its own particular pattern, and the appearance of a certain pattern proves the presence of that particular compound in the material being examined.

The power of a diffracted beam is dependent on the quantity of the corresponding crystalline material in the sample.   It is accordingly possible to obtain a quantitative determination of the relative amounts of the constituents of a mixture of solids.

X-RAY EMISSION ANALYSIS

The fluorescent emission of x-rays provides one of the most potent tools available to the analyst for the identification and measurement of heavy elements in the presence of each other and in any matrix.   The method fails with elements lighter than sodium and is only marginal below calcium.

Excitation of the sample is achieved by irradiation with a beam of primary x-rays of greater energy than the characteristic x-radiation which it is desired to excite in the sample.   Because of the need for high-energy primary radiation, a tungsten-target tube is generally employed. The radiation need not be monochromatic.

Each heavy element in the sample is excited to emit the same frequencies it would if it were made the target in a separate x-ray tube, but essentially free of the *Bremsstrahlung* continuum. Then, by the methods of x-ray diffraction, the various wavelengths can be separated and measured. A diffraction apparatus operated with a crystal grating of known spacing is termed an *x-ray spectrometer,* and is strictly analogous to grating spectrometers for visible light. Several designs are possible which are similar in principle to the monochromators described earlier. Several analyzing crystals are frequently needed to cover different ranges, and must be provided with precision interchangeable mountings. The effectiveness of a crystal at separating radiations of the elements is given by a *dispersion curve,* a plot of the angle of deviation observed on the goniometer, as a function of the atomic number Z of the fluorescing element (Fig. 10-16).

Fig. 10-16   X-ray dispersion curves for a number of crystals.
EDDT = ethylenediamine *d*-tartrate.

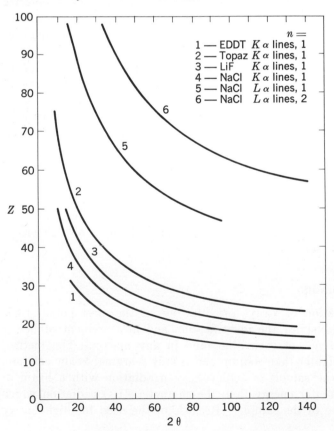

The data are usually presented in degrees of angle 2θ, which has the same significance as in Fig. 10-8. One design of spectrometer is shown in Fig. 10-17.

X-ray emission provides a convenient means of analysis for high-alloy steels and for heat-resistant alloys of the chromium-nickel-cobalt type. Analysis of low percentages is limited by the absorption of the emitted radiation of the element by the materials of the specimen. Samples in which the main constituent is an element of high atomic weight will absorb a higher percentage of the radiation than would be absorbed by a light element. Thus nickel in an aluminum alloy can be determined with greater sensitivity than is possible for nickel in steel or in a silver or lead alloy, where the absorption of the Ni $K\alpha$ radiation is high. In favorable circumstances accuracy of the order of 0.5 percent of the element present can be achieved. The limit of detectability may be as low as a few parts per million.

Figure 10-18 is a reproduction of an automatic record showing an emission spectrum of a chromium-nickel plate on a silver-copper base. The method has been applied to the determination of hafnium in zirconium and tantalum in niobium (11). It was possible, for example, to determine 1 percent tantalum in niobium with a precision of ±0.04 percent (i.e., ±4 percent of the amount present).

Emission methods have also been applied successfully in trace analysis following a preconcentration step. In one report (13) a 90-min electrolysis onto a cathode of pyrolytic graphite, and subsequent x-ray examination

Fig. 10-17 Basic geometry of x-ray emission spectrometer. (*Philips Electronics, Inc.*)

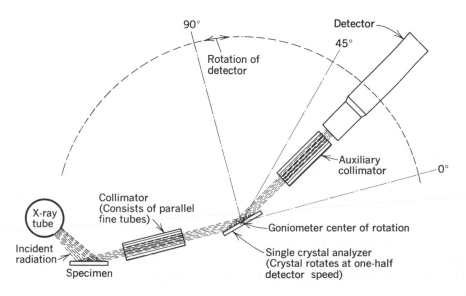

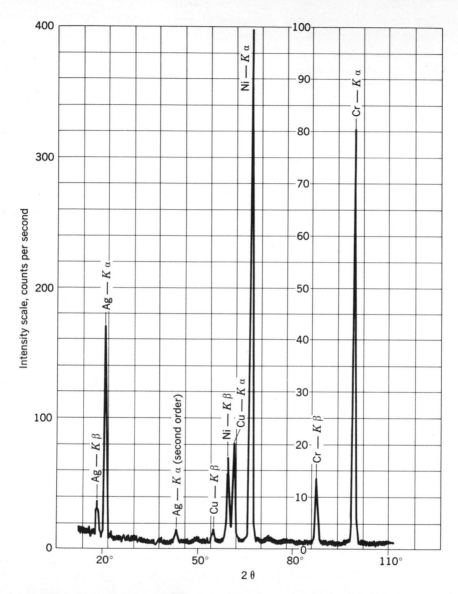

Fig. 10-18 X-ray emission spectrum of a silver-copper alloy plated with nickel and chromium. (*Philips Electronics, Inc.*)

of the cathode surface, permitted determination of various metals to concentrations of a fraction of 1 μg · ml^{-1}. Pyrolytic graphite is particularly advantageous, because it can readily be cleaved into thin layers suitable for mounting in the x-ray apparatus. In addition, carbon is of low enough atomic number to contribute almost no background.

ENERGY-DISPERSIVE X-RAY SPECTROMETERS (14–16)

Equation (2-2), $E = h\nu$, can be rewritten as

$$E\lambda = hc \tag{10-8}$$

where hc is, for all practical purposes, a constant. This formulation stresses the mathematical symmetry of energy and wavelength and suggests the possibility that spectral dispersion may be possible in terms of energy as well as wavelength. Energy dispersion depends on the availability of detectors that respond linearly to the energy content of incident photons. This requirement must be carefully distinguished from response to *total* incident energy or power, which is a function of both energy per photon and the number of photons received per unit of time. Energy detectors in the present sense are available in the x-ray region, namely, those detectors we have designated *proportional:* scintillation counters, gas counters operated at an intermediate range of voltages, and lithium-drifted silicon or germanium detectors. These types vary in their ability to resolve close x-ray lines, as shown in Fig. 10-6. In fact, it was the development of practical solid-state detectors that made possible energy-dispersive spectrometers in a useful form.

Wavelength instruments with a diffraction crystal are capable of greater resolution than their energy-dispersive counterparts, but the latter show a gain of 100 times or more in efficiency and hence sensitivity. This is because they detect all frequencies simultaneously, another instance of the Fellgett advantage mentioned in connection with interferometric and Hadamard spectrometers (Chap. 5).

It should be pointed out that the spectra produced by the two classes of spectrometer are identical, except for differences in resolving power. This fact is emphasized in Fig. 10-5 by the inclusion of two abscissa scales. The graph can be drawn linear in either variable, exactly as in the infrared, where the choice is between wavelength and wave number.

A basic energy-dispersive spectrometer with a radioactive source is shown in Fig. 10-19 (5). An x-ray tube is also usable as a source but contributes greater background. The background can be cut, no matter which radiation source is employed, by installing a fluorescent plate between the primary source and the sample. Figure 10-20 shows two configurations in which a radioactive source and sample can be utilized, with and without an intermediate fluorescent plate. Figure 10-21 depicts a complete instrument.

The signal from the solid-state detector is analyzed by a series of electronic energy discriminators that permit separate counting of signal pulses of successive energy brackets. How this operates is suggested for a six-channel analyzer in Fig. 10-22. The pulses correspond to photons

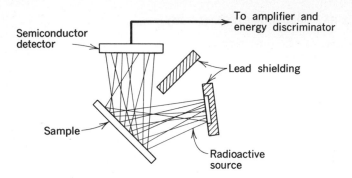

Fig. 10-19 Nondispersive x-ray analyzer with radioactive source, schematic. [*Science* (5).]

arriving at random times. The figure indicates two "peaks," corresponding approximately to channels 2 and 5. Obviously many more channels and data points would be needed to make a valid experiment. Multichannel pulse-height analyzers are available from a number of manufacturers. A favorite size has 400 channels. They are used for a variety of applications besides that described here.

Figure 10-23 shows a spectrum taken with an energy-dispersive spectrometer. Many applications for instruments of this type have been reported. Hanson (17), in a 25-page article, has described a comprehensive program for determining the elemental composition of any object made of metal, glass, or ceramic: he includes the necessary data for 71 elements. Rasberry (of the National Bureau of Standards) has reported (18) comparative data for four types of portable energy-dispersive x-ray fluorescence analyzers for in situ detection of Pb in wall paint; he found a lower limit of detectability of less than 1 mg of Pb per cm^2.

ELECTRON MICROPROBE ANALYSIS (19)

In the past several decades the methods of focusing electron beams (*electron optics*) have been developed to a high degree. It is possible to utilize a finely focused beam to excite x-rays in a solid sample as small as 1 μm^3 (10^{-18} m^3). A schematic diagram of a typical apparatus is shown in Fig. 10-24. The electron beam originates at the electron gun at the top of the diagram, and is focused by specially shaped magnets (magnetic lenses) onto the sample. (This section of the instrument is very similar to an electron microscope used in reverse.) The radiation produced in the sample is analyzed in an x-ray spectrometer. An optical microscope must be incorporated, so that the operator can locate precisely the desired spot on the sample.

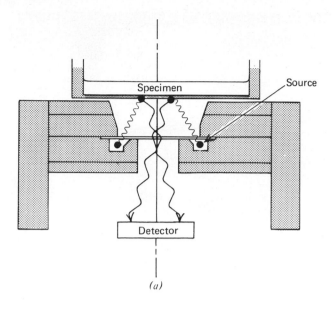

(a)

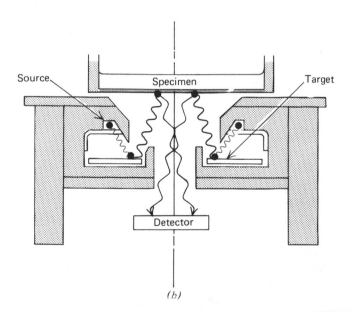

(b)

Fig. 10-20 Nondispersive x-ray emission analyzers. (*a*) The
specimen is irradiated directly by the gamma rays from the
radioactive source. (*b*) Radiation reaches the specimen only as
x-rays from a fluorescent target excited by the primary gamma
rays. (*EDAX.*)

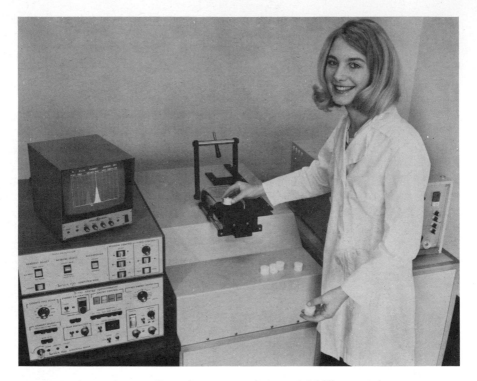

Fig. 10-21   A complete nondispersive x-ray analyzer.   A 3-kW x-ray tube and secondary fluorescent target are located below the sample holder. The multichannel analyzer reads out on the oscilloscope screen at the left. (*Kevex Corporation.*)

The detailed design is more complex than shown in the figure; it is not an easy problem to dovetail the electron system, the optical system, and the x-ray system so that they do not interfere with each other, and yet so that each can operate at its maximum efficiency.   The electron beam must, of course, be in a highly evacuated chamber, which provides further complications.   The associated electronics must include controls for the electron lenses, the electron gun, the x-ray detector and its amplifier, and the vacuum gages and pumps.   The complete instrument is therefore large, complex, and expensive.   In spite of this, it provides such a wealth of information that a number of firms make such equipment.   It is useful for phase studies in metallurgy and ceramics, for following the process of diffusion in the fabrication of transistors, for establishing the identity of impurities and inclusions, and many other purposes.

Until recently this was the only method available for analysis of extremely small objects or areas of a larger object.   With the advent of

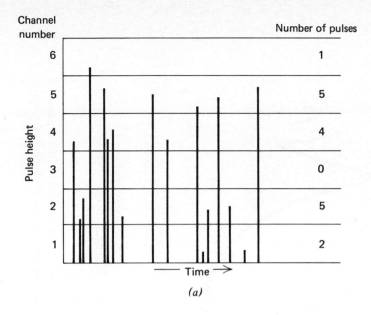

(a)

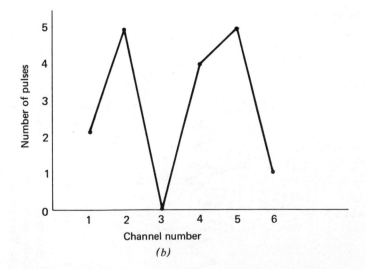

(b)

Fig. 10-22 Multichannel analyzer. (a) Hypothetical distribution of 17 pulses over 6 energy channels. (b) Channel profile.

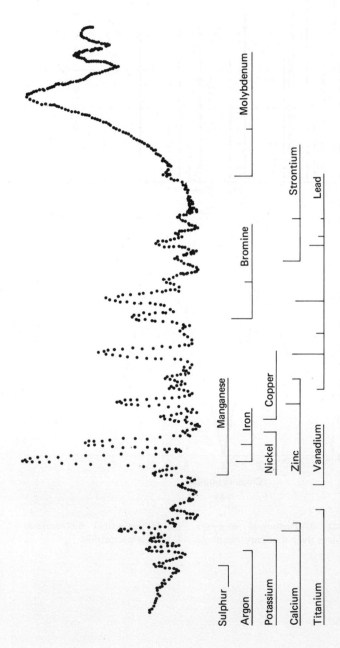

Fig. 10-23 Analysis of particulates in an urban air sample. Air was pumped through a filter, and the filter paper examined in a nondispersive analyzer energized with ^{109}Cd gamma rays. Each dot represents one channel of a multichannel analyzer. (*Kevex Corporation.*)

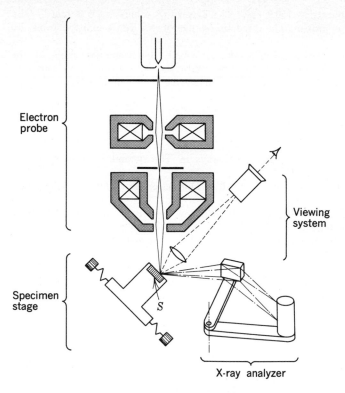

Fig. 10-24 Electron-probe microanalyzer, schematic view, showing the basic elements. [*Wiley* (19).]

the laser an alternative method has come to hand; the laser beam can be focused on a small area, and carries enough power to vaporize most materials. The resulting puff of vapor can be analyzed by optical emission or absorption photometry or even by mass spectrometry.

PROBLEMS

10-1 What are the dimensions of the "constant" C in Eq. (10-5)?

10-2 A metallic sample is irradiated in an x-ray spectrometer with radiation from a tungsten target tube. The spectrometer is equipped with a calcite crystal for which the grating space is 302.9 pm. A strong line is observed at an angle $2\theta = 34°23'$. Calculate the wavelength of the observed x-ray and, from a handbook listing, determine what element must be present. (The spectrum may be assumed to be in the first order.)

10-3 The μ_m values for cobalt at several wavelengths are given as

λ, pm	50.0	100.0	150.0	200.0	250.0	300.0
μ_m	15.5	110.0	345.0	87.0	177.0	270.0

The Ni $K\beta$ radiation occurs at 150.0 pm. The density of cobalt is 8.9 g · cm^{-3}. The intensity of Ni $K\alpha$ is approximately three times that of Ni $K\beta$ as emitted from the target.

By plotting the absorption coefficient for cobalt, determine its value at the wavelengths of Ni $K\alpha$ and Ni $K\beta$ radiations. With the data given above and that in Table 10-1, determine the ratio of Ni $K\alpha$ to Ni $K\beta$ that will penetrate a cobalt foil of 0.005-mm thickness.

REFERENCES

1. Compton, A. H., and S. K. Allison: "X-rays in Theory and Experiment," Van Nostrand, Princeton, N.J., 1935.
2. Klug, H. P., and L. E. Alexander: "X-Ray Diffraction Procedures," Wiley, New York, 1954.
3. Liebhafsky, H. A., H. G. Pfeiffer, E. H. Winslow, and P. D. Zemany: "X-Rays, Electrons, and Analytical Chemistry," Wiley, New York, 1972.
4. Friedlander, G., J. W. Kennedy, and J. M. Miller: "Nuclear and Radiochemistry," 2d ed., Wiley, New York, 1964.
5. Bowman, H. R., E. K. Hyde, S. G. Thompson, and R. C. Jared: *Science,* **151**:562 (1966); T. Hall: *Science,* **153**:320 (1960); H. R. Bowman and E. K. Hyde: *Science,* **153**:321 (1966).
6. Vollmar, R. C., E. E. Petterson, and P. A. Petruzzelli: *Anal. Chem.,* **21**:1491 (1949).
7. Griffin, L. H.: *Anal. Chem.,* **34**:606 (1962).
8. Bartlett, T. W.: *Anal. Chem.,* **23**:705 (1951).
9. Dunn, H. W.: *Anal. Chem.,* **34**:116 (1962).
10. Rudman, R.: *J. Chem. Educ.,* **44**:A7, A99, A187, A289, A399, A499 (1967).
11. Birks, L. S., and E. J. Brooks: *Anal. Chem.,* **22**:1017 (1950).
12. Parrish, W.: *Science,* **110**:368 (1949).
13. Vassos, B. H., R. F. Hirsch, and H. Letterman: *Anal. Chem.,* **45**:792 (1973).
14. Frankel, R. S., and D. W. Aitken: *Appl. Spectrosc.,* **24**:557 (1970).
15. Porter, D. E., and E. Woldseth: *Anal. Chem.,* **45**:604A (1973).
16. Kneip, T. J., and G. R. Laurer: *Anal. Chem.,* **44**(14):57A (1972).
17. Hanson, V. F.: *Appl. Spectrosc.,* **27**:309 (1973).
18. Rasberry, S. D.: *Appl. Spectrosc.,* **27**:102 (1973).
19. Wittry, D. B.: In I. M. Kolthoff and P. J. Elving (eds.), "Treatise on Analytical Chemistry," pt. I, vol. 5, chap. 61, Wiley-Interscience, New York, 1964.

Chapter 11

Electron and
Ion Spectroscopy

Several different phenomena can result from the bombardment of a substance by energetic particles or radiation. The primary process is the ejection of electrons from target atoms, leaving vacancies. Following this, relaxation (i.e., return to the normal configuration) may follow either of two competing paths: characteristic x-rays may be emitted, or secondary (Auger*) electrons may be ejected. In the preceding chapter, the analytical significance of x-ray emission was explored; this chapter will demonstrate the importance of electron emissions.

Another useful technique arises from the bombardment of a target material by such positive ions as He^+ or Ar^+. In this case, the energy content of ions rebounding after elastic collisions gives a clue as to the nature of the target atoms.

These various methods are summarized in Table 11-1.

Figure 11-1 is a generalized diagram of a heavy atom showing a series of electronic energy levels (1). The outer segment consists of a continuous band containing valence electrons in molecular orbitals. Below this band are discrete levels occupied by the core electrons, those not directly involved in chemical bonding.

* Pronounced as in French, "oh-jay."

Table 11-1   Input-output energy forms in electron and ion spectroscopy*

Energy in	Energy out		
	X-rays	*Electrons†*	*Ions*
X-rays	[X-ray fluorescence] [X-ray diffraction]	Photoelectrons, Auger electrons; ESCA, IEE, PESIS, XPS	[Ionization]
Far ultraviolet	. . . .	Photoelectrons; PES, PESOS, UPS	[Ionization]
Electrons	[X-ray emission]	Secondary emission, Auger electrons; AES, EIS, LEED, LEES	[Impact ionization]
Ions	. . . .	. . . .	Ion-scattering, ISS

* Brackets indicate phenomena outside the present discussion.
† AES, Auger electron spectroscopy; EIS, electron impact spectroscopy; ESCA, electron spectroscopy for chemical analysis; IEE, induced electron emission; ISS, ion-scattering spectroscopy; LEED, low-energy electron diffraction; LEES, low-energy electron spectroscopy; PES, photoelectron spectroscopy; PESIS, photoelectron spectroscopy of inner shells; PESOS, photoelectron spectroscopy of outer shells; UPS, ultraviolet photoelectron spectroscopy; XPS, x-ray photoelectron spectroscopy. (Many of these designations are synonymous; their multiplicity is an indication that the field has not yet become stable.)

In the upper right portion of the diagram are listed some radiation sources capable of ejecting electrons at successive levels. The far-ultraviolet radiations from singly and doubly ionized helium can only remove valence electrons to produce ions. The soft Y $M\zeta$ x-ray* can cause ejection of outer core electrons, but more energetic radiation such as Al $K\alpha$ or Cr $K\alpha$ is required to reach the inner shells.

X-RAY PHOTOELECTRON SPECTROSCOPY (XPS)

This is the area most commonly referred to as ESCA, though that designation should be of wider applicability. The energy relations in the ejection of a photoelectron by an x-ray can be expressed as

$$E_b = h\nu - (E_k + C) \tag{11-1}$$

E_b, called the *binding energy*, is defined by this relation as the difference between the energy $h\nu$ of the x-ray photon and the kinetic energy of the ejected electron as corrected by a quantity C depending on the properties of the particular spectrometer system (2). The binding energy is specific for a given electron in a given element, and can serve for the identification

* Yttrium M-zeta.

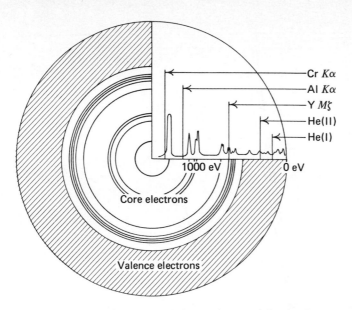

Fig. 11-1 Schematic system of electronic energy levels of an atom in a molecule. The inset shows the regions excitable by several sources of radiation, having the following wavelengths and energies:

Cr $K\alpha$	0.2294 nm	5406.0 eV
Al $K\alpha$	0.8342 nm	1487.0 eV
Y $M\zeta$	9.34 nm	132.8 eV
He$_{II}$	30.38 nm	41.0 eV
He$_{I}$	58.44 nm	21.0 eV

[*Endeavour* (1).]

of that element. Figure 11-2 shows a plot of binding energies that have been determined by photoelectron spectroscopy (3). For light elements ($Z < 30$), two well-separated energy peaks are observed, corresponding to the K and L electrons. From about 35 to 70 a triplet (L_I, L_{II}, L_{III}) is seen, while above 70 a more complex pattern of M and N electrons becomes evident.

Even though the electrons come from the inner core shells, their binding energies are affected measurably by the state of chemical combination (e.g., oxidation state), as this changes the effective force field of the nucleus. Hence E_b exhibits a sensitivity to the molecular environment of the emitting atom, a chemical shift, amounting to a spread of up to about 10 eV for many elements. Thus sulfur ($Z = 16$), whose $2p$ electrons have a normal binding energy of 165 eV (see Fig. 11-2), can yield photoelectrons corresponding to binding energies from about 160 to 168 eV. Some of

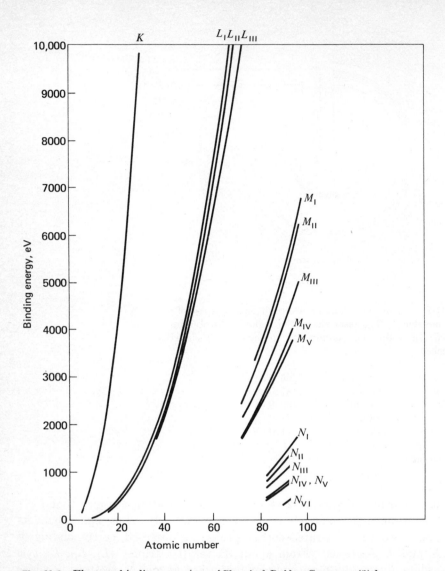

Fig. 11-2 Electron binding energies. [*Chemical Rubber Company* (3).]

these chemical shifts are summarized in Fig. 11-3, taken from Hercules (4). Similar correlation charts for carbon and nitrogen have been published (4, 5). These charts must be regarded as provisional, but doubtless more definitive versions will become available in the future.

Two examples of photoelectron spectra, showing the variation in binding energy of electrons in different atoms of the same compound, are depicted

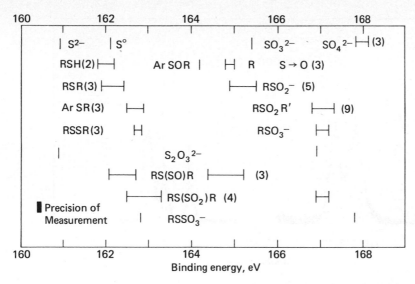

Fig. 11-3 Correlation chart for sulfur $2p$ electron binding energies and functional groups. |——| indicates range of observed energies; the number of compounds used in correlation is given in parentheses. [*Analytical Chemistry* (4).]

in Fig. 11-4. The relative areas beneath successive peaks give a measure of the numbers of atoms in the molecule with the same chemical environment. The binding energies of $1s$ electrons in methyl and methylene carbons bound only to hydrogen and other carbon atoms is very nearly identical (5), so the A peak in Fig. 11-4a can be identified as due to the first, second, and fifth carbon atoms in the formula. The other two peaks correspond to the two different carbons bound to oxygen; which is which can be decided only by correlations with other compounds through a chart for carbon chemical shifts. The two oxygen peaks in the same figure, as expected, are of equal size. The four peaks of Fig. 11-4b are all equal in area. Note that the abscissas can be specified either as the binding energy itself or as chemical shifts relative to a convenient peak taken as a reference point.

XPS, in common with other branches of electron spectroscopy, is essentially a *surface* tool, as the electrons have little chance of escape if they originate more than perhaps 5 nm below the surface of the sample. This makes it particularly valuable for a variety of applications involving surface properties. On the other hand, it can be applied to bulk samples only if the surface is representative of the bulk composition. In many materials, the surface can be etched away gradually by bombardment with ions of argon or another element, so sources of such ions are sometimes

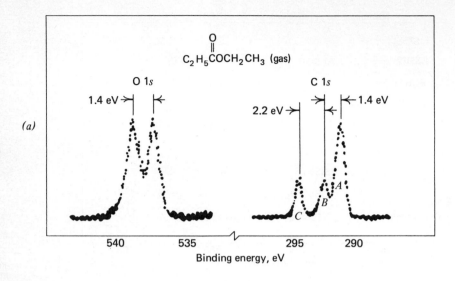

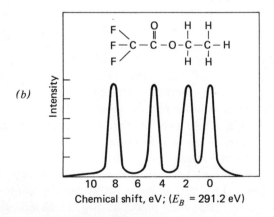

Fig. 11-4 (*a*) Photoelectron spectrum of ethyl propionate. Note the splitting of the oxygen and carbon 1*s* lines due to chemical shifts. (*b*) Carbon 1*s* electron spectrum of ethyl trifluoroacetate excited by monochromatized x-rays. The peaks correspond to the carbon atoms written above. [*Endeavour* (1).]

built into ESCA instruments. This allows the operator to examine the chemical species present at successive depths within a sample by running an electron spectrum after each short treatment with ions.

XPS can be applied to many problems of structure determination or identification of compounds in both solid and gas phases. Two represen-

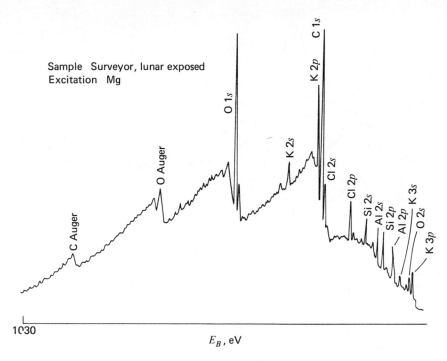

Fig. 11-5 Identification by XPS of elements in lunar soil returned by a Surveyor rocket. [*American Laboratory* (6).]

tative examples are shown in Figs. 11-5 and 11-6 (6). Details of additional examples can be found in the literature (2, 4–6).

ULTRAVIOLET PHOTOELECTRON SPECTROSCOPY (UPS)

The major difference between spectroscopy of electrons ejected by far-ultra-violet radiation and by x-rays is the restriction indicated in Fig. 11-1: Only valence electrons can be observed. This provides an opportunity to obtain direct information about bonding, oxidation states, and molecular ionization potentials. However, association of spectral peaks with individual atoms may not be possible, because of delocalization of electrons in molecular orbitals. Simple molecules can be identified through use of UPS as a fingerprint technique. Figure 11-7 shows a UPS spectrum of air with its major constituents identified (7). At the present state of the art, UPS does not seem to present an analytical tool comparable in versatility to XPS (7, 8).

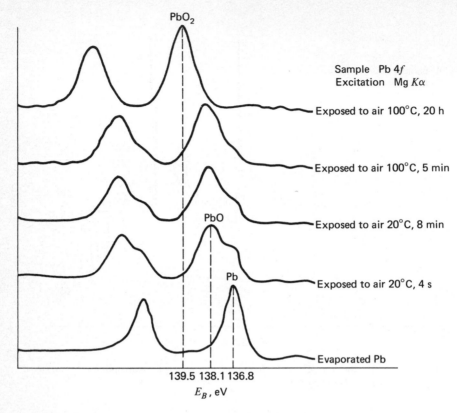

Fig. 11-6 Binding energy shifts of the lead $4f$ electrons as oxide layers are formed on a pure lead sample. [*American Laboratory* (6).]

ELECTRON-IMPACT SPECTROSCOPY (EIS)

A diagram similar to Fig. 11-1 adapted to excitation by electron bombardment would show that, just as with electromagnetic irradiation, the energy of the primary beam determines the depth of penetration into the target atom. Low-energy incident electrons generally result in the promotion of valence electrons from ground to excited states (contrasted with UPS, where ions are usually produced). It is to this phenomenon that the designation *electron-impact spectroscopy* is given. High-energy electrons induce the Auger effect, to be discussed subsequently.

In EIS, the energy required to promote a valence electron is most conveniently measured by observing the *decrease* in kinetic energy of the incident electrons as they are scattered by the molecule. In presently available equipment, only gaseous samples can be handled, and only forward-scattered electrons are detected. The resulting information concerns

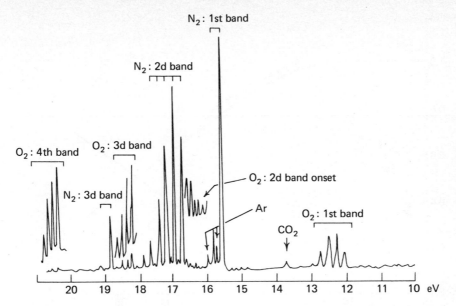

Fig. 11-7 Photoelectron spectrum of air.   Portions of the spectrum are rerun at higher sensitivity.   [*Analytical Chemistry* (7).]

the energy differences between ground and excited states, particularly with respect to vibrational and rotational levels.   The method thus complements infrared and Raman spectroscopy.   At least in theory, all levels can be observed up to the available energy; there are no selection rules to forbid some transitions.   A considerable potential for analytical usefulness is indicated; the determination of CO in air down to about 50 ppm has been reported (9), but the ultimate limit of detection is undoubtedly much lower.

AUGER ELECTRON SPECTROSCOPY (AES)

When a core electron is ejected, whether it be by action of a primary x-ray or an incident energetic electron, an electron from a higher level will drop into the vacancy.   The excess energy will be released either as a secondary (fluorescent) x-ray, or by the ejection of another electron from one of the upper or intermediate levels, with whatever kinetic energy is available from the orbital transition.   The latter is the *Auger effect*.   In Fig. 11-8 the probabilities of these two processes are plotted against the atomic number, showing that Auger spectroscopy is of the most value for light elements (4).

The kinetic energy E_A of an Auger electron from the L shell is given by

$$E_A = (E_K - E_L) - E_L \qquad\qquad (11\text{-}2)$$

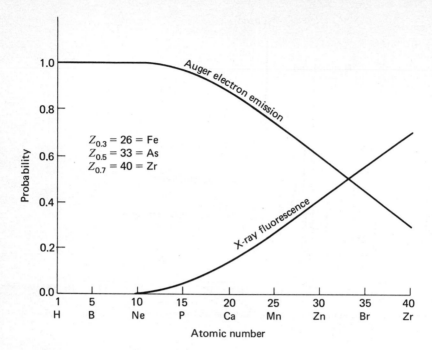

Fig. 11-8 Probability of Auger electron emission and x-ray fluorescence as a function of atomic number. [*Analytical Chemistry* (4).]

where E_K and E_L are binding energies of electrons in the K and L shells. The quantity $E_K - E_L$ is the energy released by the electron falling from the L to the K shell; the second E_L is the energy required to remove the Auger electron from the L level.   Note that the Auger energy is independent of the energy in the primary radiation, whereas by Eq. (11-1) the kinetic energy of photoelectrons changes if the source of the primary beam is altered.   Figure 11-9 shows this effect for a sample irradiated with x-rays from two different targets (6).

As Fig. 11-9 demonstrates, Auger peaks are seen in XPS spectra as well as in electron-generated spectra. The latter, however, are more satisfactory if Auger spectra alone are desired. There is bound to be considerable background due to random scattering of electrons. This is commonly discriminated against by recording the *derivative* of the signal rather than the signal itself.

Auger spectra can be used to identify elements (Fig. 11-10) (10) in the surface layers of solid samples, but they are not sensitive to the state of chemical combination; there are generally no chemical shifts.

Figure 11-11 gives an example of surface analysis by Auger spectroscopy (11). (Note the derivative nature of the curves.) The sample is

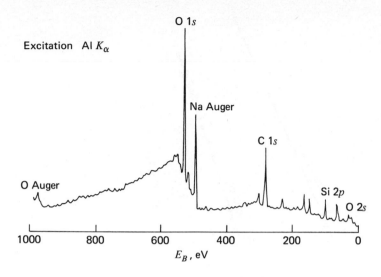

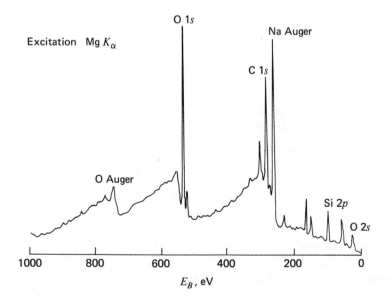

Fig. 11-9 X-ray induced photoelectron spectrum of a glass sample with Al and Mg $K\alpha$ radiation. Auger lines are quickly identified by their position changes on the binding-energy scale. [*American Laboratory* (6).]

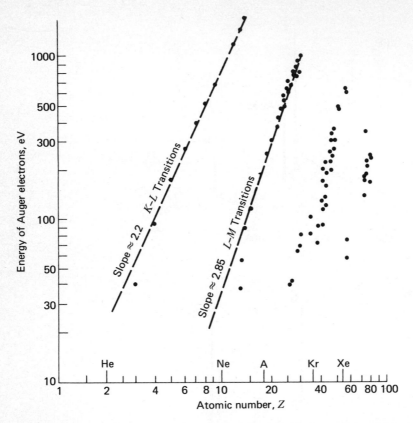

Fig. 11-10   Auger electron energies as a function of atomic number.   [*Journal of Applied Physics* (10).]

a silicon plate on which a film of Nichrome has been deposited.   Before ion treatment to etch away the surface, the largest concentration of atoms was oxygen.   After removal of a 10-nm layer, oxygen had nearly disappeared and chromium and nickel were evident, while at 20 nm these also were nearly gone and only silicon could be seen.

INSTRUMENTATION

An electron spectrometer must contain the following components: (*1*) a source of radiation with which to excite the sample; (*2*) an energy analyzer, the counterpart of a monochromator; (*3*) an electron detector; and (*4*) a high-vacuum system.   The entire system also must be shielded from the earth's magnetic field.   In addition, many instruments are provided with an ion-bombardment gun and other accessories.

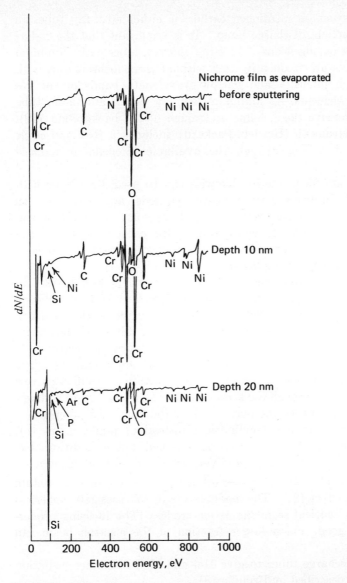

Fig. 11-11   Representative Auger spectra from various depths of a 15-nm Nichrome film on a silicon substrate.   [*Journal of Vacuum Science and Technology* (11).]

RADIATION SOURCES

The source of radiation, as mentioned earlier, is either an x-ray tube, an electron gun, or a helium discharge lamp. It is important that the radiation be energetically homogeneous. As x-ray sources, tubes with aluminum or magnesium targets are frequently used without a monochromator, with reliance on the high intensity and narrow wavelength bands of the $K\alpha$ lines of these light elements. An aluminum window on an aluminum tube acts as a filter to remove the $K\beta$ line and much of the background (12). One commercial instrument (Hewlett-Packard) includes a Rowland-circle crystal monochromator, even though the available intensity is reduced thereby.

An electron beam for excitation likewise can be used directly or with a monochromator. Electrons from a heated cathode, accelerated by an electric field, are fairly homogeneous. They have some spread, however, owing to the fact that they are emitted from the cathode with a range of kinetic energies. If it is desired to obtain a greater degree of homogeneity, an energy filter may be employed.

Several forms of electron energy filters are shown in Fig. 11-12. In the system in (a), the electrons are subject to a retarding field between the two grids; only those with sufficient energy to overcome the field can escape to the right. This is essentially a high-pass filter, in that there is a lower but not an upper limit to the energy of the emerging electrons.

In form (b), the electron beam enters at a 45° angle into the space between two parallel plane conductors; only those within a specified narrow energy band will emerge through the second slit. The energy band is determined by the distance between the plates and by the impressed potential.

If the electrodes are made cylindrical instead of plane, as in (c), not only will good energy discrimination be obtained, but in addition electrons entering the filter at a slightly divergent angle will be focused on the exit slit. The angle 127°17′ $(= \pi/\sqrt{2}$ radians) is required to obtain this double-focusing effect (8). The modification in (d) uses 180° spherical rather than 127° cylindrical segments as electrodes. The focusing properties are nearly as good, and a larger fraction of the electron beam can be accepted.

The special discharge tubes to give He_I and He_{II} resonance radiation for UPS studies are described by Brundle (8).

ENERGY ANALYZERS

The energy analyzer (monochromator) between sample and detector has a more stringent efficiency requirement than those described above (Fig. 11-12), since the electron beam is orders of magnitude less intense.

Retarding field analyzers (Fig. 11-12a) have been used, but geometric restrictions are needed to ensure that all electrons are moving normal to the grids. Magnetic deflection analyzers (Fig. 11-13a) (4) are effective,

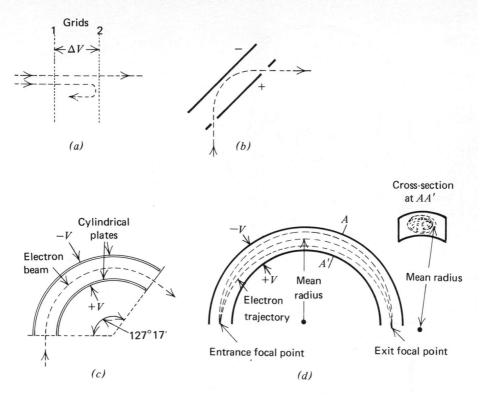

Fig. 11-12 Energy filters. (*a*) Retarding field; (*b*) parallel plates; (*c*) cylindrical; (*d*) spherical.

but less convenient to design and use than electrostatic types. The nonuniform magnetic field employed provides double focusing at $254°56'$ ($= \pi\sqrt{2}$ radians).

The most widely used monochromators utilize either cylindrical or spherical electrostatic fields. The cylindrical type is an extension of the filter of Fig. 11-12*b* (not *c*), as though the plane electrodes were bent into a cylinder about the line joining input and output slits. This is shown in Fig. 11-13*b*. From the theory of this analyzer (13) it appears that optimum focus will be obtained with the angle α between the electron beam and the axis of symmetry within a few degrees of $42°20'$. Practical designs accept electrons in an annular cone of several degrees around this value. The spherical analyzer (Fig. 11-14) is an extension of the filter of Fig. 11-12*d*, making use of a nearly complete sphere.

DETECTORS

Though not the only possible detection device, the electron multiplier is almost universally employed because of its sensitivity and convenience.

(a)

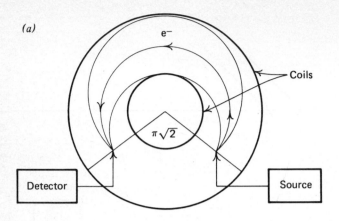

(b)

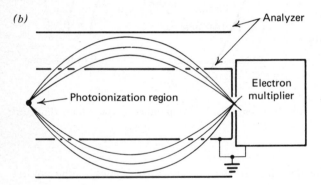

Fig. 11-13 Electron-energy analyzers. *(a)* Magnetic [*Analytical Chemistry* (4)]; *(b)* cylindrical electrostatic.

There are several varieties of electron multipliers; one is the same as the photomultiplier described in Chap. 3, but accepts electrons directly instead of from a photocathode. Just as with photomultipliers, the system of dynodes can be arranged in many different patterns. Another widely used form, the *channel electron multiplier* (Fig. 11-15) is constructed from a small glass tube, usually curved. The inner surface is coated with a high-resistance conductor, which is connected to a source of 2 to 3 kV and acts as a combination of a continuous dynode and a resistive voltage divider. This device accepts electrons at one end and emits more electrons at the other, hence acts as a current amplifier. Typically it has a gain up to 10^8, and such sensitivity that it can count individual electrons.

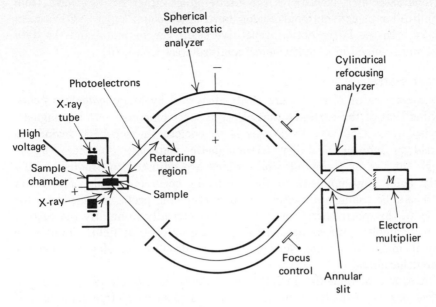

Fig. 11-14 Spherical electrostatic analyzer.

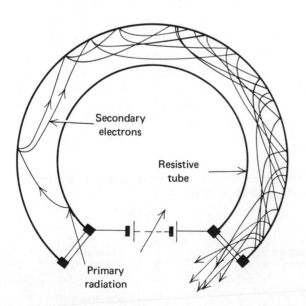

Fig. 11-15 Electron multiplication in a channel detector.

The associated electronics can be designed either to take data from the multiplier in conventional analog form or to count pulses with digital circuitry. In the latter option, repetitive scans can be made and the data stored and summed in a multichannel analyzer (see Chap. 10).

AUXILIARY SYSTEMS

The trajectories of electrons are easily distorted by stray *magnetic fields*, including that of the earth, so that electron spectrometers must be magnetically isolated. One way to do this is to enclose the sensitive regions in a shield of high-permeability ferromagnetic material. Another way is through the use of *Helmholtz coils*, which are adjusted to produce a field exactly equal and opposite to the naturally occurring field. The latter system can be made automatic. A magnetometer probe is placed in the vicinity of the spectrometer and connected electrically to modify the current in the Helmholtz coils as needed to maintain a constant field. Thus even transient magnetic events, such as the operation of an elevator, can be rendered harmless.

A source of difficulty in XPS and UPS is the *sample charging* effect. As electrons are expelled from the sample, a positive charge remains. If the sample is electrically conductive and in contact with the metal parts of the spectrometer, the positive charge cannot accumulate, but if the sample is an insulator, the positive charge will quickly build up to the point where electrons are prevented from leaving. In this event, the observed signal is diminished, and scattered electrons, which are always present to a greater or lesser extent, are attracted to the sample, so that an equilibrium is attained. Particularly in instruments designed to reduce electron scattering, this charging effect can be very serious. It is overcome by including in the instrument a source of low-energy electrons to flood the sample. This can actually result in additional useful information. Figure 11-16 shows two portions of a spectrum of vanadium diboride that has

Fig. 11-16 Electron spectrum of vanadium diboride, showing the effect of flooding with low-energy electrons. [*Hewlett-Packard Company* (14).]

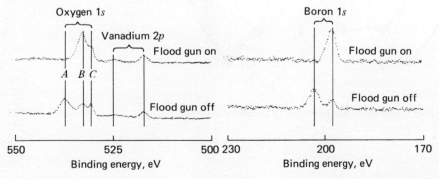

been exposed to air (14).   The 1s electrons of boron show two peaks when the flood gun is turned off, but only one (the area equaling the sum of the other two) with the gun on.   This is interpreted to mean that there is only one kind of boron present, but that about two-thirds of its atoms are electrically insulated from the bulk of the sample.   The vanadium peaks (2p) show no change with the gun on or off, but the changes in the 1s peaks of oxygen are taken to show that the peak marked C is due to atoms linked to vanadium, while the A and B oxygens are bound to boron.

The *high-vacuum system* is conventional, with either an oil-diffusion pump or an ion pump backed up by a mechanical forepump.   In some instruments, particularly those for gaseous samples, differential pumping is required, with two sets of pumps, so that the pressure in the analyzer can be held as low as possible (10^{-8} torr $\cong 1$ μPa is typical), while that in the sample area is perhaps 100 times greater.

Figures 11-14, 11-17, and 11-18 show diagrammatically three commercial electron spectrometers, with contrasting energy analyzers.

ION-SCATTERING SPECTROSCOPY (ISS)

In the bombardment of a solid sample by a beam of positive ions, elastic collisions with target atoms occur, and the ions are scattered thereby. Depending on the kinetic energy of the ions, the target atoms may be driven away from the solid surface, or they may not.   In the latter case, the recoil energy is absorbed by the bulk sample.   Removal of surface atoms in this manner is known as *sputtering*, and is the mechanism for producing a clean surface for ESCA studies, as already noted.

It is possible to collect the sputtered atoms and determine their masses in a small, dedicated mass spectrometer.   Of interest in the present discussion, because of its similarity to ESCA, is the examination of the scattered ions.   The energy E of a scattered ion is less than the energy E_0 of the ion before impact.   The relation is (for a scattering angle of 90°):

$$E = E_0 \frac{M_s - M_0}{M_s + M_0} \qquad (11\text{-}3)$$

where M_s and M_0 are, respectively, the masses of the surface atom and the incident ion.   This is valid only if $M_0 < M_s$.   The energy E is most sensitive to small differences in M_s if M_0 is only slightly smaller than M_s, so it is advantageous to use ions from a variety of gases.   The noble gases, He and Ar especially, are most frequently selected in the interests of avoiding side effects.

An instrument for ISS is diagrammed in Fig. 11-19.   Ions are formed by bombarding gas atoms with electrons.   The positive ions are accelerated

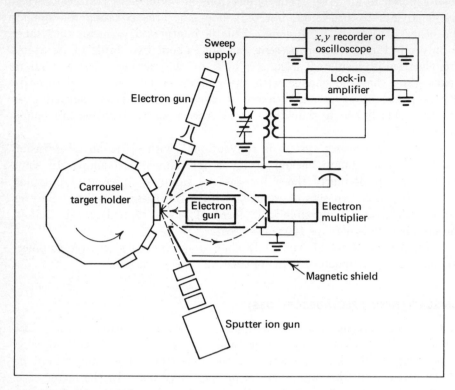

Fig. 11-17 Schematic layout of electron optics and electronics for an Auger electron spectrometer. (*Physical Electronics Industries, Inc.*)

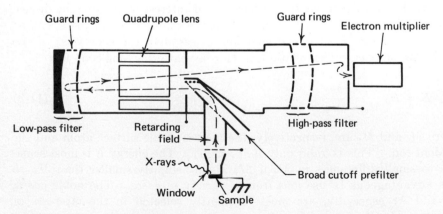

Fig. 11-18 Electron optics of an ESCA spectrometer. (*E. I. Du Pont de Nemours & Co.*)

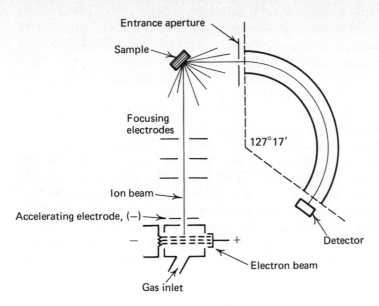

Fig. 11-19 Ion-scattering spectrometer. [*Redrawn from Surface Science* (16).]

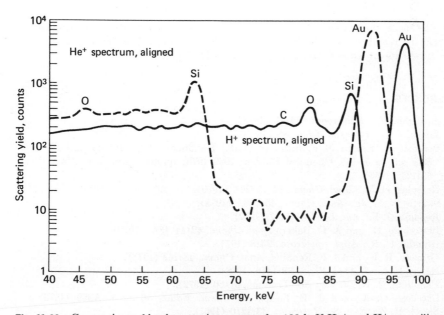

Fig. 11-20 Comparison of backscattering spectra for 100-keV He$^+$ and H$^+$ on a silicon surface coated with a film of gold. [*Surface Science* (16).]

and focused on the sample at an angle of 45°. Ions are scattered in all directions, but only those in a selected small solid angle are admitted to the 127° electrostatic analyzer. The detector can be a solid-state (Si) device, or a channel electron multiplier, since energetic ions will cause release of electrons within the channel. Detailed descriptions of two instruments can be found in the literature (15, 16).

A representative scattered-ion spectrum is shown in Fig. 11-20 (16). The sample was crystalline silicon with a thin gold film. Note that with H^+ ions the several peaks are crowded toward the high-energy end of the spectrum relative to the analogous peaks with He^+ ions, as would be predicted qualitatively by Eq. (11-3). The designation *aligned* refers to the fact that, with these small ions, it is possible to orient the sample so that the ion beam penetrates a significant distance through aligned channels in the crystal structure. [The actual energies recorded in Fig. 11-20 are slightly lower than Eq. (11-3) predicts, because the apparatus collected ions scattered at an angle of 120° from the forward direction, rather than 90°.]

PROBLEMS

11-1 Cr $K\alpha$, the most energetic source listed in Fig. 11-1, is also the *least* energetic of the x-rays included in Table 10-1. Can you suggest why shorter-wavelength x-rays (such as Co $K\alpha$ or Mo $K\alpha$) are not often utilized in ESCA?

11-2 Reconcile the statement that the energy of Auger electrons is independent of the primary radiation and the change in position of Auger lines in Fig. 11-9. The wavelengths are Al $K\alpha$ = 832.0 pm and Mg $K\alpha$ = 987.0 pm.

REFERENCES

1. Siegbahn, K.: *Endeavour,* **32:**51 (1973).
2. James, T. L.: *J. Chem. Educ.,* **48:**712 (1971).
3. Bearden, J. A., and A. F. Burr: In "CRC Handbook of Chemistry and Physics," 50th ed., p. E-185, Chemical Rubber, 1969–1970; quoted from *Rev. Mod. Phys.,* **39:**125 (1967).
4. Hercules, D. M.: *Anal. Chem.,* **42**(1):20A (1970).
5. Swartz, W. E., Jr.: *Anal. Chem.,* **45:**788A (1973).
6. Rendina, J. F.: *Am. Lab.,* **4**(2):17 (1972).
7. Betteridge, D., and A. D. Baker: *Anal. Chem.,* **42**(1):43A, (1970).
8. Brundle, C. R.: *Appl. Spectrosc.,* **25:**8 (1971).
9. Grojean, R. E., and J. F. Rendina, *Anal. Chem.,* **43:**162 (1971).
10. Harris, L. A.: *J. Appl. Phys.,* **39:**1419 (1968).
11. Palmberg, P. W.: *J. Vac. Sci. Technol.,* **9:**160 (1972).
12. Lucchesi, C. A., and J. E. Lester: *J. Chem. Educ.,* **50:**A205, A269 (1973).
13. Sar-el, H. Z.: *Rev. Sci. Instrum.,* **38:**1210 (1967).
14. Kelly, M. A., and C. E. Tyler: *Hewlett-Packard J.,* **24**(11):2 (1973).
15. Goff, R. F., and D. P. Smith: *J. Vac. Sci. Technol.,* **7:**72 (1970).
16. Buck, T. M., and G. H. Wheatley: *Surf. Sci.,* **33:**35 (1972).

Chapter 12

Magnetic Resonance Spectroscopy

A radically different type of interaction between matter and electromagnetic forces can be observed by subjecting a sample simultaneously to two magnetic fields, one stationary and the other varying at some radiofrequency. At particular combinations of fields, energy is absorbed by the sample, and the absorption can be observed as a change in the signal developed by a radiofrequency detector and amplifier.

This energy absorption can be related to the magnetic dipolar nature of spinning nuclei. Quantum theory tells us that nuclei are characterized by a spin quantum number I which can have positive values of $n/2$ (in units of $h/2\pi$; h is Planck's constant), where n can be 0, 1, 2, ... If $I = 0$, the nucleus does not spin, and hence cannot be observed by the methods here considered; this applies to ^{12}C, ^{16}O, ^{32}S, and others. Maximum sharpness of absorption peaks occurs with nuclei for which $I = \frac{1}{2}$, including among others ^1H, ^{19}F, ^{31}P, ^{13}C, and ^{29}Si. The first three of these are easily observed, because they constitute substantially 100 percent of the corresponding elements in natural abundance, whereas the others occur in small proportions.

The spinning nuclei simulate tiny magnets, and so interact with the externally impressed dc magnetic field H. It might be supposed that they would all line up with the field like so many compass needles, but instead their rotary motion causes them to precess like a gyroscope in a gravitational field. According to quantum mechanics, there are $2I + 1$ possible orientations, and hence energy levels, which means that the proton, for example, has two such levels. The energy difference between them is given by

$$\Delta E = \frac{\mu H}{I} \tag{12-1}$$

where μ is the magnetic moment of the spinning nucleus. Energy will be absorbed from the radiofrequency field at frequency ν if the condition $h\nu = \Delta E$ is fulfilled. This characteristic frequency is called the *Larmor frequency*. The above relations can be combined into the expression

$$\frac{\omega}{H} = \frac{2\pi\mu}{hI} \tag{12-2}$$

in which ω, the angular frequency of precession, is set equal to $2\pi\nu$.

The ratio ω/H is a fundamental constant characteristic of any nuclear species that has a finite value of I. This is called the *gyromagnetic ratio* (sometimes the *magnetogyric ratio*) and has the symbol γ.

At a frequency of 100 MHz (one of the commonly used values) the energy difference ΔE is about 10^{-2} cal $\cdot$ mol^{-1}, which means that the source of radiofrequency need not be very powerful, but the detector must be quite sensitive.

Another effect of the imposed alternating field at the Larmor frequency is to cause all the spinning nuclei to precess *in phase*. Thus we have a multitude of nuclear oscillators, which according to electromagnetic theory must radiate energy; since they are in phase with each other, they will act as a *coherent* source. Their radiation can be picked up by another coil in the neighborhood of the sample, positioned with its axis mutually perpendicular to those of the oscillator coil and the fixed field. Hence two types of nuclear magnetic resonance (NMR) spectrometers are possible, the single-coil instrument, in which absorption is measured, and the two-coil variety, which measures resonant radiation.

INTRODUCTORY EXPERIMENT

Let us consider the single-coil NMR spectrometer of Fig. 12-1, containing a glass sample tube filled with distilled water. The radiofrequency oscillator is set at 5 MHz, and the magnetic field varied at a constant rate from 0 to about 1 T,* while the output of the detector is monitored. The

---

* The SI unit of magnetic flux density (field strength) is the *tesla* (T) which corresponds to 10^4 gauss.

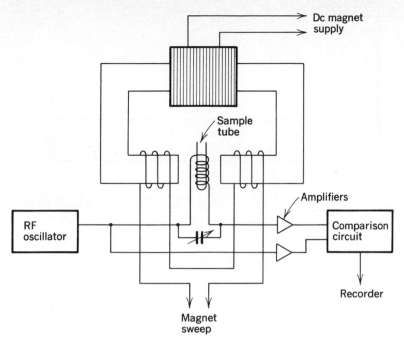

Fig. 12-1 An NMR spectrometer, single-coil type, schematic.

resulting graph will resemble Fig. 12-2, with resonance maxima corresponding to each isotope of each element present for which $I > 0$.   (It is customary to plot NMR spectra as peaks above a base line, no matter whether
measured as absorption or as resonant emission.)   The indicated field
values correspond to the gamma ratios of the various isotopes at $\nu = 5$
MHz.   The copper peaks result from the wire of the radiofrequency coil.
Clearly this provides the possibility of effective qualitative analysis.
Quantitative measurements are also possible, through integration of the
areas beneath the peaks.

An instrument capable of performing an experiment of this type is
called a *wide-line* or *low-resolution* NMR spectrometer.   Its great potential
as an analytical tool has not been widely exploited, no doubt partly because
of its relatively high cost.   It has found rather limited use in investigations
of the physical environment of nuclei, including certain crystal parameters
in solid samples.

HIGH-RESOLUTION NMR

In practice it is found that NMR is of maximum chemical utility if restricted to the study of the fine structure in the resonance of a single

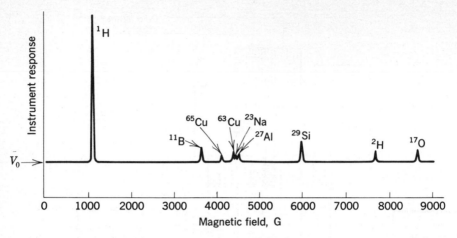

Fig. 12-2 NMR spectrum from a sample of water in a borosilicate glass container, taken on a low-dispersion instrument with a frequency of 5 MHz. (*Varian Associates.*)

nuclear species; an instrument for this purpose is called a *high-resolution* NMR spectrometer. The majority of instruments have parameters suited only to detecting the resonance of hydrogen nuclei (protons). This means that both the magnetic field and the frequency can be held constant, except for a necessary variation in one of them by a few parts per million, as we shall see presently.

In the preceding discussion it has been assumed that the nucleus under consideration is actually subjected to the applied magnetic field with its full measured intensity. If this were strictly true, then only a single peak would be seen, resulting from resonance with all the protons in the sample, as in a wide-line instrument. This is only a rough approximation of the truth, however. A high-resolution spectrometer can evidence two distinct types of structure in the NMR absorption due to proton resonance, known respectively as the chemical shift and spin-spin coupling.

THE CHEMICAL SHIFT

Every nucleus is surrounded by a cloud of electrons in constant motion. Under the influence of the magnetic field these electrons are caused to circulate in such a sense as to oppose the field. This has the effect of partially shielding the nucleus from feeling the full value of the external field. It follows that either the frequency or the field will have to be changed slightly to bring the shielded nucleus into resonance. It is customary in many instruments to accomplish this by an adjustment of the magnetic field through an auxiliary winding carrying varying direct current which sweeps the field over a narrow span (less than 100 μT per tesla of field). The electronic circuitry is such that the value of the added field is converted to its frequency equivalent for presentation to the recorder.

The value of the shift depends on the chemical environment of the proton, since this is the source of variations in shielding by electrons; it is thus called the *chemical shift*. Although the chemical shift is measured as a field or frequency, it is in reality a *ratio* of the necessary change in field to the applied field, or of the necessary change in frequency to the standard frequency, and hence is a dimensionless constant, usually designated by δ, and specified in parts per million (ppm).

Since we cannot observe the resonance of a test tube full of protons without any shielding electrons, there is no absolute standard with which to compare shifts. Therefore an arbitrary comparison standard must be adopted. For organic materials, whenever solubilities will permit, carbon tetrachloride (no protons) is used as solvent, and a small quantity of tetramethylsilane (TMS), $(CH_3)_4Si$, is added as internal standard. This material is chosen because not only are all its hydrogen atoms in an identical environment, but they are more strongly shielded than the protons in any purely organic compound. The position of TMS on the chemical shift scale is arbitrarily assigned the δ value of 0. Some authors prefer to give TMS the value 10 and denote the shift by τ, where $\tau = 10 - \delta$, as this results in small positive values for nearly all other samples. Greater shielding corresponds to an upfield chemical shift. That is, the field must be increased to compensate for the shielding; δ decreases with increased shielding, and τ increases. In the older literature (1960 is "old" in this field), δ was often referred to the proton resonance in water, benzene, or another substance which has only a single peak.

It is possible to construct a diagram (Fig. 12-3) of approximate ranges of δ or τ for protons in various chemical environments (1). The exact values depend to a large degree on substituent effects, solvent, concentration, hydrogen bonding, etc., but are reproducible for any given set of conditions.

SPIN-SPIN COUPLING

The second type of structure frequently observed in NMR spectra is due to the interaction of the spin of a proton with that of another proton or protons attached to (usually) an adjoining carbon. The interaction involves the spins of the bonding electrons of all three bonds (H—C, C—C, and C—H), but we need not be concerned with the detailed mechanism. If the protons are in equivalent environments, the interaction will not be manifest, but otherwise the maxima at each chemical shift position will be split into a close multiplet.

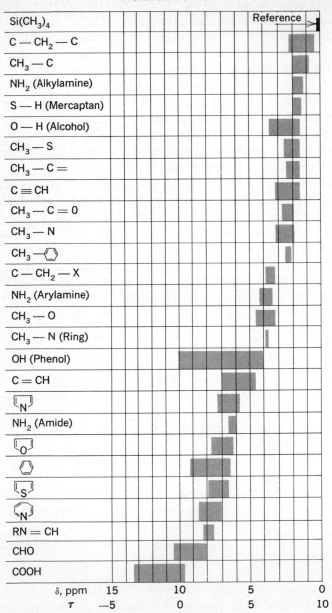

Fig. 12-3 Chart showing ranges of proton chemical shifts for various molecular environments. [*Chemical and Engineering News* (1).]

As an example, consider the spectrum of ethyl iodide in deuterochloroform solution with added TMS (Fig. 12-4). In (a) the spectrum is shown as determined at high resolution. The areas enclosed beneath the multiplets at δ 3.2 and 1.8 are in the ratio of 2:3; the first is produced by the two methylene protons, the second by the three protons of the methyl group. The maximum at δ = 0 is the TMS reference peak. The methylene resonance is split into a quadruplet in the approximate ratio of 1:3:3:1; these magnitudes are the result of the possible relative orientations of the spins of the three methyl protons: there is an equal probability of their being all lined up *with* the field, which we may denote by $\left[\substack{\leftarrow\\\leftarrow\\\leftarrow}\right]$, or *against* it, $\left[\substack{\rightarrow\\\rightarrow\\\rightarrow}\right]$, but it is three times as probable that two will face one way and one the other $\left[\substack{\rightarrow\\\leftarrow}\right]\left[\substack{\leftarrow\\\rightarrow}\right]\left[\substack{\leftarrow\\\rightarrow}\right]$, and similarly $\left[\substack{\leftarrow\\\rightarrow}\right]\left[\substack{\rightarrow\\\rightarrow}\right]\left[\substack{\rightarrow\\\leftarrow}\right]$. One direction of orientation will result in slightly greater shielding, the other in slightly less, hence the 1:3:3:1 ratio. By analogous reasoning it can be shown that the two methylene protons should be expected to cause splitting of the methyl resonance in a 1:2:1 ratio.

The spacings between the components of both multiplets are all equal and designated by a *coupling constant J* which has units of frequency. J is typically between 1 and 20 Hz. It will be noted that the intensity ratios 1:3:3:1 and 1:2:1 are not exactly followed. Those peaks of each group which are closer to the other group are larger in proportion; the nearer together the chemical shifts, the more marked is this effect.

In Fig. 12-4b is shown a trace produced by an integrator which is built into the spectrometer. The step heights are proportional to the areas

Fig. 12-4 NMR spectra of ethyl iodide dissolved in deuterochloroform (CDCl₃). (a) Normal spectrum at high resolution; (b) integral spectrum. (*Courtesy of R. F. Hirsch.*)

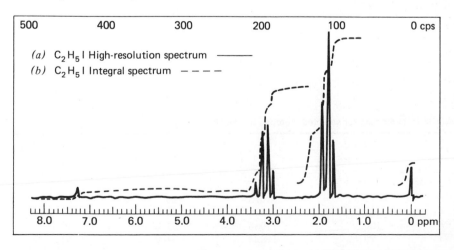

(a) C₂H₅I High-resolution spectrum ———
(b) C₂H₅I Integral spectrum — — — —

beneath the corresponding peaks of the NMR spectrum. The integral step corresponding to an entire multiplet gives a quantitative measure of the number of protons contributing to that resonance.

INSTRUMENTATION FOR NMR

The requirements of the design of high-resolution NMR spectrometers are quite severe. The magnetic field must be uniform over a large enough region to cover the area subtended by a sample probe. The diameter of the pole pieces must be at least four times that of the area which is required to be uniform. If the closest peaks to be resolved are separated by unit J value, in a 60-MHz spectrometer, this field must be homogeneous to within 1 part in 60 million. Even the most carefully machined magnet cannot equal these requirements, so sets of specially shaped auxiliary windings called *shimming coils* are provided and powered with adjustable direct current to counteract any residual inhomogeneity. The field observed by the nuclei under study may still have some lack of uniformity originating in the sample itself and in its container; this is largely averaged out by spinning the sample within the coils by means of a small air turbine.

The radiofrequency excitation is derived from a highly stable crystal-controlled oscillator and fed to the probe coils and the receiver through coaxial cables. The resolution obtainable increases with the frequency, but so does the magnetic field requirement and hence the cost. The least expensive high-resolution instruments commonly manufactured operate at 60 MHz ($H_0 = 1.409$ T for protons); the next higher class at 100 MHz ($H_0 = 2.350$ T); premium instruments are available at 270 MHz (6.342 T).

In many instruments, provision is made for operation at lower frequencies to match the stated magnetic field to other nuclei. Table 12-1 gives the frequencies for a number of nuclei for which $I = \frac{1}{2}$, together with other pertinent data.

FREQUENCY LOCK

An automatic circuit to maintain a constant field-to-frequency ratio is essential to a high-resolution NMR spectrometer. It functions by a fre-

Table 12-1   NMR data for selected isotopes for which $I = \frac{1}{2}$

Isotope	Natural abundance	γ, $rad \cdot T^{-1} \cdot s^{-1}$	$\nu(H_0 = 1.409\ T)$, MHz	$\nu(H_0 = 2.350\ T)$, MHz
^1H	99.98	2.68	60.0	100.0
^{13}C	1.11	0.675	15.1	25.2
^{19}F	100	2.520	56.5	94.2
^{31}P	100	1.086	24.3	40.5
^{203}Tl	29.52	1.528	34.2	57.1

quency-controlling feedback circuit that locks into some specific nuclear resonance, and continuously adjusts the radiofrequency oscillator to keep the reference signal maximized. Without this automatic control, the performance of the instrument would be dependent on the magnetic surroundings, and would be adversely affected by extraneous magnetic events in the vicinity.

In most instruments the reference signal is taken from a nucleus in the sample itself (*internal lock*), often TMS, which then serves both this control function and the reference for calibration of chemical shifts. Some instruments provide for an *external lock*, in which the reference material is located in a separate probe, a short distance removed from the analytical sample. The external lock has the advantage that control is not interrupted during change of samples, and the disadvantage that reference and analytical samples may not feel exactly equal magnetic fields; it provides greater convenience at the expense of precision. Another choice, especially for work with nuclei other than ^1H, is between a *homonuclear* and a *heteronuclear lock*. In the latter the lock is established on a nuclear species other than the one under study; in ^{13}C NMR, for example, the lock can be taken on deuterium nuclei in a deuterated solvent such as $CDCl_3$. The ^2H nucleus, with $I = 1$, gives a somewhat broadened peak without noticeable chemical shifts, hence is suitable for a reference frequency.

DOUBLE RESONANCE

The spin-spin coupling previously described is sometimes of considerable help in the identification of resonances, but in relatively complex molecules it can complicate the spectrum to the point where it becomes impossible to elucidate. Of great assistance in such cases is a *spin decoupler*. This amounts to an auxiliary oscillator which can produce an alternating current at a selectable frequency and impose the corresponding field on the sample with considerable intensity. If this added signal is tuned to the resonant frequency of one set of coupled protons while the contribution of the other set to the spectrum is under observation, it will be found that the multiplet caused by the coupling will collapse, leaving a single sharp peak. In Fig. 12-4a, for example, if the auxiliary frequency is set at 110 Hz downfield (from TMS, that is, $\delta = 1.83$), the quadruplet at $\delta = 3.20$ will change to a singlet at the same location and with the same area. Likewise, if the decoupling field is adjusted to 190 Hz ($\delta = 3.20$), the triplet at $\delta = 1.83$ will collapse to a singlet. (For such a simple spectrum as that illustrated, this is hardly worth the trouble.)

This decoupling is caused by the rapid equilibration of the undesired protons between their two energy states, so that the protons being observed cannot distinguish the separate states and hence cannot be split.

Another type of double resonance, called the *nuclear Overhauser effect*

(NOE), depends on disruption of the normal relaxation mechanism (3). *Relaxation* refers to the process by which a nucleus, having absorbed energy, returns to its low-energy state. The most important of such processes involves dipole-dipole interaction between nuclear magnetic dipoles. This interaction is strongly dependent on the distance of separation of the nuclei concerned, an inverse sixth-power relation. It can be shown theoretically that, if one of these nuclei is saturated by radiation from an auxiliary oscillator, the relaxation of the other nucleus is diminished, resulting in a larger NMR signal. If the two nuclei are both protons, the enhancement can be by a factor of 1.5; if a ^{13}C resonance is observed while an adjacent proton is irradiated, the signal can be increased by nearly three times. This effect increases the signal-to-noise ratio to a useful degree, but more important, it can help pinpoint which peaks are due to nuclei in close proximity to each other. Thus it can distinguish conformational or other stereoisomers. As an example, consider the compound (4):

$$
\begin{array}{c}
H \qquad COOH \\
\diagdown \quad \diagup \\
C \\
\diagup \; 2 \; \diagdown \\
\underset{6}{CH_3}{-}\underset{5}{CH}{=}\underset{3}{C} \qquad \underset{1}{N}{-}H \\
\diagdown \; 4 \; \diagup \\
C \\
\diagup \quad \diagdown \\
H \qquad H
\end{array}
$$

Irradiation of the methyl group produced an enhancement of the two protons at position 4 but not that at 2, showing that the methyl is much closer to the number 4 carbon, hence *trans* with respect to the carboxyl.

APPLICATIONS OF PROTON NMR

The theory already discussed is sufficient indication of the great utility of the method in the *qualitative* identification of pure substances. Atlases are available of NMR spectra, comparable to those of optical absorption spectra, for comparison of unknowns with authentic samples already studied. For substances of unknown structure, NMR provides a valuable diagnostic tool. Chemical shifts and spin coupling and decoupling observations are all useful in this connection.

In the first place, it is easy to establish how many different environments for hydrogen atoms exist in the molecule and, with the aid of the integrator, how many atoms are located in each. Then, through study of fine structure (multiplicity) of peaks, it is possible to determine which hydrogen types are closest to each other. Spin decoupling and NOE observations are invaluable in this phase of the study.

As a *quantitative* tool, NMR has the advantage that a pure sample of the substance sought is not necessary. This contrasts with most optical spectra, and with such methods as gas chromatography, in which an authentic pure sample is essential. In NMR, a pure reference compound is still required as an internal standard but it can be any compound that has an easily identifiable spectrum not overlapping with the sample. An example (2) is a determination of the degree of esterification of pentaerythritol, $C(CH_2OH)_4$, by a complex mixture of acids represented by R—COOH. A weighed portion of the mixed esters was combined with a weighed amount of benzyl benzoate, and gave the NMR spectrum depicted in Fig. 12-5. The sharp peak at $\delta = 5.35$ is due to the single —CH_2— group of benzyl benzoate, and its integral serves to calibrate the spectrum in terms of area per proton. The three broad peaks at $\delta = 1.65, 4.0$, and 4.4 correspond to —OH, —CH_2—OH, and —CH_2—OCOR (italicized hydrogens), respectively. The —OH resonance cannot be used because of the overlap of several sharp peaks originating in the protons of the R groups, but the other two peaks are usable. Since the relation between integrated area and number of protons has already been established, it is a simple matter to compute the concentrations of the —CH_2OH and —CH_2—OCOR functional groups as milliequivalents per gram of sample. It should be noted, in Fig. 12-5, that the analytical peaks are broad because of the contributions to each of a considerable number of molecular species with differing R groups, so that the chemical shifts are only approximately the same. Note also that the relatively large noise evident in these peaks does not show up in the integrator traces. This is characteristic of integrators; noise is equally probable in positive and negative senses, so that area contributions tend to cancel whereas the signal is always positive. The experimental results quoted in the reference showed standard deviations of the order of 2 to 10 percent, considered quite acceptable (2).

Mixtures of compounds can be analyzed in favorable cases with excellent precision. An example is shown in Fig. 12-6. A mixture approximating that which might be found in some types of petroleum was prepared from 1,2,3,4-tetrahydronaphthalene (tetralin), naphthalene, and *n*-hexane. The aromatic hydrogens showed overlapping multiple resonances in the region marked (*a*) in the figure. The hydrogens on carbon atoms adjoining an aromatic ring (alpha to the ring) appeared at (*b*), and the purely aliphatic hydrogens at (*c*). Tetralin has 4 aromatic protons, while naphthalene has 8; only tetralin has alpha protons, of which there are 4; finally, tetralin has 4 protons which are essentially aliphatic, and hexane contributes all 14 of its protons in the same region. A set of three simultaneous equations can be solved for the mole fractions of the three compounds,

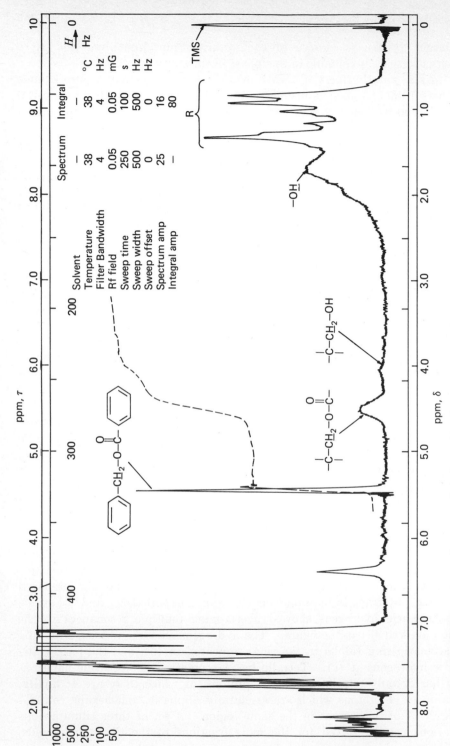

	Spectrum	Integral	
Solvent	—	—	
Temperature	38	38	°C
Filter Bandwidth	4	4	Hz
Rf field	0.05	0.05	mG
Sweep time	250	100	s
Sweep width	500	500	Hz
Sweep offset	0	0	Hz
Spectrum amp	25	16	
Integral amp	—	80	

Fig. 12-5 NMR spectrum of pentaerythritol esterification products. [*American Laboratory* (2).]

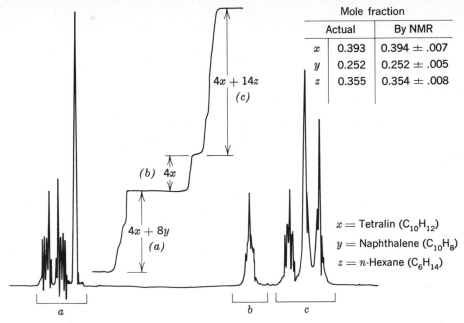

	Mole fraction	
	Actual	By NMR
x	0.393	0.394 ± .007
y	0.252	0.252 ± .005
z	0.355	0.354 ± .008

x = Tetralin ($C_{10}H_{12}$)
y = Naphthalene ($C_{10}H_8$)
z = n-Hexane (C_6H_{14})

Fig. 12-6 Analysis of a hydrocarbon mixture by NMR. (See text for details.) (*Varian Associates.*)

leading to the data in the table inset in the figure. These are in extremely good agreement with the composition of the mixture as calculated from the weights.

NMR SHIFT REAGENTS

The chemical shifts of protons located near an atom with a lone electron pair (Lewis base) can be markedly altered by coordinating the electron pair with a lanthanide atom. For this purpose an organic complex of Eu or Pr is used as reagent. An example is

known as $Eu(DPM)_3$. The effect of the paramagnetic Eu(III) atom is to shift downfield the resonances of nearby protons, but with only slight effect on more distant protons. Figure 12-7 shows an example. This effect

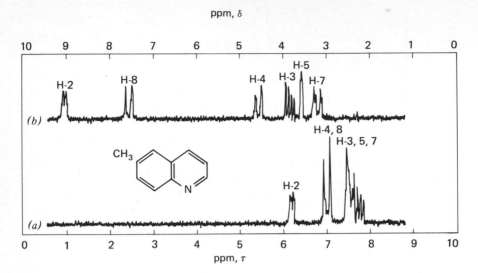

Fig. 12-7 NMR spectra of 6-methylquinoline, over the range 5 to 15 δ. (*a*) CDCl₃ solution; (*b*) same with 0.3 mmol Eu(DPM)₃ added. Proton assignments are marked. (*Perkin-Elmer Corporation.*)

can be very useful in separating overlapping peaks and clarifying locations of sets of protons within a molecule.

NMR OF OTHER ELEMENTS

Of the 18 elements other than hydrogen that have natural isotopes with spin $I = \frac{1}{2}$, the only ones with sensitivities high enough for practical utility at the present state of instrumentation are: ^{19}F (83.3 percent sensitivity), ^{203}Tl and ^{205}Tl (13.5), ^{31}P (6.63), ^{129}Xe (0.55), ^{115}Sn, ^{117}Sn, and ^{119}Sn (0.45), ^{195}Pt (0.34), ^{123}Te and ^{125}Te (0.22), ^{207}Pb (0.193), ^{111}Cd and ^{113}Cd (0.134), ^{77}Se (0.062), ^{29}Si (0.037), and ^{13}C (0.018). The figures are percentages of the proton sensitivity for equal field strengths, and take into account the natural abundances of the magnetic isotopes.* Of these, only ^{19}F, ^{31}P, and ^{13}C are utilized to any great extent. The principles are, of course, the same as for hydrogen. The chemical shifts in general increase with atomic number, covering a span of about 600 ppm for ^{19}F and ^{13}C, 1000 for ^{31}P, and up to about 25,000 (i.e., 2.5 percent) for ^{207}Pb. This means

---

* These sensitivities can be rationalized as follows (for ^{13}C): The intrinsic sensitivity of a nucleus can be shown theoretically to be proportional to γ^3. Given $\gamma_{1H} = 2.68$, $\gamma_{13C} = 0.675$, and the natural abundance of $^{13}C = 1.11$ percent, the sensitivity for ^{13}C is $(0.675/2.68)^3(1.11) = 0.0177$ percent.

that useful information can be obtained with instruments of lower resolution than required for protons (5).

FLUORINE 19

NMR of ^{19}F offers no particular difficulties, and can be carried out by the same techniques as proton spectroscopy at the appropriate field-frequency combinations (see Table 12-1) (6). An interesting possibility is the preparation of fluorine-containing derivatives of organic compounds as a means of avoiding solvent and other interferences present in ^{1}H NMR. For example, Leader (7) has used hexafluoroacetone as a reagent to form easily identified derivatives with alcohols and amines. He gives chemical shifts for adducts with 125 compounds.

PHOSPHORUS 31

Except for its lower sensitivity, ^{31}P shows magnetic properties similar to those of ^{1}H and ^{19}F (8). A representative application (9) is the quantitative analysis of condensed phosphates, hydroxymethylphosphines, and thiophosphates. The method was in excellent agreement with prior methods, and was much faster (no alternative method for thiophosphates was known to the authors) (9).

CARBON 13

The sensitivity for this isotope is so low that unaided standard NMR spectrometers are not adequate for its study. Repetitive scans with signals stored in an electronic memory and automatically averaged (a multichannel analyzer) can be used to advantage. The method of choice at present involves Fourier-transform instruments, to be described in the next section. These are expensive, but give perhaps 100 times the sensitivity of a standard instrument with a signal averager (10).

^{13}C spectra show chemical shifts that are more sensitive to details of structure than are proton shifts. Differences between structural and stereoisomers can be seen easily. Since most carbon atoms in organic compounds are attached to hydrogens, ^{13}C–^{1}H spin-spin interactions are pronounced, though because of the low abundance of ^{13}C they are not important in proton spectroscopy (11). Hence spin decoupling is essential. This can be accomplished with a wide-band oscillator, to decouple *all* protons simultaneously, allowing all ^{13}C multiplets to collapse to singlets. Other double-resonance methods permit retention of coupling information concerning directly attached protons, but elimination of that due to protons on adjacent or further removed carbons. With this technique, called *off-resonance decoupling*, a nonprotonated ^{13}C gives a singlet, ^{13}CH a doublet, ^{13}CH$_2$ a triplet, and ^{13}CH$_3$ a quartet. More details and a wealth of examples can be found in the literature (10, 11).

FOURIER-TRANSFORM (FT) NMR

The efficiency of the spectroscopic examination of a sample in NMR, just as in infrared, can be greatly improved by exciting all possible resonances simultaneously rather than successively as in scanning. In the infrared this means irradiating the sample with undispersed (white) radiation, changing the effective time scale with an interferometer, and then converting information from a time to a frequency scale. In NMR, the sample is likewise flooded with radiation of all possible frequencies in the range of interest, followed by time-to-frequency (Fourier) conversion (12, 13). An interferometer is unnecessary, since the signals are already on a time scale that can be handled directly.

The most convenient way of obtaining a wide range of frequencies is by means of a short burst or pulse of radiofrequency energy. A pulse of this type is the equivalent of a band of frequencies centered around the oscillator (carrier) frequency, the bandwidth depending on the duration of the pulse; a 10-μs pulse covers a frequency range of about 10^5 Hz.

Following each pulse the excited nuclei, which are precessing in the fixed magnetic field, continue to emit signals at their characteristic Larmor frequencies as they relax back to their ground states. This emission following the excitation pulse is known as *free-induction decay* (FID). These emissions can be picked up by a probe coil. Each pulse contains the information of one complete spectrum, collected in an interval between pulses of 1 or 2 s. These data are digitized by sampling at a high rate, and stored in an electronic memory for subsequent averaging over many pulses.

The FID of a single resonant frequency takes the form of an exponentially decaying sine wave of frequency $\nu = |\nu_L - \nu_C|$, where ν_L is the Larmor frequency, and ν_C is the carrier frequency. With multiple reso-

Fig. 12-8   FID pattern for $^{13}CH_3I$.   [*JEOL (U.S.A.), Inc.*]

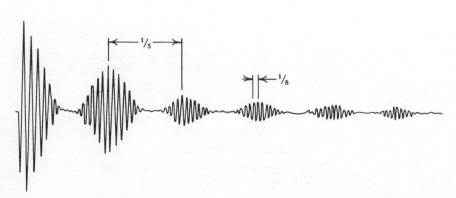

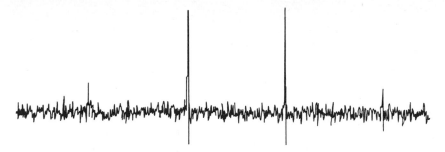

Fig. 12-9 Normal NMR spectrum of $^{13}CH_3I$. [*JEOL (U.S.A.), Inc.*]

nances, the FID patterns rapidly become more complicated. Figure 12-8 shows the FID for $^{13}CH_3I$. The conventional NMR spectrum of this compound (Fig. 12-9) shows a 1:3:3:1 quartet separated by 135 Hz (in a 2.3-T field), the ^{13}C–^{1}H spin coupling constant. The FID shows well-defined interference beats at 135 Hz. Because of this effect, this type of presentation is called an *interferogram*. Figure 12-10 illustrates how hopelessly complex an interferogram can be, even for a relatively simple molecule, 3-ethylpyridine.

The information contained in the interferogram can be converted to a useful form by a computer programmed to take the Fourier transform. The Fourier-transform spectrum of $^{13}CH_3I$ is shown in Fig. 12-11 for comparison with the non-Fourier-transform spectrum of Fig. 12-9. The improvement to be expected is of the order of 100 times.

Fourier-transform spectrometers are manufactured by several companies, who place emphasis on their use in ^{13}C studies. They are much more expensive than their simpler counterparts, even without a computer. The required computer can be a minicomputer, built in as a part of the spectrometer, or it can be shared with other instruments.

Fig. 12-10 FID or interferogram for 3-ethylpyridine. [*JEOL (U.S.A.), Inc.*]

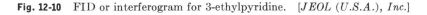

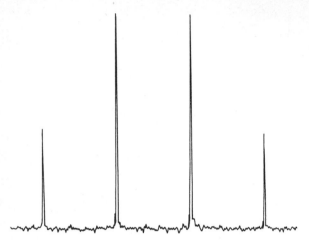

Fig. 12-11 NMR spectrum of $^{13}CH_3I$ resulting from Fourier transformation of the interferogram of Fig. 12-8. [*JEOL (U.S.A.), Inc.*]

ELECTRON SPIN RESONANCE (ESR)*

Since electrons always possess a spin, they also have a magnetic moment, and hence the basic magnetic resonance theory applies to electrons as well as to spinning nuclei. The gyromagnetic ratio for an electron turns out to be 1.76×10^3 rad $\cdot$ T^{-1} $\cdot$ s^{-1} (as compared with 2.68 for the proton). Since this is an inconvenient number to work with, it is customary to rewrite Eq. (12-1) as

$$E = hf = \frac{\mu H}{I} = g\beta H \tag{12-3}$$

where β is the Bohr magneton, a constant with the magnitude 9.2732×10^{-24} J $\cdot$ T^{-1}, and g is called the *splitting factor*. The quantum number I has the value $\frac{1}{2}$ for the electron, so that it has just two energy states, as does the proton. The value of g is 2.0023 for free electrons, and varies from this by a few percent for free radicals, transition-metal ions, and other bodies containing unpaired electrons. If electrons are paired, their magnetic moments effectively cancel each other and they are not observable.

The unpaired electrons have a great tendency to show fine structure in their resonances due to coupling with spinning nuclei in their vicinity. The principles are similar to those involved in the spin coupling of protons, but are often much more complex, frequently to the point where the source

* This subject is also known as electron magnetic resonance (EMR) and electron paramagnetic resonance (EPR).

of each component of the fine structure cannot be identified. The ethyl free radical, for example, produces an array of four triplets. Spin decoupling can sometimes help.

ESR INSTRUMENTATION

Most of the work in ESR is carried out at a constant frequency in the neighborhood of 9.5 GHz with the corresponding field of about 340 kT. Instruments are available, however, which operate as high as 35 GHz, and others are designed for lower frequencies, in the megahertz region. It is usually considered simpler design to maintain constant frequency, generated by a klystron oscillator, and vary the field to scan the resonance peaks.

The most obvious difference between the instrumentation for ESR and NMR is that, because of the high frequencies usually employed, the power is much more effectively conducted from one point to another by rigid waveguides than by flexible coaxial cables. The sample cell is inserted through an orifice in the waveguide at a point where the magnetic (rather than the electric) vector of the electromagnetic wave is undergoing maximum amplitude fluctuations. The efficiency of transfer of energy to the sample is generally low (1 to 30 percent), but even so the method is quite sensitive. Under typical conditions as little as 10^{-10} to 10^{-8} mole percent of free radicals in a 1-g sample can be detected.

It is customary to record ESR spectra in the form of the first derivative of the normal absorption spectrum, as greater resolution is attainable. Quantitative measurements must then be made on the second integral of the ESR curve.

APPLICATIONS OF ESR

Just as in NMR, a standard reference substance is convenient in ESR.   The most widely used is the 1,1-diphenyl-2-picrylhydrazyl free radical:

which is a chemically stable material with the splitting factor $g = 2.0036$. It cannot be used as an *internal* standard with other free radicals, as there is only slight variation in g values, and the standard cannot be distinguished from the substance studied; they can be recorded consecutively, however, with all parameters kept constant. A tiny chip of ruby crystal cemented permanently to the sample cell has been recommended for a stan-

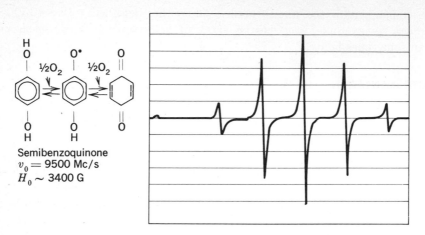

Fig. 12-12   ESR spectrum of the semiquinone intermediate in the oxidation of hydroquinone.   (*Varian Associates.*)

dard; ruby contains a trace of $Cr(III)$ entrapped in its crystal lattice and shows a strong resonance ($g = 1.4$).

Free radicals can be studied readily by ESR, even in very low concentrations.   Figure 12-12 shows the ESR spectrum of a quinone-hydroquinone redox system.   It proves conclusively that a semiquinone free-radical anion exists as an intermediate.   The five-line pattern is a consequence of the magnetic spin interaction between the odd electron and the four hydrogens on the ring; statistics show that the intensities should stand in the ratios 1:4:6:4:1.   Figure 12-13 gives the results of a kinetic study of the formation and decay of the semiquinone.   The ESR instrument was tuned to the point of maximum signal and observed over a period of time.   The sample cell was designed to permit two solutions (hydroquinone and oxygen) to mix together and immediately flow through the observation area.   The flow rate was faster than the rate of production of the radical but, as soon as the flow was stopped, its production, followed by its further reaction, could be monitored easily.   The half-time of the formation reaction under one set of conditions was 0.15 s.

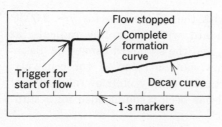

Fig. 12-13   Kinetic study of the formation and decay of the semiquinone. (*Varian Associates.*)

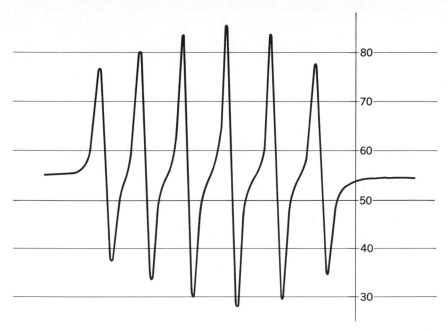

Fig. 12-14 ESR spectrum of the Mn^{2+} ion, 10^{-4} M solution. (*Perkin-Elmer Corporation.*)

Cells have been devised which permit the formation of radicals in situ by irradiation with ultraviolet, gamma, or x-rays, or by electrolytic redox reactions. The great sensitivity of the method makes possible observations of transient species which could be detected in no other way. For example, the fivefold resonance of the ethyl radical becomes visible when ethanol is subjected to x-radiation.

Another important field of application of ESR is in the estimation of trace amounts of paramagnetic ions, particularly in biological work. Figure 12-14 shows the six resonances of the Mn^{2+} ion as observed in aqueous solution. The multiplicity is given by $2I + 1$; for Mn, $I = \frac{5}{2}$, which accounts for the six peaks. This ion can be measured to $10^{-6}M$ or less, with increased instrumental sensitivity, before becoming lost in the background noise.

PROBLEMS

12-1 Justify the (*a*), (*b*), and (*c*) assignments of Fig. 12-6 by comparison with Fig. 12-3. Measure as accurately as you can the heights of the three steps in the integral curve, and calculate the amounts of x, y, and z.

12-2 Account for the small peak at $\delta = 7.32$ in Fig. 12-4*a*.

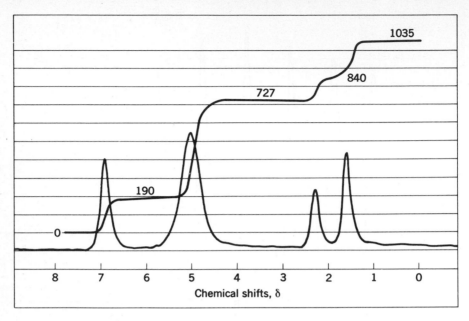

Fig. 12-15 NMR spectrogram at moderate resolution of an emulsion of cyclohexane, toluene, and water. (*Varian Associates.*)

12-3 Figure 12-15 shows the NMR spectrogram of a mixture of cyclohexane, toluene, and water. By comparison with Fig. 12-3, identify the four peaks. From the integral curve, estimate the composition of the mixture.

12-4 Corsini et al. (14) used ESR evidence to determine which of the following two formulas is correct for the complex obtained between copper and 8-quinolinethiol (C_9H_7NS):

(*a*) $Cu(I)(C_9H_7NS)(C_9H_6NS^-)$
(*b*) $Cu(II)(C_9H_6NS^-)_2$

Show that the ESR approach should be expected to distinguish between these structures.

REFERENCES

1. Bovey, F. A.: *Chem. Eng. News,* **43**:98 (1965).
2. Ward, G. A.: *Am. Lab.,* **2**(3):12 (1970).
3. Kennewell, P. D.: *J. Chem. Educ.,* **47**:278 (1970).
4. Isono, K., K. Asahi, and S. Suzuki: *J. Am. Chem. Soc.,* **91**:7490 (1969).
5. Sharp, R. R.: In I. L. Simmons and G. W. Ewing (eds.), "Applications of the Newer Techniques of Analysis," p. 123 ff., Plenum, New York, 1973.
6. Paudler, W. W.: "Nuclear Magnetic Resonance," p. 22, Allyn and Bacon, Boston, 1971.
7. Leader, G. R.: *Anal. Chem.,* **45**:1700 (1973).
8. Ref. 6, p. 24.

9. Colson, J. G., and D. H. Marr: *Anal. Chem.,* **45:** 370 (1973).
10. Stothers, J. B.: *Appl. Spectrosc.,* **26:**1 (1972).
11. Levy, G. C., and G. L. Nelson: "Carbon-13 Nuclear Magnetic Resonance for Organic Chemists," Wiley-Interscience, New York, 1972.
12. Becker, E. D.: *Appl. Spectrosc.,* **26:**421 (1972).
13. Netzel, D. A.: *Appl. Spectrosc.,* **26:**430 (1972).
14. Corsini, A., Q. Fernando, and H. Freiser, *Talanta,* **11:**63 (1964).

Chapter 13

Introduction to Electrochemical Methods

A major series of analytical methods is based on the electrochemical properties of solutions. Consider a solution of an electrolyte contained in a glass vessel and in contact with two metallic conductors. It is possible to connect this cell to an outside source of electric power and, unless the voltage is too low, this will cause a current to flow through the cell. On the other hand, the cell itself may act as a source of electrical energy and produce a current through the external connections. These effects for any specific cell may depend in both nature and magnitude on the composition of the solution, on the materials of which the electrodes are made, on mechanical features, such as electrode size and spacing and presence or absence of stirring, on the temperature, and on the properties of the external electrical circuit.

We shall consider first the properties of various types of electrodes and their interactions with solutions, and then summarize the many possible ways of utilizing them.

ELECTRODES

For some applications an *inert electrode* is required solely to make electrical contact with the solution, without entering into a chemical reaction with any component. A noble metal, usually platinum, sometimes gold or silver, is most suitable, though in some instances an electrode of carbon will give good results.

In other situations, an *active electrode* is appropriate, that is, an electrode made of an element in its uncombined state, which will enter into a chemical equilibrium with ions of the same element in the solution. The potential of the electrode will be related to the concentration of the corresponding ion. Among the commonly used electrodes of this class are silver, mercury, and hydrogen. A gas electrode (such as the hydrogen electrode) consists of a platinum wire or foil to conduct electricity with the gas bubbling over its surface; this combination acts as though it were made of the gaseous element alone. In principle, an electrode can be constructed for any element which is capable of existence in the form of simple ions, but in practice there are many restrictions. The more active elements are seldom employed as electrodes because of the obvious difficulty in preventing direct chemical attack. Hard metals such as chromium and iron tend to have inhomogeneous and nonreproducible surfaces, which impair their utility.

Other "electrodes," such as the calomel and glass electrodes, are in reality combinations of inert or active elementary electrodes with appropriate compounds, which are sometimes fabricated into convenient units for insertion into a cell. They are more correctly termed *half-cells* and will be treated as such.

THE CELL REACTION

Whenever a direct current passes through an electrolytic cell, an oxidation-reduction reaction takes place. At one electrode, defined as the *anode*, oxidation occurs with transfer of electrons from the reduced species to the electrode; at the *cathode*, reduction takes place and electrons are transferred from the electrode to the oxidized species. The primary function of the external circuit is to convey the electrons from anode to cathode. The electric circuit is completed by ionic conduction through the solution.

A generalized redox reaction can be written as

$$r\mathrm{A_{red}} + s\mathrm{B_{ox}} + \cdots \rightleftharpoons p\mathrm{A_{ox}} + q\mathrm{B_{red}} + \cdots$$

where the subscripts "red" and "ox" refer, respectively, to the reduced and oxidized forms of substances A and B. For simplicity, we shall restrict the discussion to the case in which A and B are the only substances oxidized or reduced (i.e., eliminate the "$+ \cdots$").

The equilibrium constant K is defined as

$$K = \frac{(\mathrm{A_{ox}})_{eq}{}^{p}(\mathrm{B_{red}})_{eq}{}^{q}}{(\mathrm{A_{red}})_{eq}{}^{r}(\mathrm{B_{ox}})_{eq}{}^{s}} \tag{13-1}$$

in which the parentheses denote molar activities, and the subscript "eq" emphasizes the fact that these are equilibrium quantities. We can also define a quantity Q, the activity quotient, as

$$Q = \frac{(A_{ox})_{act}^{p}(B_{red})_{act}^{q}}{(A_{red})_{act}^{r}(B_{ox})_{act}^{s}} \tag{13-2}$$

The notation "act" signifies that these quantities are the actual values in an experiment, not necessarily the equilibrium values. Now it can be shown from thermodynamic considerations that the change in free energy (maximum available work at constant temperature and pressure) is given by

$$\Delta G = RT \ln Q - RT \ln K \tag{13-3}$$

where R is the universal gas constant ($8.316 \ J \cdot mol^{-1} \cdot deg^{-1}$) and T is the Kelvin temperature. In electrochemical reactions, the free energy is related to electrical quantities through the expression

$$\Delta G = -nFE_{cell} \tag{13-4}$$

where E_{cell} is potential of the cell in volts, F is the Faraday constant, approximately 96,500 C per equivalent, and n is the number of electrons transferred for one formula unit of the reaction. Thus

$$E_{cell} = -\frac{\Delta G}{nF} = -\frac{RT}{nF} \ln Q + \frac{RT}{nF} \ln K \tag{13-5}$$

By substitution of the defined values of K and Q, followed by rearrangement of the logarithmic terms, we can show that

$$E_{cell} = \left[\frac{RT}{nF} \ln \frac{(A_{ox})_{eq}^{p}}{(A_{red})_{eq}^{r}} - \frac{RT}{nF} \ln \frac{(A_{ox})_{act}^{p}}{(A_{red})_{act}^{r}} \right]$$
$$- \left[\frac{RT}{nF} \ln \frac{(B_{ox})_{eq}^{s}}{(B_{red})_{eq}^{q}} - \frac{RT}{nF} \ln \frac{(B_{ox})_{act}^{s}}{(B_{red})_{act}^{q}} \right] \tag{13-6}$$

Now let us define

$$E_{A}^{\circ} = \frac{RT}{nF} \ln \frac{(A_{red})_{eq}^{r}}{(A_{ox})_{eq}^{p}}$$

and similarly for E_{B}°, which gives

$$E_{cell} = \left[E_{B}^{\circ} - \frac{RT}{nF} \ln \frac{(B_{red})_{act}^{q}}{(B_{ox})_{act}^{s}} \right] - \left[E_{A}^{\circ} - \frac{RT}{nF} \ln \frac{(A_{red})_{act}^{r}}{(A_{ox})_{act}^{p}} \right] \tag{13-7}$$

By this procedure, we have separated into two terms the effects of the two substances A and B on the potential of the cell. We can carry this one step further by defining what we may call *half-cell potentials*:

$$E_{\mathrm{A}} = E_{\mathrm{A}}^{\circ} - \frac{RT}{nF} \ln \frac{(\mathrm{A_{red}})_{\mathrm{act}}^{r}}{(\mathrm{A_{ox}})_{\mathrm{act}}^{p}} \tag{13-8}$$

$$E_{\mathrm{B}} = E_{\mathrm{B}}^{\circ} - \frac{RT}{nF} \ln \frac{(\mathrm{B_{red}})_{\mathrm{act}}^{q}}{(\mathrm{B_{ox}})_{\mathrm{act}}^{s}} \tag{13-9}$$

Then, from Eq. (13-7),

$$E_{\mathrm{cell}} = E_{\mathrm{B}} - E_{\mathrm{A}} \tag{13-10}$$

Equations such as (13-8) and (13-9) were first introduced by Walther Nernst, and are frequently referred to as *Nernst equations*.

Just as the expression for the E of the cell has been broken down into two portions, so the chemical equation for the cell reaction can be separated into two portions, called *half-reactions:*

$$r A_{\mathrm{red}} \rightleftharpoons p\mathrm{A_{ox}} + ne^{-}$$
$$s\mathrm{B_{ox}} + ne^{-} \rightleftharpoons q\mathrm{B_{red}}$$

where e^{-}, as usual, symbolizes an electron.

It is convenient, particularly for purposes of tabulation, to write all half-reactions with the electrons on the same side. We will follow recent practice and write them as *reductions*:

$$p\mathrm{A_{ox}} + ne^{-} \rightleftharpoons r\mathrm{A_{red}} \tag{13-11}$$

$$q\mathrm{B_{ox}} + ne^{-} \rightleftharpoons s\mathrm{B_{red}} \tag{13-12}$$

To obtain the complete cell reaction, the two half-reactions must be multiplied through by suitable numerical factors to make the n's equal. One half-reaction is then *subtracted* algebraically from the other.

The potentials E_{A} and E_{B} are, of course, associated with these half-reactions. The corresponding quantities E_{A}° and E_{B}° are called the *standard potentials* for the half-reactions, and in principle can be evaluated by setting up a half-cell in which the several activities are so chosen that their ratio as it occurs in the Nernst equation is unity and the logarithmic term becomes 0.

No valid method has been discovered for determining an *absolute* potential of a single electrode, since a measurement always requires a second electrode. Therefore it is necessary to choose an electrode to be assigned arbitrarily to the zero position on the scale of potentials. The *normal hydrogen electrode* (NHE) has been selected for this purpose. This is an electrode in which hydrogen gas at a partial pressure of 1 atm is bubbled over a platinized platinum foil immersed in an aqueous solution in which the activity of the hydrogen ion is unity. The NHE is defined by international agreement to show zero potential at all temperatures.

A graphical display of the Nernst equation for a number of metals and hydrogen is presented in Fig. 13-1. The potentials are referred for

Fig. 13-1 Electrode potentials of various metals and hydrogen as functions of the ionic activities. For the significance of the dashed curves, see text.

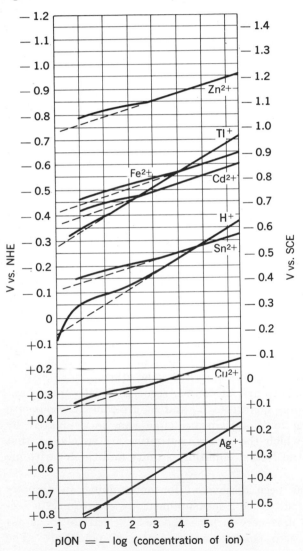

convenience to both the NHE and the saturated calomel electrode (SCE), which will be defined later. The abscissas must be on a logarithmic basis and are conveniently treated in terms of the negative logarithms of concentration or activity. If the scale of abscissas is taken as *activities*, then the straight, dashed lines give true potentials; if, on the other hand, the scale is read as *concentrations*, the solid curves will give more nearly correct values. The deviations between straight and curved portions represent the effect of the activity corrections, and indeed potential measurements provide one of the most useful methods of determining activity coefficients. It will be noted that activities and concentrations become indistinguishable at about 10^{-3} or 10^{-4} F.

The particular elements whose potentials are included in Fig. 13-1 are those which most closely follow the Nernst relation. Others tend to show considerable deviation due to complexation and other effects. For transition metals, particularly, standard potentials are less useful because the determination of activities usually requires data which are not available. For this reason *formal potentials* are often substituted. Let us rewrite Eq. (13-8) showing activity coefficients γ:*

$$
\begin{aligned}
E_A &= E_A^\circ - \frac{RT}{nF} \ln \frac{[A_{red}]^r \gamma_{A(red)}^r}{[A_{ox}]^p \gamma_{A(ox)}^p} \\
&= E_A^\circ - \frac{RT}{nF} \ln \frac{\gamma_{A(red)}^r}{\gamma_{A(ox)}^p} - \frac{RT}{nF} \ln \frac{[A_{red}]^r}{[A_{ox}]^p} \qquad (13\text{-}13) \\
&= E_A^{\circ\prime} - \frac{RT}{nF} \ln \frac{[A_{red}]^r}{[A_{ox}]^p}
\end{aligned}
$$

The quantity $E^{\circ\prime}$ is defined as the formal potential. The composition of the medium in which the measurements were made must be specified carefully whenever formal potentials are employed. The values in Table 13-1, taken from the compilation by Meites (1), show how greatly the formal potentials of a system can vary with the nature of the medium. The first entry is the accepted *standard* potential.

The standard and formal potentials for half-cells are such important and useful constants that a great many of them have been determined with precision and are available in tables.

In many situations it is convenient or even mandatory to separate physically the two electrodes of a cell, always maintaining electrolytic contact between them. This contact may be made in the pores of a porous cup or fritted-glass barrier, or it may be through the medium of a salt bridge. Such a separation of the electrodes is necessary when the electro-

---

* We will drop the subscript "act" from Eq. (13-8); square brackets denote concentrations, while parentheses continue to signify activities.

Table 13-1 Formal potentials of the system Fe(III)/Fe(II) (at 25°C)

Medium	$E^{°\prime}$, V versus NHE
Standard	+0.771
1 F HClO$_4$	0.735
0.5 F HCl	0.71
1 F H$_2$SO$_4$	0.68
5 F HCl	0.64
10 F HCl	0.53
2 F H$_3$PO$_4$	0.46
0.5 F Na$_2$ tartrate (pH 5–6)	0.07
1 F K$_2$C$_2$O$_4$ (pH 5)	0.01
10 F NaOH	−0.68

lytes of the two half-cells are incompatible, or when the redox reaction would occur directly without transfer of electrons via the external circuitry if the electrolyte of one half-cell were to come into contact with the other electrode.

The boundary between the two electrolytes constitutes an additional source of emf, the *liquid-junction potential*, arising from unequal diffusion of cations in one direction and anions in the other. There is no way in which this extraneous potential can be entirely eliminated, but it can be minimized by the use of a *salt bridge* containing a concentrated solution of KCl, NH$_4$NO$_3$, or other salt, the two ions of which have about the same mobilities.

Thus it happens that we frequently employ half-cells that not only have their own half-reactions and half-cell potentials, but also have independent physical existence. It must always be remembered, however, that *one* half-cell is of no use; any application must involve at least two.

There are several types of half-cells of particular importance. In the general half-reaction, such as Eq. (13-11), one or both substances may be in solution, one may be insoluble (such as a free metal), one may be an ion in equilibrium with a slightly soluble salt in the presence of excess solid phase, one or both may be participants in other equilibria, such as complex formation, with some substance which does not enter directly into the redox reaction. A few examples follow.

1. *A Metal in Equilibrium with Its Ions (Class I Electrodes)*

$$\mathrm{Zn^{2+} + 2e^- \rightleftharpoons Zn} \qquad E° = -0.763 \text{ V}$$

$$\mathrm{Cu^{2+} + 2e^- \rightleftharpoons Cu} \qquad E° = +0.337 \text{ V}$$

$$\mathrm{Ag^+ + \ e^- \rightleftharpoons Ag} \qquad E° = +0.799 \text{ V}$$

The oxidized form A_{ox} is the cation, the reduced form A_{red} the free metal. The $E°$ values quoted are in volts relative to the NHE at 25°C. The electrolyte for these half-cells is ordinarily a solution of a salt of the metal with the anion of a strong mineral acid, its selection dictated by solubilities and complexing tendencies: the sulfates, nitrates, and perchlorates are often appropriate. Equation (13-8) takes the form

$$E_A = E_A° + \frac{RT}{nF} \ln (A_{ox})$$

because the activity of a pure solid element is always taken as unity. The quantity (A_{ox}) refers to the activity of the simple cationic species; in the presence of a complex-forming substance this will be less, often much less, than the total amount of the metal in solution.

2. *A Metal in Equilibrium with a Saturated Solution of a Slightly Soluble Salt (Class II Electrodes)*

$$AgCl_{(s)} + e^- \rightleftharpoons Ag + Cl^-_{(a=1)} \qquad E° = +0.2222 \text{ V}$$

$$Hg_2Cl_{2(s)} + 2e^- \rightleftharpoons 2Hg + 2Cl^-_{(a=1)} \qquad E° = +0.2676 \text{ V}$$

Half-cells of this type are widely used as *reference electrodes*, which in effect constitute secondary standards to replace the inconvenient NHE. For such service the activity (or concentration) of the anion is established at a selected value by the addition of a solution of a soluble salt with the same anion; in the examples cited, a solution of potassium chloride is usually chosen. The properties required of a practical reference electrode include ease of fabrication, reproducibility of potential, and low temperature coefficient. A few common reference electrodes are

Saturated calomel electrode (SCE):

$$Hg_2Cl_{2(s)} + 2e^- \rightleftharpoons 2Hg + 2Cl^-_{(sat'd\ KCl)} \qquad E = +0.246 \text{ V}$$

Normal calomel electrode (NCE):

$$Hg_2Cl_{2(s)} + 2e^- \rightleftharpoons 2Hg + 2Cl^-_{(1\ N\ KCl)} \qquad E = +0.280 \text{ V}$$

Normal silver–silver chloride electrode:

$$AgCl_{(s)} + e^- \rightleftharpoons Ag + Cl^-_{(1\ N\ KCl)} \qquad E = +0.237 \text{ V}$$

3. *A Metal in Equilibrium with Two Slightly Soluble Salts with a Common Anion (Class III Electrodes)*

$$Ag_2S_{(s)} \rightleftharpoons 2Ag^+ + S^{2-}$$

$$CdS_{(s)} \rightleftharpoons Cd^{2+} + S^{2-}$$

This half-cell can serve as a measure of the activity of Cd^{2+}. It is a requirement that the second salt (CdS) be slightly more soluble than the first (Ag_2S). One widely applicable electrode which can be placed in this class is that involving the equilibria between EDTA, Hg^{2+} ion, and the ion of a di-, tri-, or tetravalent metal; the slightly dissociated complexes play the same roles as the slightly soluble sulfide salts in the above example.

4. *Two Soluble Species in Equilibrium at an Inert Electrode*

$$Ce^{4+} + e^- \rightleftharpoons Ce^{3+} \qquad E° = +1.61 \text{ V}$$

$$2Hg^{2+} + 2e^- \rightleftharpoons Hg_2^{2+} \qquad E° = +0.920 \text{ V}$$

$$Fe^{3+} + e^- \rightleftharpoons Fe^{2+} \qquad E° = +0.771 \text{ V}$$

The only function of the inert electrode is to transport electrons to or from the ions in the solution. The $E°$ values refer to conditions such that the activities of the two ionic species are equal, which does not necessarily mean that the total concentrations of oxidized and reduced forms of the element are equal. In the presence of complex formers, especially, the tendency of the metal in the two oxidation states to form complexes may not be the same, so that an $E°$ value based on total concentrations rather than activities of free ions can be either larger or smaller than the quoted values.

SIGN CONVENTIONS

We have chosen to write half-reactions as reductions, which gives $E°$ values which are negative with respect to the NHE for metals which are more powerful reducing agents than hydrogen, and positive $E°$ values for those which are less powerful (see the Appendix for a more complete list):

$$Cu^{2+} + 2e^- \rightleftharpoons Cu \qquad E° = +0.337 \text{ V}$$

$$2H^+ + 2e^- \rightleftharpoons H_2 \qquad E° = \quad 0 \text{ V}$$

$$Zn^{2+} + 2e^- \rightleftharpoons Zn \qquad E° = -0.763 \text{ V}$$

The $E°$ values so stated are known as *standard electrode potentials*. This is consistent with experiment, for if a cell is constructed with electrodes of copper and zinc, each in contact with its own ions, the copper is the one which is found to be positive, and the zinc negative, as observed with any ordinary voltmeter or potentiometer. If we write a half-reaction as an oxidation, then the sign of the associated potential must be reversed, but it should *not* be called the *electrode potential*. There are two conflicting conventions for determining the sign of the potential of an electrode.

The convention followed here is in accord with the recommendations of the International Union of Pure and Applied Chemistry (IUPAC) meeting in Stockholm in July, 1953. A complete account of the several sign conventions and of the IUPAC recommendations can be found in a paper by McGlashen (2).

REVERSIBILITY

The terms *reversible* and *irreversible* are used with several different meanings, according to the context. In a purely chemical sense, a reaction is irreversible if its products either do not react with each other, or do so in such a way as to give products other than the original reactants. As an electrochemical example, consider a cell made up of zinc and silver–silver chloride electrodes in dilute hydrochloric acid. If this cell is short-circuited by an external connection, the half-reactions which occur are

$$Zn \rightarrow Zn^{2+} + 2e^-$$
$$2AgCl + 2e^- \rightarrow 2Ag + 2Cl^-$$

and the entire cell reaction is

$$Zn + 2AgCl \rightarrow 2Ag + Zn^{2+} + 2Cl^-$$

However, if this cell be connected to a source of electricity at a high enough voltage to force current through it in the reverse direction, the reactions will be

$$2H^+ + 2e^- \rightarrow H_2$$
$$2Ag + 2Cl^- \rightarrow 2AgCl + 2e^-$$

and the overall reaction is

$$2Ag + 2H^+ + 2Cl^- \rightarrow 2AgCl + H_2$$

Thus it is seen that the silver–silver chloride electrode is reversible while the zinc–hydrochloric acid half-cell is not.

In a *thermodynamic* sense, a reaction is reversible only if an infinitesimal change in driving force will cause a change in direction, which is equivalent to saying that the system is in thermodynamic equilibrium. This implies that the reaction is fast enough to respond instantaneously to any small change in an independent variable. Thermodynamic reversibility is an ideal state which real systems may approximate more or less closely. If an electrochemical reaction is rapid enough that the departure from equilibrium is negligible, it can be considered reversible. A given reaction may be effectively reversible when observed by one technique (as potential measurement with no current flowing) and yet deviate noticeably from reversibility when studied under slowly changing conditions, as in polarography, and become "totally" irreversible when subjected to rapid changes, as in certain fast-scan or ac procedures.

POLARIZATION

An electrode (and hence a cell) is said to be *polarized* if its potential shows any departure from the value which would be predicted from the Nernst equation. This circumstance may occur, for example, when an arbitrary potential is impressed upon an electrode, or when a significant amount of current is drawn through it. Changes in potential due to actual changes in the concentrations of ions at the electrode surfaces are sometimes referred to as a form of polarization, *concentration polarization*. This expression is not recommended, however. The activities concerned should always be those of the ions at the electrode surface.

OVERVOLTAGE

According to thermodynamic definitions, any half-cell is operating irreversibly if appreciable current is flowing. The actual potential of a half-cell under such conditions cannot be calculated; but it is always greater than the corresponding reversible potential as computed from the Nernst equation (i.e., more negative for a cathode, more positive for an anode). The difference between the equilibrium potential and the actual potential is known as *overvoltage*. Overvoltage (3) can then be thought of as the extra driving force necessary to cause a reaction to take place at an appreciable rate. Its magnitude varies with current density, temperature, and with the materials taking part in the reaction. Of especial importance is the overvoltage required to reduce H^+ ion (or water) to hydrogen gas, a process which, in the absence of overvoltage, would take place at 0 V (for activity of H^+ ion equal to unity, the NHE). Figure 13-2 gives representative values of hydrogen overvoltage on a number of cathodes, all in 1 F hydrochloric acid solution (4).

Overvoltage can be of real advantage in some circumstances. Cations of metals such as iron and zinc, for example, can be reduced to the free metals at a mercury cathode, even though their standard potentials are more negative than the NHE, because the high overvoltage of hydrogen on mercury prevents its liberation. At a platinum cathode these ions cannot be reduced from an aqueous solution, as the potential cannot exceed that required for hydrogen liberation.

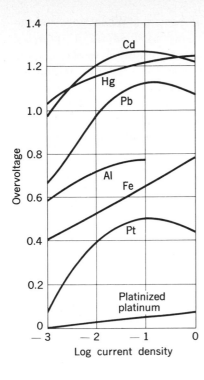

Fig. 13-2 Hydrogen overvoltage on various metals as a function of current density (amperes per square centimetre). The electrolyte is 1 F HCl. [*Wiley* (4).]

ELECTROANALYTICAL METHODS

1. POTENTIOMETRY This is a direct analytical application of the Nernst equation through measurement of the potentials of nonpolarized electrodes under conditions of zero current.

2. CHRONOPOTENTIOMETRY According to this method, a known constant current is passed through the solution and the potential appearing across the electrodes is observed as a function of time. The potential holds a nearly constant value for a period of time proportional to the concentration of an electroactive species. The related measurement of current changes following application of a constant potential is *chronoamperometry*.

3. VOLTAMMETRY AND POLAROGRAPHY These are methods of studying the composition of dilute electrolytic solutions by plotting current-voltage curves. In the usual procedure the voltage applied to a small polarizable electrode (relative to a reference electrode) is increased negatively over a span of 1 or 2 V, and the resulting changes in current through the solution noted. *Voltammetry* is the general name for this method; the term *polarography* is usually restricted to applications of the dropping mercury electrode.

4. CONDUCTIMETRY In this analytical method two identical inert electrodes are employed, and the conductance (reciprocal of resistance) between

them is measured, usually with an ac-powered Wheatstone bridge. Specific effects of the electrodes are eliminated as far as possible.

5. COULOMETRY   This is a method of analysis which involves the application of Faraday's laws of electrolysis, the equivalence between quantity of electricity and quantity of chemical change.

6. CONTROLLED POTENTIAL SEPARATIONS   It is frequently possible to achieve quantitative separations by means of electrolytic oxidation or reduction at an electrode the potential of which is carefully controlled.   The quantity of separated substance may often be measured coulometrically or gravimetrically.

Interesting correlations of many of these electroanalytical methods have been presented in the literature (5–7).   These papers will be more readily appreciated *after* detailed study of the several methods individually.

REFERENCES

1. Meites, L.: In L. Meites (ed.), "Handbook of Analytical Chemistry," table 5-1, p. 5-6, McGraw-Hill, New York, 1963.
2. McGlashen, M. L.: *Pure Appl. Chem.*, **21**(1):1 (1973).
3. Bokris, J. O'M.: *J. Chem. Educ.*, **48**:352 (1971).
4. Daniels, F., and R. A. Alberty: "Physical Chemistry," 3d ed., p. 266, Wiley, New York, 1966.
5. Kolthoff, I. M.: *Anal. Chem.*, **26**:1685 (1954).
6. Reilley, C. N., W. D. Cooke, and N. H. Furman: *Anal. Chem.*, **23**:1226 (1951).
7. Reinmuth, W. H.: *Anal. Chem.*, **32**:1509 (1960).

Chapter 14

Potentiometry

As discussed in the preceding chapter, the Nernst equation gives a simple relation between the relative potential of an electrode and the concentration of a corresponding ionic species in solution. Thus measurement of the potential of a reversible electrode permits calculation of the activity or concentration of a component of the solution.

As an example, suppose we construct a cell with a silver electrode dipping into a solution of silver nitrate, the latter connected by a salt bridge to a saturated calomel reference electrode. It is required to find the concentration of the silver salt solution. The cell assembly is diagrammed in Fig. 14-1. The beaker on the left in the figure contains the unknown solution, and the tube on the right contains the saturated calomel electrode. The central beaker is for the purpose of connecting the potassium chloride solution of the SCE with the salt bridge from the silver solution. The salt bridge is filled with an agar gel which contains ammonium nitrate for electrolytic conduction. (This arrangement prevents contamination of the SCE by silver and also reduces the liquid-junction potential to negligible proportions.)

If a potential-measuring instrument is connected to the two electrodes, the silver will be found to be positive; the SCE, negative. Suppose the

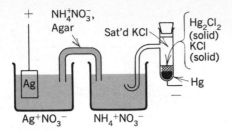

Fig. 14-1 Galvanic cell assembly with silver and saturated calomel electrodes, schematic.

meter indicates a 0.400-V difference in potential between electrodes.   We can write

$$E_{Ag} = E°_{Ag} + \frac{RT}{nF} \ln (Ag^+)$$

$$E_{SCE} = +0.246 \text{ V}$$

and

$$E_{cell} = E_{Ag} - E_{SCE} = E°_{Ag} - E_{SCE} + \frac{RT}{nF} \ln (Ag^+)$$

Solving for log (Ag^+), and noting that $n = 1$, we have (at 25°C):

$$\log (Ag^+) = \frac{E_{cell} - E°_{Ag} + E_{SCE}}{2.303(RT/nF)}$$

$$= \frac{0.400 - 0.799 + 0.246}{0.0591} = -2.59$$

$$(Ag^+) = \text{antilog} (-2.59) = 2.57 \times 10^{-3} F$$

Note that in the first step, writing $E_{cell} = E_{Ag} - E_{SCE}$, we always subtract the more negative from the more positive.   Note also that the quantity $2.303(RT/F)$ has the value 0.0591 at 25°C; this is a convenient figure to remember.

For another example, consider a cell consisting of a platinum electrode dipping into a solution nominally 0.1 F in ferrous sulfate.   The potassium chloride salt bridge of a calomel electrode is inserted directly into the same solution, since no harmful interaction can occur.   The calomel electrode is the SCE; the temperature is 25°C.   A potential difference of 0.395 V is observed.   It is desired to find the percentage of iron(II) which has been converted by air oxidation to iron(III).   The half-cell potentials are:

$$E_{\text{Pt}} = E^{\circ}_{\text{Fe}^{3+}/\text{Fe}^{2+}} + \frac{RT}{nF} \ln \frac{(\text{Fe}^{3+})}{(\text{Fe}^{2+})}$$

$$E_{\text{SCE}} = +0.246 \text{ V}$$

$$E_{\text{cell}} = E_{\text{Pt}} - E_{\text{SCE}}$$

$$= E^{\circ}_{\text{Fe}^{3+}/\text{Fe}^{2+}} - E_{\text{SCE}} + 0.0591 \log \frac{(\text{Fe}^{3+})}{(\text{Fe}^{2+})}$$

$$\log \frac{(\text{Fe}^{3+})}{(\text{Fe}^{2+})} = \frac{E_{\text{cell}} - E^{\circ}_{\text{Fe}^{3+}/\text{Fe}^{2+}} + E_{\text{SCE}}}{0.0591}$$

$$= \frac{0.395 - 0.771 + 0.246}{0.0591} = -2.20$$

$$\frac{(\text{Fe}^{3+})}{(\text{Fe}^{2+})} = \text{antilog} \ (-2.20) = 6.3 \times 10^{-3} = 0.63 \text{ percent}$$

which shows that 0.63 percent of the iron(II) was converted to iron(III).

THE CONCENTRATION CELL

If two identical electrodes are placed in beakers containing solutions similar in every way except concentration (connected by a salt bridge), the potential between the electrodes is related to the *ratio* of the two concentrations.   For example, in a chloride analysis, one silver–silver chloride electrode may be dipped into a solution of unknown chloride concentration $[\text{Cl}^-]_x$, the other being dipped into a standard solution, concentration $[\text{Cl}^-]_s$.   Activity corrections can be neglected even at relatively high concentrations, as their effect will be similar in the numerator and denominator of the logarithmic term.

Suppose the standard is 0.1000 F.   Its single electrode potential is

$$E_s = +0.222 - 0.0591 \log 0.1000 = +0.281 \text{ V}$$

If the unknown solution is more concentrated, say 0.1500 F, then

$$E_x = +0.222 - 0.0591 \log 0.1500 = +0.271 \text{ V}$$

The measured potential difference is found, as usual, by subtracting the more negative from the more positive:

$$E_{\text{cell}} = E_s - E_x = +0.010 \text{ V}$$

On the other hand, if the concentration of the unknown is less than that of the standard, say 0.0680 F, then

$$E_x = +0.222 - 0.0591 \log 0.0680 = +0.291 \text{ V}$$

and

$$E_{\text{cell}} = E_x - E_s = +0.010 \text{ V}$$

As $E°$ is the same for both electrodes, its value cancels, and need not be included in the calculations. Since any measured potential (such as the +0.010 V above) may correspond to either of two unknown concentrations, the concentration cell may give rise to ambiguous interpretations. This may be avoided by noting carefully the relative signs of standard and unknown.

The concentration cell provides a highly sensitive analytical tool (1) because most sources of error arising within the cell cancel out in a method which is a comparison of two nearly identical solutions. This means that the overall precision of the method may very well be limited by the measuring instrument. As in any potentiometric method, the sensitivity (at 25°C) is $0.0591/n$ volts for a change of a factor of 10 in concentrations. If the measurement were made relative to a universal reference electrode (i.e., *not* a concentration cell), a measuring instrument with a range of perhaps as much as 2 V would be required, whereas with a concentration cell a 20-mV instrument can be employed. The latter permits 100 times the precision of the former.

ION-SELECTIVE MEMBRANE ELECTRODES

A whole class of highly useful electrodes that show varying degrees of specificity and selectivity utilize a *membrane* to confine an inner reference solution and reference electrode, and at the same time make electrolytic contact with the outer (test) solution (2). A few typical constructions are shown in Fig. 14-2. The membrane in each of these electrodes acts by an ion-exchange mechanism.

THE GLASS ELECTRODE

Electrode (*a*) is the familiar glass electrode for pH measurement. It consists of a small bulb of pH-sensitive glass containing a buffered chloride solution and an internal reference electrode, usually silver–silver chloride or calomel. The glass is a partially hydrated aluminosilicate containing sodium or calcium ions and often small amounts of lanthanide ions. It is permeable to hydrogen ions in the sense that an ion can join an oxygen site on the lattice at the same time that another, on the opposite surface, leaves, thus maintaining electrical neutrality in the glass. An equilibrium is easily established with hydrogen ions in inner and outer solutions, and a potential is produced:

$$E = K + \frac{RT}{F} \ln \frac{(H^+)_{inner}}{(H^+)_{outer}} \tag{14-1}$$

K in this expression includes the difference between the characteristic potentials of inner and outer reference electrodes (which might be identical), the liquid-junction potential, and a small unpredictable contribution known as the *asymmetry potential*, perhaps due to different physical strains in the two surfaces of the glass.

Fig. 14-2 Ion-selective electrodes. (*a*) Glass; (*b*) liquid ion exchange; (*c*) ionic crystal membrane.

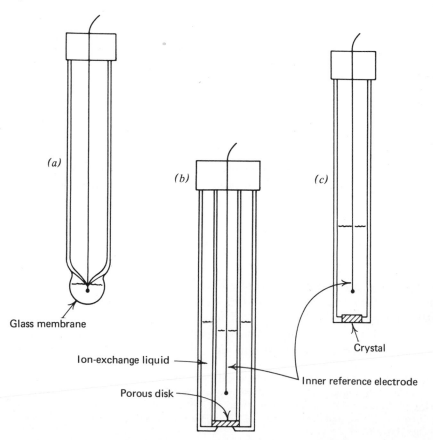

(*a*)

(*b*)

(*c*)

Glass membrane

Ion-exchange liquid

Porous disk

Crystal

Inner reference electrode

By introducing the pH definition,* Eq. (14-1) can be rewritten (at 25°C) as

$$E = K + 0.0591 \, (\text{pH}_{\text{outer}} - \text{pH}_{\text{inner}}) \qquad (14\text{-}2)$$

In any given electrode the pH of the inner solution is constant, so we can combine it with K and drop subscripts, to give

$$E = K' + 0.0591 \, \text{pH} \qquad (14\text{-}3)$$

In a practical pH meter, the constant K' is taken into account in the calibration procedure, whereby the electrode assembly is immersed in a standard buffer and the indicator arbitrarily made to show the correct value.

The glass electrode is characterized by a "pH of zero potential," where the pH of inner and outer solutions is the same. If this is at pH 4, for example, then an external solution at pH 4, should, by Eq. (14-2), give $E = K$, which will be very small if inner and outer electrodes are identical.

Combination glass and reference electrodes are fabricated by many manufacturers. In most designs the reference electrode makes contact with its salt-bridge solution (KCl or other) which is held in an annular container surrounding the glass electrode. The salt solution acts as an electrostatic shield for the high-resistance glass electrode. The combination electrode has become quite popular because of its convenience. The cost is somewhat less than for separate glass and reference electrodes.

One shortcoming of the glass pH electrode is that an erroneously high reading is obtained in the presence of large concentrations of alkali metal ions, the *sodium error*. The previous equations are no longer adequate. The more complete relation, known as the Nernst-Eisenman equation, is

$$E = \text{constant} + \frac{RT}{F} \ln \, [(\text{H}^+) + k_1(\text{Na}^+) + k_2(\text{K}^+) + \cdots] \qquad (14\text{-}4)$$

where the k's are *selectivity coefficients*. An acceptable pH electrode should have very small k values, 10^{-3} or less. These coefficients are not strictly constants, and cannot be used to calculate corrections for observed pH values. They are useful in determining the suitability of various electrodes for particular analytical situations.

* Although the pH scale is defined as the negative logarithm of the activity of hydrogen ions, it has been found expedient to set up a practical or working scale based on a series of buffer solutions prepared in a prescribed manner. A full discussion of both theoretical and practical aspects of the pH concept can be found in the monograph by Bates (3).

Eisenman (4) and others have shown that the selectivity coefficients are related to exchange constants K for an ion-exchange process taking place at the glass surface, and to the diffusion mobilities of the ions within the glass. For example, if consideration is limited to H^+ and Na^+ ions, the coefficient k_1 is given by

$$k_1 = \frac{(H^+)_{sol}(Na^+)_{surf}}{(H^+)_{surf}(Na^+)_{sol}} \frac{U_{Na^+}}{U_{H^+}}$$

where the subscripts denote activities in solution and adsorbed on the glass surface; U_{Na^+} and U_{H^+} are the respective mobilities in the glass.

Glass electrodes can be made with high sensitivity toward Na^+ and Ag^+ ions and only moderate pH response. This type of electrode can be used to measure either Na^+ or Ag^+ ion if the other is absent or held constant, and if the solution is adequately buffered.

LIQUID-MEMBRANE ELECTRODES

Electrodes sensitive to a variety of cations or anions can be prepared with a "membrane" consisting of a liquid ion-exchange material (4). This type of electrode may be constructed as shown in Fig. 14-2b. A small disk of hydrophobic filter material forms the barrier between inner and outer electrolytes. The disk is in contact at its perimeter with an organic solvent, immiscible with water, held in the annular space. Dissolved in this solvent is a salt of the desired ion with a counterion of relatively high molecular weight and much greater solubility in the organic phase than in water. The solvent is pulled into the pores of the filter disk by capillarity, where it makes electrical contact with both aqueous solutions. Equilibrium is thus established between the common ion in the membrane and the solutions. The potential of the inner electrode follows a Nernst relation, just as does that of the glass electrode. Some examples of electrodes with liquid membranes are listed in Table 14-1.

It has been reported recently that liquid ion-exchange electrodes can operate without an internal reference solution. The ion-exchange material is incorporated into a polymer film placed directly in contact with a platinum wire (5) or a carbon rod (6). The resulting electrodes appear to be electrical equivalents of the more elaborate versions, and certainly less delicate and expensive.

DOUBLE-MEMBRANE ELECTRODES

The versatility of ion-selective membrane electrodes can sometimes be increased by the use of a second membrane. The partial pressure of CO_2 dissolved in blood plasma or other fluid is routinely measured with a glass electrode covered with a thin film of Teflon or other gas-permeable polymer. Trapped between the film and the glass is a layer of aqueous $NaHCO_3$

Table 14-1 Representative ion-selective electrodes (2)

Liquid membranes

Ion measured	Exchange site*	Principal selectivity constants
Ca^{2+}	$(RO)_2PO_2^-$	H^+ 10^7; Zn^{2+} 3.2; Fe^{2+} 0.80; Pb^{2+} 0.63; Cu^{2+} 0.27; Ni^{2+} 0.08; Sr^{2+} 0.02; Mg^{2+} 0.01; Ba^{2+} 0.01; Na^+ 0.0016
NO_3^-	NiL_3^{2+}	ClO_4^- 10^3; I^- 20; ClO_3^- 2; Br^- 0.9; S^{2-} 0.57; NO_2^- 0.06; Cl^- 0.006; SO_4^{2-} 0.0006
ClO_4^-	FeL_3^{2+}	OH^- 1.0; I^- 0.012; NO_3^- 0.0015; Cl^- 0.00022; SO_4^{2-} 0.00016

Solid-state membranes

Ion measured	Membrane	Principal interferences
F^-	LaF_3	OH^-
Cl^-	$AgCl(Ag_2S)$	Br^-, I^-, S^{2-}, NH_3, CN^-
Br^-	$AgBr(Ag_2S)$	I^-, S^{2-}, NH_3, CN^-
I^-	$AgI(Ag_2S)$	S^{2-}, CN^-
SCN^-	$AgSCN(Ag_2S)$	Br^-, I^-, S^{2-}, NH_3, CN^-
S^{2-}, Ag^+	Ag_2S	Hg^{2+}
CN^-	$AgI(Ag_2S)$	I^-, S^{2-}
Cu^{2+}	$CuS(Ag_2S)$	Hg^{2+}, Ag^+
Pb^{2+}	$PbS(Ag_2S)$	Hg^{2+}, Ag^+, Cu^{2+}
Cd^{2+}	$CdS(Ag_2S)$	Hg^{2+}, Ag^+, Cu^{2+}

* The ligand L is a substituted 9,10-phenanthroline moiety.

solution, about 0.01 M. In use, CO_2 diffuses through the plastic film, in an amount determined by its partial pressure in the sample. The resulting pH change in the $NaHCO_3$ solution is sensed by the glass electrode.

The inherent specificity of enzymes can sometimes be used to advantage in electrodes (8). A second membrane consisting of a gel layer containing urease, for example, sensitizes an ammonium-ion glass electrode to urea. The enzyme-catalyzed hydrolysis of urea produces ammonium ion at a *rate* proportional to the urea concentration. The response is shown in Fig. 14-3. It will take several minutes for a steady potential to be attained, but a valid analysis can be obtained in less than 1 min by determining the initial slope. Other possible enzyme electrodes are reviewed in reference 9.

SOLID-STATE MEMBRANE ELECTRODES

Another important class of electrodes makes use of a wafer or pellet of crystalline material as the membrane. A prime example is the Orion*

* Orion Researches, Cambridge, Mass.

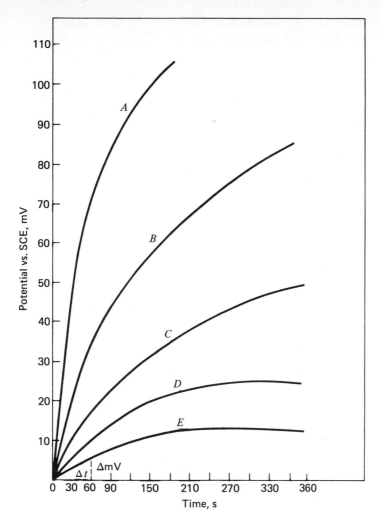

Fig. 14-3 Response curves of potential (versus SCE) against time for the urease-catalyzed hydrolysis of urea. Urea concentration, 0.01 M; pH, 7.0; buffer, 0.1 M tris. Units of urease are: *A*, 0.04; *B*, 0.016; *C*, 0.008; *D*, 0.004; and *E*, 0.002. [*Analytical Chemistry* (7).]

fluoride electrode in which the membrane is a single crystal of lanthanum fluoride, LaF_3. Structurally this takes the form of Fig. 14-2c. The crystal lattice is such that fluoride ions can move about freely within an immobile framework of lanthanum ions. The response is completely specific to fluoride, as no other ions are accepted into the crystal. The electrode shows true Nernstian response from approximately 10 to 10^{-6} M fluoride activity;

it is limited in extremely dilute solutions only by the finite solubility of the crystal. As with any anion-sensing electrode, the sign of the logarithmic term is negative:

$$E = K - \frac{RT}{F} \ln (F^-) \tag{14-5}$$

The fluoride electrode has become a particularly important analytical tool, because prior methods for determining this ion involved time-consuming indirect methods of limited accuracy.

Comparable electrodes can be fabricated from a polycrystalline pressed pellet of silver sulfide, responsive to either Ag^+ or S^{2-} ions down to the order of 10^{-19} or 10^{-20} M.

Chloride, bromide, iodide, and thiocyanate electrodes can be made by pressing in a pellet a mixture of Ag_2S and the silver salt of the desired anion AgX. The silver ions can move in the solid matrix but, as these salts are more soluble than the sulfide, the resulting equilibrium involves the X^- anion and its counterpart in the solutions. The Nernst equation takes the form

$$E = K + \frac{RT}{F} \ln (Ag^+) = K + \frac{RT}{F} \ln \frac{K_{sp\ (AgX)}}{(X^-)}$$

$$= K' - \frac{RT}{F} \ln (X^-)$$

where K' includes K and the log of K_{sp}.

An analogous series of electrodes can be made by incorporating the sulfide of a second metal into the Ag_2S pellet. In the case of lead, for example, the applicable Nernst equation for this class III electrode is

$$E = K + \frac{RT}{2F} \ln (Pb^{2+})$$

Representative solid-state membrane electrodes are listed in Table 14-1.

The effects of extraneous ions on solid-state membrane electrodes contrast with such effects on liquid (and glass) ion-exchange types. In the case of liquid membranes, the effects are best described, as we have seen, in terms of selectivity constants. Interferences with solid-state electrodes, on the other hand, are primarily related to solubility products. For example (5), an AgBr (Ag_2S) electrode will cease responding to bromide ions and become effectively a thiocyanate electrode if the activity of SCN^- exceeds that of Br^- by the ratio of the K_{sp}'s, as then the AgBr on the surface of the electrode will all be converted to AgSCN. Lesser amounts of SCN^-, however, will have no deleterious effect.

These electrodes based on solid-state membranes are among the most convenient and useful detectors for cations and anions in solution that are available. It is to be hoped that comparable electrodes will be developed for many other ions.

Another type of selective electrode, invented by E. Pungor and manufactured in Hungary, consists of a silicone rubber membrane with an insoluble salt such as silver iodide imbedded in it like a "filler." The loaded membrane must be soaked for several hours in a solution of the ion to which it is sensitized. These electrodes are not widely used in the United States.

One may question why a membrane electrode sensitized to a metal is preferred over the direct use of a wire of that metal to sense its ions in solution. There are several reasons for this. For one thing, many metals are not stable in the presence of water (e.g., Na, Ca). At the other end of the activity scale, metals such as Ag act as inert sensors of the redox potential of ionic couples, which often interferes with their use to measure their own ions. Many of the metals of intermediate activity do not readily enter into ionic equilibrium because of interfering surface conditions.

REFERENCE ELECTRODES

The requirements for the reference electrode are likely to be more severe with the newer ion-selective electrodes than in pH measurements, because of the greater accuracy often desired (10). It is frequently sufficient to measure pH to the nearest hundredth of a pH unit (± 6 mV), whereas almost invariably measurements of ions other than H^+ need to be an order of magnitude better. Hence small variations in the reference electrode that would be negligible in pH measurement become important.

One of the most common reference electrode configurations for pH work utilizes as the site of the liquid junction a fiber of asbestos or similar material sealed through the glass containing tube. The inner electrolyte, usually saturated KCl, is maintained at a higher level than the test solution, so that there is a slight outward flow of KCl, preventing contamination of the inner solution. The inner electrode may be Hg/Hg_2Cl_2 or $Ag/AgCl$, the latter being the more common. There is some tendency for the fiber to become partially clogged, preventing adequate flow and affecting the potential, so it is desirable to use an electrode with a sleeve-type junction, providing greater flow.

POTENTIOMETRIC TITRATIONS

A great variety of titration reactions can be followed potentiometrically. The only requirement is that the reaction involve the addition or removal of some ion for which an electrode is available. The potential of the indicating electrode may well change by a relatively small amount during an experiment—100 mV is typical—while the observed potential difference between indicator and reference electrodes is considerably greater. In such a case, the precision of measurement can be improved through the use of a concentration cell. The two vessels of the cell will contain identical electrodes reversible to the ion being titrated. In the reference side is placed a solution identical to that expected at the equivalence point. Then, when the titration in the indicator vessel reaches equivalence, the potential will be zero. The increased precision results from the possibility of using the measuring instrument in its most sensitive range.

CONSTANT-POTENTIAL TITRATIONS

Another approach to potentiometric titration is to measure the amount of titrant needed to maintain the indicator electrode at a constant potential. The titration curve then becomes a plot of volume of standard solution added as a function of time. This procedure has been employed rather extensively in the field of enzymology.

For example, the enzyme cholinesterase acts to decompose acetylcholine, producing acetic acid in the process. The enzyme is highly sensitive to pH, and the medium must be held very close to pH 7.4 for the reaction to proceed optimally. A bicarbonate buffer has been used in the past, and the carbon dioxide liberated by the acetic acid measured manometrically. This is awkward, and in addition has been attacked on the grounds that physiological conditions are not reproduced.

The present method makes use of a pH meter connected through a servo system to control a motor-driven syringe buret supplying NaOH solution at a constant rate, as needed to maintain constant pH. A pH meter and associated equipment used in this manner are called a *pH-stat*. Figure 14-4 shows an example of a curve by which the cholinesterase activity of a sample of animal tissue was determined (9). A 0.3-g portion of tissue was homogenized in physiological saline solution (0.9 percent NaCl), and the pH was then adjusted to 7.4 (at point P on the curve) and held for about 10 min to establish a base line showing that no acid was being liberated spontaneously. Then an excess of acetylcholine iodide (Ach·I) was added, whereupon the inflow of NaOH commenced. For this experiment the cholinesterase activity was determined from the slope to be 4.96×10^{-6} mol·g^{-1}·min^{-1}.

This technique does not seem to have been applied outside of biochemical work, but is certainly of general applicability.

INSTRUMENTATION

The Nernst equation, the fundamental relation in the potentiometry of cells, is strictly valid only if no current passes through the cell. This

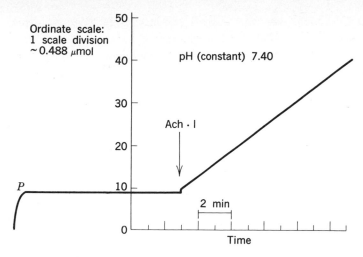

Fig. 14-4 Titration of cholinesterase at constant pH. [*Acta Pharmacologica et Toxicologica* (11).]

imposes restrictions on the design of measuring instruments. There is no potential-measuring device that ensures truly zero current, but this condition can be approached closely enough for analytical purposes. The classical method is by the use of a precision potentiometer, an instrument wherein the voltage of the cell is counterbalanced with an accurately measured fraction of the voltage from a stable internal source. Potentiometers are seldom used in modern analytical laboratories.

The most prevalent instrument for measuring cell potentials is an electronic voltmeter known as a *pH meter* (or *specific-ion meter*). It contains an amplifier designed to sense the impressed signal voltage and produce a proportional meter deflection or a numerical (digital) indication, or the response of a recorder pen. The resistance of a glass electrode cell may be as high as 10^8 Ω and, if the error is to be held to 0.1 percent, the input resistance of the amplifier must not be less than 10^{11} Ω. Until the last few years, this was achieved only with selected vacuum tubes, but now the development of field-effect transistors (FETs) has allowed the design of pH meters with all solid-state components, increasing reliability and reducing cost. The details of electronic circuitry will be discussed in Chap. 26.

There are many manufacturers of pH and specific ion-meters, each offering many models. Some of these are battery-operated, some plug into ac power lines. They fall roughly into three categories, with high, medium, and relatively low precision and accuracy (and price). The first group is primarily designed for research purposes, the second for general laboratory use, the third for field use where small size and rugged construction are more important than a high degree of precision.

A line-operated laboratory pH meter of the deflection type must have three controls on its panel, and may have a fourth. These are (*1*) a switch with "standby" and "operate" positions; (*2*) a calibration or standardization adjustment, which amounts to a zero offset; with it one adjusts the instrument to read the correct value when the electrodes are immersed in a standard buffer; (*3*) a temperature compensator which permits alteration of the sensitivity to account for the temperature dependence of the Nernst potential. Some pH meters also have a scale selector which allows the instrument to cover the whole pH range (usually 0 to 14) or to fill the scale with a selected portion of that range, perhaps 2 or 3 pH units; this type is called an *expanded-scale* pH meter.

Electrodes with glass membranes are the only ones with especially high resistance. Any electronic meter capable of measuring their potentials will be equally satisfactory with any other electrode systems mentioned in this chapter.

AUTOMATIC TITRATORS

In industrial analytical laboratories and for some types of research problems, automatic operation is widely applied to potentiometric titrations. There are two classes of titrators, those in which the instrument plots a complete titration curve on chart paper, and those that act to close an electrically operated buret valve exactly at the equivalence point. Those of the first type consist in essence of a pH meter connected to a strip-chart recorder. The chief difficulty arises from the need for delivery of titrant at a constant rate, which is not possible from a conventional gravity-flow buret. This can be overcome by substituting a constant-flow pump or a motor-driven syringe.

An example is the Sargent-Malmstadt titrator.* This employs a special capacitive electronic circuit to doubly differentiate automatically the variations in potential that are fed into it from the electrodes. If the potential as a function of volume, $E = f(V)$, is given by curve a of Fig. 14-5, then the derivative dE/dV is represented by curve b, and the second derivative d^2E/dV^2 by curve c. In the titrator, advantage is taken of the fact that the positive peak in curve c comes very slightly *before* the equivalence point. The first *decrease* in the second-derivative current energizes a relay that closes the buret stopcock. The mechanical inertia in the moving parts causes a delay which just offsets the advance warning given by the second-derivative curve, so that the flow of titrant is effectively stopped at just the right moment.

* Sargent-Welch Scientific Company, Skokie, Ill.

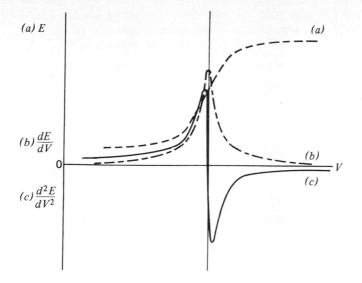

Fig. 14-5 A potentiometric titration curve. (*a*) The usual form; (*b*) the first derivative; (*c*) the second derivative.

PROBLEMS

14-1 A voltaic cell is formed with a lead electrode in 0.015 F lead acetate against a cadmium electrode in 0.021 F cadmium sulfate. The two solutions are joined by a salt bridge containing ammonium nitrate. What is the potential of the cell at 25°C if the activity coefficients in the two solutions are considered equal?

14-2 What is the pH of a solution at 25°C if a potential of 0.703 V is observed between a hydrogen electrode and an NCE placed therein?

14-3 Exactly 25.00 ml of 0.1000 F silver nitrate was pipetted into each of two beakers. Identical silver electrodes were inserted, and the two solutions joined by a potassium nitrate salt bridge. The potential difference between the electrodes was, as expected, zero. A 10.00-ml aliquot portion of a lead nitrate solution was added to one beaker and 10.00 ml distilled water to the other, whereupon a potential of 0.070 mV appeared between the silver electrodes. If the lead nitrate solution had been prepared by dissolving a 10.00-g sample of metallic lead in nitric acid and diluting to a volume of 100 ml, calculate the percentage of silver impurity in the lead. Estimate the precision of the analysis if the potentiometer has a maximum uncertainty of ±1 μV.

14-4 A cell is set up as follows: silver electrode, unknown solution, salt bridge, saturated KCl solution, $Hg_2Cl_{2(s)}$, mercury electrode. (*a*) Which electrode is the reference electrode, and which the indicator? (*b*) What is the purpose of the salt bridge, and what electrolyte should it contain? (*c*) If the potential of the cell is found to be 0.300 V, with the silver electrode more positive than the mercury, what is the concentration of silver ion in the unknown?

14-5 A cell is assembled with two platinum wires as electrodes dipping into separate beakers which are connected by a salt bridge, each beaker containing 25.00 ml of a mixture of Fe^{2+} and Fe^{3+} ions, each ion 1 F. Now 1.00 ml of a solution of a reducing agent is added to one side, whereupon the potential difference changes from 0 to 0.0260 V. Compute the normality of the solution of reducing agent.

14-6 What is the maximum concentration of Cu^{2+} ion which can exist in solution in contact with metallic zinc?

14-7 Derive an equation for the potential of a CO_2 electrode as a function of the partial pressure of CO_2 in a sample.

14-8 The fluoride electrode can be used in class II to measure lanthanum ions, La^{3+}. Derive the relation between potential and La^{3+}-ion activity.

14-9 An aqueous solution (pH 5.0) is to be examined for its free fluoride-ion concentration. A 100-ml portion of the solution, measured with a fluoride electrode, is found to give a reading of 120 mV against a suitable reference electrode. Exactly 1.00 ml of a 0.100 M solution of KF is added to the test solution with stirring, whereupon the potential changes to 132 mV. Calculate the fluoride-ion concentration in the sample. Assume 25°C.

REFERENCES

1. Furman, N. H.: In J. H. Yoe and H. J. Koch, Jr. (eds.), "Trace Analysis," chap. 9, Wiley, New York, 1957.
2. Durst, R. A. (ed.): Ion-Selective Electrodes, *Natl. Bur. Stand. Spec. Publ.* 314 (1969).
3. Bates, R. G.: "Determination of pH: Theory and Practice," 3d ed., Wiley, New York, 1973.
4. Eisenman, G.: "Glass Electrodes for Hydrogen and Other Cations," Dekker, New York, 1967.
5. Ross, J. W., Jr.: In R. A. Durst (ed.), Ion-Selective Electrodes, chap. 2, *Natl. Bur. Stand. Spec. Publ.* 314 (1969).
6. James, H., G. Carmack, and H. Freiser: *Anal. Chem.,* **44**:856 (1972).
7. Ansaldi, A., and S. I. Epstein: *Anal. Chem.,* **45**:595 (1973).
8. Guilbault, G. G., R. K. Smith, and J. G. Montalvo, Jr.: *Anal. Chem.,* **41**:600 (1969).
9. Weetall, H. H.: *Anal. Chem.,* **46**:602A (1974).
10. Caton, R. D., Jr.: *J. Chem. Educ.,* **50**:A571 (1973); **51**:A7 (1974).
11. Jensen-Holm, J., H. H. Lausen, K. Milthers, and K. O. Møller: *Acta Prarmacol. Toxicol.,* **15**:384 (1959).

Chapter 15

Voltammetry, Polarography, and Related Methods

In the previous chapter we dealt with potentials of unpolarized electrodes, which means that no current could be allowed to flow. We shall now investigate phenomena accompanied by the passage of considerable current; in particular, we will plot curves showing the magnitude of the current as a function of the voltage on the electrode of interest (the *working electrode*). This general method of studying the composition of a solution is called *voltammetry*. The term *polarography* is used chiefly for voltammetry at the dropping mercury electrode, but the distinction is not always followed.

The passage of current necessitates a modification of the reference electrode or, better, the addition of a third electrode to the cell. The use of the same electrode as reference and current carrier is undesirable in principle. The electrode must be so constructed as to have low resistance, in order not to introduce excessive error; a 10-mV error will result from only 10 μA flowing through 1000 Ω. In addition the metal-solution interface must have a much larger area than the working electrode, so that the current density and any resulting polarization effects will be less. This can be satisfactory for currents less than about 10 μA with a working electrode of micro dimensions. The fiber-tip reference electrodes often used with a pH meter are too high in resistance for this application.

The three-electrode cell is greatly to be preferred in general voltam-metry. The third (auxiliary) electrode can be a simple wire of platinum or silver, or a mercury pool. The reference electrode, since it does not carry current, can be of any convenient physical form.

The two forms of voltammetric assemblies are compared in Fig. 15-1. (Actual instruments will be described later, following theoretical considerations.) A variable voltage source is connected in series with a microammeter and the current-carrying electrodes. The actual potential at the working electrode is measured by a suitable (usually electronic) voltmeter, relative to the reference electrode.

The working electrode in voltammetry is inert, responding to whatever electroactive species may be present in solution. The choice of electrode depends largely on the range of potentials it is desired to investigate. For potentials more positive than the reference electrode (assumed to be the SCE), the best choice is platinum. Mercury cannot be made more positive than about $+0.25$ V versus SCE because of the ease of its anodic dissolution. Platinum is limited in the positive direction only by the oxidation of water $(2H_2O \rightarrow O_2 + 4H^+ + 4e^-)$, which occurs at about $+0.65$ V. On the other hand, for negative potentials platinum can only be used to about -0.45 V, at which potential hydrogen is liberated $(2H^+ + 2e^- \rightarrow H_2$ or $2H_2O + 2e^- \rightarrow H_2 + 2OH^-)$, while mercury, due to its high overvoltage for hydrogen, can be utilized as far as -1.8 V in acid or about -2.3 V in basic media. Figure 15-2 shows graphically these limiting potentials, and also indicates the conventions regarding the sign of the current and method of plotting voltammetric curves.

Fig. 15-1 Two- and three-electrode systems for voltammetry. Wkg, working electrode; Ref, reference electrode; Aux, auxiliary electrode.

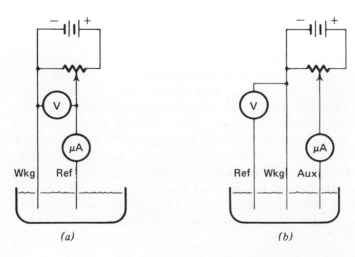

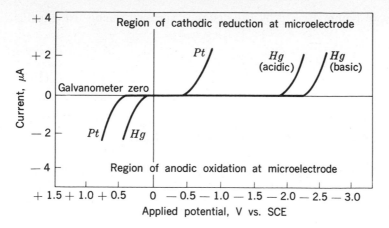

Fig. 15-2 Conventions for plotting voltammetric curves. The curves labeled Pt or Hg indicate the approximate potential limits attainable at these electrodes.

DIFFUSION-LIMITED CURRENT

Consider a plane, inert microelectrode in a deaerated, unstirred solution containing 0.1 M KCl and 10^{-3} M $PbCl_2$. Assume an Ag auxiliary electrode and a saturated calomel reference electrode with a KCl salt bridge. Suppose that a switch is closed in an external circuit, suddenly applying 1.0 V across the electrodes, with the microelectrode negative. This potential is great enough to reduce Pb^{2+} ion to the metal, but not to reduce H_2O (at pH 7). K^+ ion would of course require a much higher potential, and there is nothing else present that is reducible. At the anode, an amount of Ag equivalent to the reduced Pb^{2+} is converted to AgCl, a process that has no measurable effect on the cell.

Several things will happen: (1) K^+ and Pb^{2+} ions will start to move in the electric field toward the microelectrode, and Cl^- ions in the opposite direction. K^+ ions, not being reducible, will almost instantaneously form a sheath (about one ion thick) around the electrode, which will result in almost completely neutralizing the field, so far as the bulk of the solution is concerned. The Pb^{2+} ions will not contribute to this sheath, because they *are* reducible; every Pb^{2+} ion that approaches the electrode will immediately be discharged and deposited on the surface. But since the field has been neutralized by K^+ ions, the only way the Pb^{2+} ions can reach the vicinity of the electrode is by *diffusion*. (Because of the requirement of overall electrical neutrality, the reverse movement of Cl^- ions will have dropped to the low level required to match the reduction of Pb^{2+} ions.)

The rate of movement of any species because of diffusion is proportional to the concentration gradient, the difference between the concentrations at any two points divided by the distance between them.   In calculus terms,

$$\frac{dC}{dt} = D \frac{d^2C}{dx^2} \tag{15-1}$$

where C refers to the concentration of the diffusing species, and D is a constant of proportionality called the *diffusion coefficient*.   This is *Fick's law* (1).

Application of Fick's law to the electrolysis problem we are considering leads to the relation

$$i = nFA \left(\frac{D}{\pi t}\right)^{\frac{1}{2}} C \tag{15-2}$$

in which i is the electrolysis diffusion current (in microamperes) flowing at time t (seconds) from the start of an experiment, n is the number of electrons involved in the electrode reaction, F is the Faraday constant (approximately 9.65×10^4 C per equivalent),* A is the area (in square centimetres) of the electrode, D is Fick's diffusion coefficient (centimetres squared per second), and C is the bulk concentration of electroactive species (in millimoles per liter).   The concentration is assumed to be zero at the electrode surface.   Both i and n are taken as positive for cathodic reductions and negative for anodic oxidations.   It is important to note the proportionality between the current and concentration; however, since the current falls off with the square root of time, rather than assuming a fixed value, this equation is not a convenient basis for analytical work.†   A more practical relation will be derived in a later section.

THE DROPPING-MERCURY ELECTRODE (DME)

The most widely used microelectrode is mercury in the form of a succession of droplets emerging from a very fine-bore glass capillary (Fig. 15-3). This has several major advantages to offset the inconvenience of handling mercury.   One is the high hydrogen overvoltage of a mercury cathode. Another advantage is the fact that the electrode surface is continually being renewed, and hence cannot become fouled or poisoned.   In comparison with solid electrodes, the DME has the advantage that the increasing area of the electrode during the lifetime of a drop more than offsets the

---

* Note that C is the accepted symbol (abbreviation) for coulomb, whereas C is here defined as a concentration.

† Analysis by means of current-time curves at constant potential is called *chronoamperometry*.   It has some value in studies of the electron-transfer kinetics in irreversible systems (2).

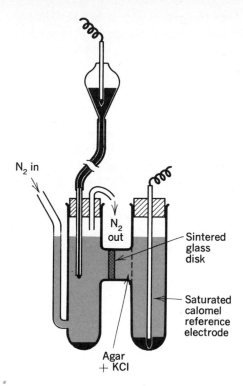

N₂ in

N₂ out

Sintered glass disk

Saturated calomel reference electrode

Agar + KCl

Fig. 15-3   Polarographic H cell.

decreasing current observed with an electrode of fixed size; this, as we will see shortly, makes quantitative analysis much more practicable.

A variety of electrolysis cells for polarography have been described. The most familiar for two-electrode polarography, the H cell (Fig. 15-3), is provided with a built-in reference electrode. The liquid junction is located in the pores of a sintered-glass disk backed up with a section of salt-impregnated agar gel to minimize convection and diffusion between the two compartments. The diameter of the cross arm and of the mercury pool is made large enough to keep the resistance low.

This same vessel is often used for three-electrode work, with a wire of Pt or Ag inserted near the DME capillary. However, the bulky H cell is unnecessary, and a cell improvised from a small beaker will be quite satisfactory unless very small samples are to be examined.

Provision must be made for deaerating the solution prior to making measurements. This is accomplished by bubbling nitrogen or other inert gas through a side arm (as in Fig. 15-3) or through a small, fritted-glass gas-dispersion tube.

The capillary for the dropping electrode is made from a section several centimetres long cut from glass tubing of 0.03- to 0.05-mm internal diameter. It must be handled with great care to prevent any water or aqueous solution from entering the capillary, since it is practically impossible to clean it out if it once becomes fouled. After use the capillary should be removed from the cell and rinsed with distilled water while the mercury is still flowing from the tip. It can then be clamped in the air (protected from dust), and the mercury reservoir lowered sufficiently to stop the flow. It should *not* be left immersed in water.

There is some tendency for minute amounts of aqueous solution to penetrate into the capillary between the glass and the mercury column, even during operation. This causes anomalous variations in current, visible in high-sensitivity operation. Cooke and coworkers (3) have discussed this phenomenon and how to avoid it.

To derive an equation for the DME corresponding to Eq. (15-2), we make the assumption that the flow of mercury is constant, and the approximation that the drop is spherical in shape right up to the moment of separation. This leads to the expression

$$i = 708.2nD^{1/2}m^{2/3}t^{1/6}C \qquad\qquad (15\text{-}3)$$

where m is the *rate* of flow of mercury (in milligrams per second). The numerical coefficient includes geometrical factors, the Faraday constant, and the density of mercury. If the current is plotted as a function of time, a fluctuating curve is obtained, as in Fig. 15-4 (4), with a period of a few seconds. The drop time τ can be varied by changing the head of

Fig. 15-4 Current-time characteristics at the dropping mercury electrode.

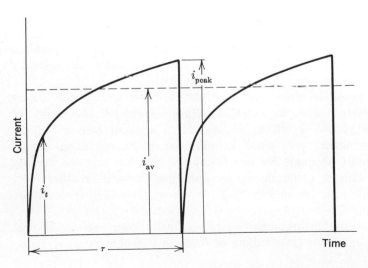

mercury; it varies slightly with the applied potential, being greatest at about
—0.5 V versus SCE.*

This fluctuating current is inconvenient to read. The classical procedure is to estimate the *average* value $\bar{\imath}$, indicated in the figure. Integration of Eq. (15-3) over the lifetime of one drop τ, and then dividing by the time interval, gives for the average current:

$$\bar{\imath} = \frac{1}{\tau} \int_0^\tau i_d \, dt = \frac{6}{7} i_d \tag{15-4}$$

$$\bar{\imath} = 607.0 \, nD^{1/2} m^{2/3} \tau^{1/6} C \tag{15-5}$$

Equations (15-3) and (15-5) are forms of the well-known Ilkovič equation.

Current fluctuations due to falling mercury drops can be eliminated or minimized by a number of instrumental techniques involving either (*a*) rapid scanning so that the whole experiment takes place within the life of a single drop, or (*b*) some type of electronic gating system such that the current is sampled only for a fraction of a second just before each drop falls. These methods will be described later.

Although the temperature does not enter explicitly into the Ilkovič equation, it is nevertheless important, as each factor of the equation (except n) is to some extent temperature-dependent. The chief effect is through the temperature coefficient of the diffusion constant D. The value of the diffusion current increases at the rate of 1 or 2 percent per degree in the vicinity of room temperature. Therefore the temperature of the electrolysis cell should be controlled to within a few tenths of a degree for high-precision measurements.

The diffusion coefficient D varies with the viscosity of the medium, and hence such a variation is found in the diffusion current. If other factors are held constant, the current should be inversely proportional to the square root of the relative viscosity. This relation is found to be valid in the absence of colloidal material, but the proportionality fails when the viscosity is increased through addition of gelatin or other hydrophilic colloid.

The diffusion current also varies from its normal value if the concentration of supporting electrolyte is less than 25 or 30 times that of the reducible substance. This effect results from the fact that under these conditions the reducible ions carry an appreciable fraction of the current; this fraction is known as the *migration current*. The migration current is due to the electrostatic attraction or repulsion between the DME and

---

* This is the potential at which a mercury-water interface shows a maximum interfacial tension, known as the *electrocapillary maximum*. This is also the potential that an isolated mercury electrode assumes when placed in the solution.

the ions. Hence the observed diffusion current is increased slightly for the reduction of cations and diminished for the reduction of anions if the concentration of supporting electrolyte is lowered, while it is unchanged for reduction of nonionic species. This situation invalidates the Ilkovič equation, since the transport of reducible species to the electrode is no longer governed only by diffusion.

VOLTAGE-SCANNING POLAROGRAPHY

Up to this point we have considered current-time curves produced at a constant impressed potential. Now we will take up the effect of changing the potential. This can be done stepwise or continuously. The stepwise procedure is generally manual. The operator sets the potential with a circuit like one of those of Fig. 15-1, waits perhaps a minute for a steady state to be attained, records the average current, and then moves to the next desired potential.

To facilitate determinations of entire current-voltage curves, automatically recording instruments, generally known as *polarographs,* have been developed. In the simplest of these, the potential applied to the cell is continuously increased, usually in the negative direction (i.e., the DME negative to the SCE), and the current is recorded. As the potential is linear with time, and the recording paper moves with constant speed, the resulting curve, called a *polarogram,* can be labeled in terms of current and potential units.

Consider a cell, such as that of Fig. 15-3, provided with DME indicator and SCE reference electrodes and filled with an oxygen-free solution which is 0.1 M in KCl and 0.001 M in $CdCl_2$. The polarogram obtained by the usual procedure resembles that shown in Fig. 15-5, which is idealized for clarity of discussion.

The curve divides itself into three regions. In region A the potential is too low to permit reduction of any of the substances known to be present. The small current which does flow is called the *residual current* and can be explained as the sum total of currents due to the reduction of traces of impurities (possibly iron, copper, or oxygen), and a so-called *charging current* which results from the fact that the mercury-solution interface, with its sheath of unreducible ions, acts like a capacitor of continually increasing area. The residual current is small and is reproducible if care is taken to eliminate reducible impurities.

In the vicinity of B (−0.5 V in the above example), the current starts to increase above the value of the residual current alone. This added current removes Cd^{2+} ions from the layers of electrolyte in contact with the surface of the electrode, by reduction to the metal; it is replaced by diffusion from the body of the solution.

At C the current shows a saturation effect which is caused by the

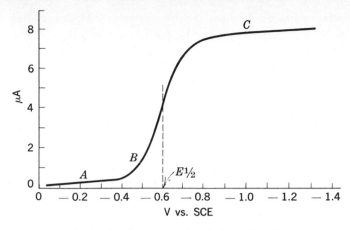

Fig. 15-5 Idealized polarogram of cadmium chloride (0.001 M) in 0.1 M potassium chloride.

total depletion of Cd^{2+} ions in the near vicinity of the DME. More Cd^{2+} ions continually reach the electrode by diffusion, and are reduced immediately on arrival. The rate of diffusion is determined solely by the difference between the bulk concentration of the solution and zero, the concentration at the electrode surface. The value of the limiting current at C, called the *diffusion current i_d*, is directly dependent on the concentration of the reducible species; it must, of course, be corrected for the residual current.

If the negative potential is increased beyond C in Fig. 15-5, the current will rise slowly and uniformly (parallel to the residual current curve) up to about -2 V, at which point it again increases rapidly, corresponding to the reduction of hydrogen ions or water.

THE SHAPE OF THE POLAROGRAPHIC WAVE

The equation for the current as a function of potential can be derived from the Nernst equation, provided that the reaction at the microelectrode proceeds reversibly. We will carry out this derivation for the important special case of the reduction of a simple cation to a metal soluble in mercury. Since solutions studied polarographically are almost always very dilute, we can assume that the activity coefficient of the cation is not appreciably different from unity. The same assumption applies to the activity coefficient of the metal in the amalgam.

Let us write the half-reaction in the form

$$M^{n+} + ne^- \rightarrow M_{(Hg)}$$

where $M_{(Hg)}$ denotes the metallic M dissolved in mercury.   The potential of the electrode must then be that given by the Nernst equation:

$$E_{DME} = E^\circ - \frac{RT}{nF} \ln \frac{[M]_{Hg}}{[M^{n+}]_{aq}} \tag{15-6}$$

Since the current i is limited by diffusion,* it follows that

$$i = k([M^{n+}]_{aq} - [M^{n+}]_{aq}{}^0)$$

where the superscript signifies conditions at the mercury surface.   For the limiting current i_d, $[M^{n+}]_{aq}{}^0$ becomes very small, and we have

$$i_d = k[M^{n+}]_{aq}$$

The concentration of M in the amalgam is proportional to the current, or

$$i = k'[M]_{Hg}$$

The constant k' is identical with k, except that the diffusion constant D is replaced by D', a similar diffusion constant for M within the amalgam, so that the ratio $k/k' = \sqrt{D/D'}$.

These several equations can be combined to give

$$E_{DME} = E^\circ - \frac{RT}{2nF} \ln \left(\frac{D}{D'}\right) - \frac{RT}{nF} \ln \frac{i}{i_d - i} \tag{15-7}$$

At the point where $i = \frac{1}{2}i_d$, the last term drops out, and the potential is designated as $E_{\frac{1}{2}}$, the *half-wave potential*:

$$E_{\frac{1}{2}} = E^\circ - \frac{RT}{2nF} \ln \left(\frac{D}{D'}\right) \tag{15-8}$$

Hence,

$$E_{DME} = E_{\frac{1}{2}} - \frac{RT}{nF} \ln \left(\frac{i}{i_d - i}\right) \tag{15-9}$$

Equation (15-8) shows that the half-wave potential, which is an easily measured quantity, is simply related to the standard potential E°.   The diffusion constants D and D' are usually not greatly different, so that $E_{\frac{1}{2}}$ is always nearly equal to E° (in the absence of complexing agents; see below).

* Currents, in this discussion, are assumed to have been corrected for the residual current blank.

Equation (15-9) gives the form of the polarographic wave in terms of the parameters i_d and $E_{1/2}$, and provides a convenient method for establishing the value of n. The most direct measure of n is the slope of the tangent to the curve at the half-wave point. Another, more precise, method is to plot values of $\log i/(i_d - i)$ against the potential $-E_{DME}$. The equation predicts a straight line with slope given by $2.303RT/nF$, or (at 25°C) $0.0591/n$. The point on this curve corresponding to $i = \frac{1}{2}i_d$ will give a precise measure of $E_{1/2}$.

The preceding derivation is concerned with the reduction of a simple (aquo) ion. In the presence of a complex former, the half-wave potential (Eq. 15-8) is shifted to more negative values in accordance with the relation

$$E_{1/2} = E° + \frac{RT}{nF} \ln K_c - \frac{pRT}{nF} \ln [X] \tag{15-10}$$

where K_c is the instability constant of the complex, $[X]$ is the concentration of complexing agent, and p is the number of moles of X which combine with 1 mol of metal M. (This equation is based on the assumption that the diffusion coefficients involved are nearly equal.) Equation (15-9) is essentially unaltered by the presence of a complex, though the value of i_d for a given concentration may change slightly.

For the case of reduction to a species which is insoluble in both water and mercury, e.g., iron and chromium metals, the equations must be modified somewhat. This calculation has little practical importance, because in all known examples such a reduction is irreversible, which rules out application of the Nernst equation. Analyses based on such reductions may be just as useful and correct as though the process were reversible, but they must be treated empirically.

Another case which is important is the reduction of one water-soluble species to another, for example, the reduction of ferric to ferrous ions. If we return to the notation employed in Chap. 13 [Eq. (13-11), etc.], we may write as a general equation

$$A_{ox} + ne^- \rightleftharpoons A_{red}$$

If both forms of A are present in solution, the potential of the DME is given by

$$E_{DME} = E_{1/2} - \frac{RT}{nF} \ln \frac{i - i_{d(a)}}{i_{d(c)} - i} \tag{15-11}$$

and the half-wave potential is

$$E_{1/2} = E° - \frac{RT}{2nF} \ln \frac{D_{ox}}{D_{red}} \tag{15-12}$$

where $i_{d(a)}$ represents an *anodic* diffusion current due to oxidation of A_{red} at the DME, $i_{d(c)}$ is the *cathodic* diffusion current corresponding to the reduction of A_{ox}. D_{ox} and D_{red} are the diffusion coefficients of the respective forms, which may safely be assumed very nearly equal to each other. Thus $E°$, the standard potential for the redox couple, is very nearly equal to $E_{1/2}$. If either A_{ox} or A_{red} is absent from the solution, the corresponding i_d becomes zero in Eq. (15-11), but the value of $E_{1/2}$ remains unchanged.

The above discussion was given for the most part in terms of cathodic reduction taking place at the DME, as this is the most widely applicable condition. However, the same discussion holds true where the electrode process is anodic oxidation.

Figures 15-6 to 15-8 show a few polarograms to illustrate the above points. Figure 15-6 shows waves of one reducible species at several concentrations, together with the residual curve (5).

Figure 15-7 is a polarogram (6) showing the sequential reduction of five cations with various properties: (*1*) Ag^+ is reduced so easily that its wave cannot be formed completely, though the corresponding diffusion plateau is well defined; (*2*) Tl^+ shows a reversible reduction with $n = 1$; (*3*) Cd^{2+} is also reversible, but the slope shows that $n = 2$; (*4*) Ni^{2+} shows a more drawn-out curve, though $n = 2$, which indicates irreversibility; and (*5*) Zn^{2+} resembles Cd^{2+} in showing a reversible wave with $n = 2$.

Figure 15-8 shows the anodic oxidation of ferrous ion (curve *c*) compared with the reduction of ferric ion (curve *a*) (6). Curve *b* is obtained when both forms are present in equivalent concentrations. The vertical lines indicate the observed positions of the half-wave potentials, which should be identical.

Fig. 15-6 Polarograms showing the construction of a calibration curve. To 10 ml of 1 M NH_3 and 1 M NH_4Cl (curve 1) have been added successive 0.05-ml increments of 0.05 M Cd^{2+} (curves 2 to 8); each curve starts at −0.2 V; each scale division corresponds to 200 mV. [*Academic* (5).]

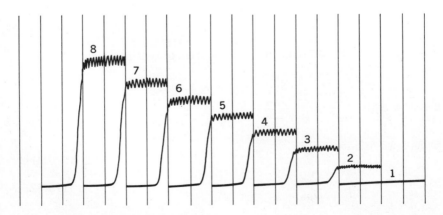

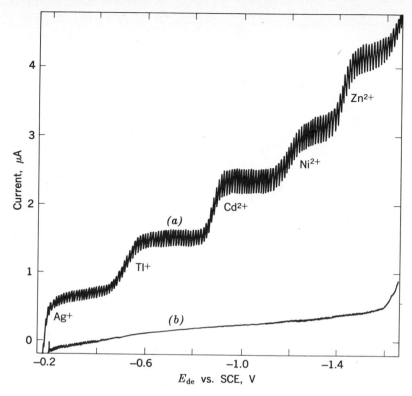

Fig. 15-7 Polarograms of (a) a solution containing 0.1 M each of Ag(I), Tl(I), Cd(II), Ni(II), and Zn(II) in 1 M NH$_3$ and 1 M NH$_4$Cl with 0.002 percent Triton X-100, and (b) the supporting electrolyte alone. [*Wiley-Interscience* (6).]

MAXIMA

Frequently, as the applied potential is raised, the current, after increasing as an ion is discharged, fails to level off but decreases again, leaving a maximum in the curve. Such a maximum may be only a slight hump, or it may be a very sharp peak, exceeding the true wave height by a factor of 2 or more. This phenomenon appears to be related to the tangential streaming motion of solution past the surface of the drop.

Maxima can usually (but not always) be eliminated by the addition of an organic surfactant. Gelatin or certain dyes, such as methyl red, are sometimes effective. The nonionic detergent Triton X-100* has been found to be particularly useful as a maximum suppressor, and is widely used. A stock solution of 0.2 percent is convenient; 0.1 ml of this for each 10 ml of solution in the polarographic cell is usually satisfactory. Care must be taken not to use too much suppressor, or the desired wave

* Rohm and Haas Company, Philadelphia, Pa.

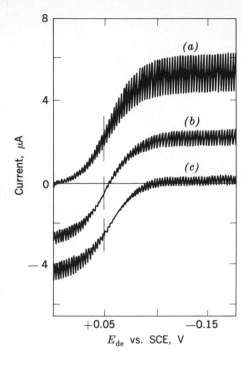

Fig. 15-8 Polarograms of (*a*) 1.4 mM Fe(III), (*b*) 0.7 mM each of Fe(II) and Fe(III), and (*c*) 1.4 mM Fe(II); the supporting electrolyte is saturated oxalic acid containing 0.0002 percent methyl red. [*Wiley-Interscience* (6).]

may be distorted or suppressed along with the maximum. The half-wave potential may be shifted a few millivolts from its normal value.

OXYGEN INTERFERENCE

Dissolved oxygen is reducible at the microelectrode in many media. A typical example of the effect of oxygen on a polarogram is shown in Fig. 15-9 (6). Two waves are observed. The first ($E_{1/2} = -0.05$ V versus SCE) is caused by the reduction of oxygen to hydrogen peroxide; the second ($E_{1/2} = -0.9$ V) corresponds to the reduction of oxygen to water. The first of these two waves often shows an intense, sharp maximum in the absence of a suppressor and in a dilute supporting electrolyte (not visible in the figure). This wave can be utilized in the analytical determination of dissolved oxygen, but more often appears as an interferent of high nuisance value. So in most polarographic work, provision must be made for removal of oxygen. In alkaline solutions this can often be easily accomplished by adding a small amount of potassium sulfite which reduces the oxygen quantitatively. In any case the oxygen can be removed by flushing the solution with a nonreducible gas such as nitrogen. Using a simple constricted glass tube as bubbler, this may take 20 to 30 min. The time required may be cut to 2 to 3 min by the substitution of a fritted-glass gas disperser.

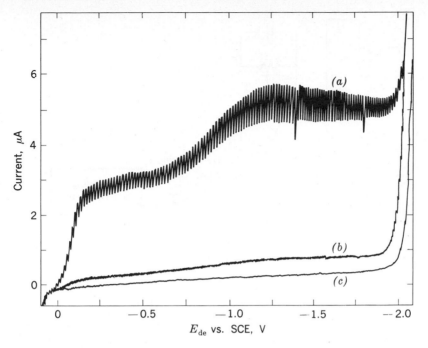

Fig. 15-9 Polarograms of 0.1 M KCl. (*a*) Saturated with air, showing the double wave of oxygen; (*b*) partially deaerated; (*c*) after complete deaeration. [*Wiley-Interscience* (6).]

It has been found (7) that Freon-12 (CCl_2F_2) is at least equally effective as nitrogen, and much more convenient. It is available in pint-size pressurized cans from refrigerator repair shops.* The stream of gas must be discontinued while data are being taken, because of the undesirable stirring action. Mercury should be excluded from the solution until after the deaeration, since it may be oxidized by the dissolved oxygen in certain media. This means that the capillary should not be inserted into the cell prior to the removal of oxygen.

INSTRUMENTATION

A satisfactory manual polarograph using one of the circuits of Fig. 15-1 can be readily assembled from components commonly available in the laboratory. The indicating instrument can be an electronic microammeter with multiple ranges.

Nearly all commercially available polarographs are designed for automatic recording. In many models (Fig. 15-10) a motor-driven voltage

*Adequate ventilation should be provided when Freon is used to avoid excessive breathing of this gas.

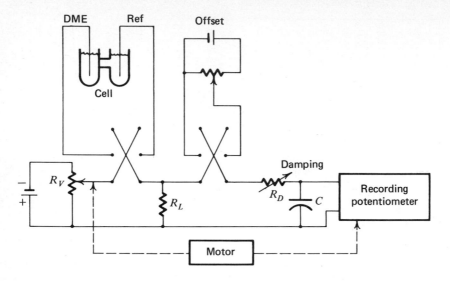

Fig. 15-10 Conventional two-electrode polarographic circuit. R_V is the motor-driven voltage source from which the working current flows through the cell and the load resistor R_L.

divider R_V provides a steadily increasing potential (*ramp*) to the cell. The resulting current produces a proportional voltage drop in a series resistor R_L, and this voltage is recorded on a strip-chart recorder. Where the recorder is an integral part of the instrument, the divider R_V and the recorder chart are driven by the same motor. An *offset* (sometimes called *bias*) control permits easy zero setting and scale expansion. A low-pass filter (R_D and C) permits a degree of damping, i.e., diminution of fluctuations due to falling mercury drops.

Automation of a three-electrode polarograph requires a different approach, making use of *operational amplifiers* (Fig. 15-11). Operational amplifiers will be discussed in considerable detail in Chap. 26. For present purposes, their salient features are the following: (*1*) no current can flow into or out of the amplifier inputs (marked A and B); (*2*) the amplifier operates in such a way as to maintain *equal* potentials at its two inputs.

An amplifier connected with only a capacitor from its output back to its input acts as an integrator with respect to time. Amplifier no. 1 in Fig. 15-11 is such an integrator and if E_1, R_1, and C_1 are constant its output E_2 will be:

$$E_2 = -\frac{E_1}{R_1 C_1}\int_0^t dt = -\frac{E_1}{R_1 C_1}t \qquad (15\text{-}13)$$

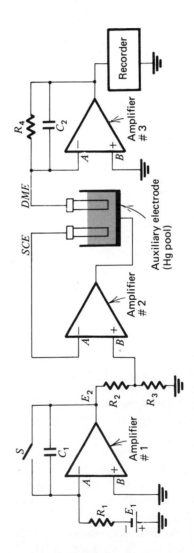

Fig. 15-11 Three-electrode polarograph, with operational amplifier circuitry.

This indicates that the output of this amplifier will increase linearly with time, with a slope determined only by circuit constants. Because of the negative sign, if E_1 is negative, then E_2 will be positive-going; this constitutes the ramp function.

An appropriate fraction of the ramp (determined by the magnitudes of R_2 and R_3) is applied to the B input of amplifier no. 2, which is thus kept at the ramp potential above ground. The only mode of response available to amplifier no. 2 is to emit current from its output to the auxiliary electrode. This current cannot go to the SCE, because the SCE connects only to the input of an amplifier, and hence cannot carry current; therefore the current must flow through the DME and the resistor R_4. Amplifier no. 3 is forced to put out just enough current to equal that passing through the cell, in order to ensure equality of potential of its two inputs. Because of this equality, the DME is essentially at ground potential at all times. The SCE, for a similar reason, is at ramp potential above ground. The result is that the no. 2 amplifier allows precisely the required amount of current to flow to maintain the SCE-DME potential difference at the desired linearly increasing voltage, with the DME *nega-tive*-going. Hence the recorder will give the true polarogram. The function of capacitor C_2 is to lengthen the time constant to give a desired amount of damping; R_4 relates to the scale of the recorder. The switch S across the integrating capacitor is used to start the experiment; the moment of opening of S corresponds to time zero. Range switching and zero offset are easily supplied, but omitted from the figure for clarity.

An additional advantage of a three-electrode polarograph is that it can be used with high-resistance, nonaqueous, or mixed-solvent systems, as well as with the more common low-resistance aqueous solutions. For this application the tip of the reference electrode or its salt bridge must be placed as close as possible to the working electrode, so as not to sense any significant part of the voltage drop across the solution.

SAMPLING CIRCUITS

The curve shown in Fig. 15-4 represents the current flowing as the result of an electron-transfer reaction, i.e., a faradaic reduction process. There is another component of current that must be considered, due to the charging of the *double-layer capacitance*. As mentioned earlier, when a negative potential (greater than the electrocapillary maximum) is first applied to the electrode, a cloud of nonreducible cations forms around the DME. The motion of these ions to form the cloud is accompanied by an equal flow of electrons within the metal electrode, and these together constitute a transient current, fully comparable to the flow of charge into a capacitor. (The external ion cloud and internal electron cloud are referred to as the *double layer*.)

As with any capacitor, current can only flow into the device if either the voltage is varying, or the "plates" of the capacitor are changing in area A or separation. In the case of the DME, the separation is the thickness of the double layer (considered to be constant), so the governing equation for the capacitive current I_c is

$$I_c = K \frac{dA}{dt} \frac{dE}{dt} \tag{15-14}$$

where the constant K includes the dielectric constant and the thickness. We can consider two cases: (1) the change of voltage (the ramp) is slow enough to be negligible during a drop period, and (2) the measurement is made near the end of a drop life, when the area is changing slowly.

It can be shown mathematically that, whereas the faradaic current increases as the one-sixth power of time, starting at zero following the birth of a new drop, the charging current is relatively large at the start and then decreases, following a minus one-third power curve. These relations are shown in Fig. 15-12.

It can be seen from these curves that the effect of charging current will be least and the faradaic current greatest in the short time interval just prior to the drop fall. A very productive approach to the increase in signal-to-noise ratio lies in a gating or timing arrangement, whereby the measurements are taken only during the last part of the growth of the drop, as indicated in Fig. 15-12, when the charging current has decayed to a negligible value and when the faradaic current is changing only slowly.

Fig. 15-12 Growth of the faradaic current and decay of the capacitive current within the lifetime of a mercury drop. The period between the dashed lines is the optimum for current measurement.

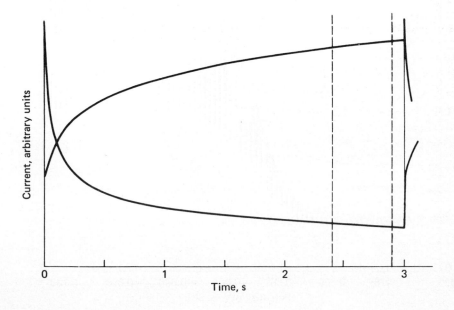

The gating system can be synchronized with the DME either by detecting the fall of the drop by some capacitive or photoelectric sensor, or by dislodging the drop mechanically at a predetermined time. This can be done by gently tapping the capillary with a small magnetically operated hammer. For example, in Fig. 15-12, a strictly reproducible timing circuit can be designed to knock off the drop at 3-s intervals and to permit the recorder to operate only between 2.4 and 2.9 s in the lifetime of each drop. During the 2.5 s when the current is *not* being sampled, the recorder will simply draw a horizontal straight-line segment. Figure 15-13 shows the result, a curve composed of small, finite steps, much easier

Fig. 15-13 Comparison of controlled drop time polarography (*a*) without and (*b*) with sampling circuitry. Curve *b* displaced upward for clarity. The solution is 10^{-3} *M* $CdCl_2$ in 1 *M* KNO_3. (*Courtesy William J. Mergens.*)

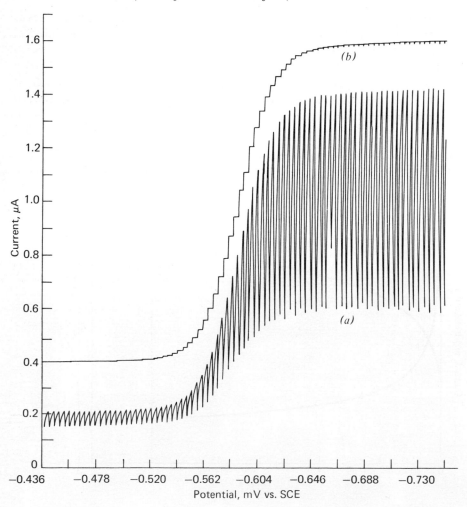

to measure quantitatively than the unmodified polarogram. This technique is sometimes called *Tast* polarography (from the German *Tast*, touch).

RAPID-SCAN POLAROGRAPHY

It is quite possible to sweep the applied voltage fast enough to obtain a complete polarogram within the last half-second of the drop life (i.e., within the indicated time slot in Fig. 15-12). The curve obtained by this method does not resemble a conventional polarogram, but shows a peak of characteristic shape, as in Fig. 15-14, which is a rapid-scan polarogram of a solution with two reducible species (8).

 The reason for the peak shape is that the slow process of diffusion is unable to supply reducible material to the electrode fast enough to keep up with the rapidly increasing potential, so that a steady state is never attained. It can be shown mathematically that E_s, the *summit potential*, is related to the half-wave potential:

$$E_s = E_{1/2} - 1.1 \frac{RT}{nF} \tag{15-15}$$

which at 25°C becomes

$$E_s = E_{1/2} - \frac{0.028}{n}\,\text{V} \tag{15-16}$$

 The value of current at the summit for a reversible system is given by an equation derived independently by Randles and Ševčik, which is analogous to the Ilkovič equation but includes the derivative dE/dt:

$$i_s = kn^{3/2}m^{2/3}t^{2/3}D^{1/2}\left(\frac{dE}{dt}\right)^{1/2}C \tag{15-17}$$

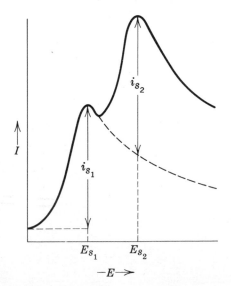

Fig. 15-14 Rapid-scan polarogram of a mixture of two reducible species. The background current for the second peak must be extrapolated from the first peak. [*Academic* (8).]

The linear relation between i_s and the concentration C is to be noted. The constancy of the ratio of i_s to $(dE/dt)^{1/2}$ is a good test for the degree of reversibility. The Randles-Ševčik equation is valid only when both oxidized and reduced species are soluble (in water or mercury).

CYCLIC VOLTAMMETRY

This is a modification of the rapid-scan technique, wherein the direction of scanning is reversed following the reduction of interest. To accomplish this, a *triangular wave* voltage is applied to the electrolytic cell (Fig. 15-15a) rather than the simple ramp function. A typical curve obtained by this method is shown in Fig. 15-15b (8). The entire process takes place in a second or less near the end of the lifetime of a mercury drop. When the voltage scan is first applied, the current will start near the origin (A), and only residual current will flow until the potential is negative enough to effect the reduction of zinc(II), whereupon a maximum appears, exactly analogous to one of those in Fig. 15-14. At point D the *direction*

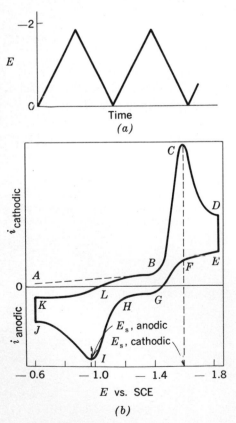

Time
(a)

E vs. SCE

(b)

Fig. 15-15 Cyclic voltammetry. (a) Triangular-wave excitation function. (b) Voltammogram obtained with 10^{-3} F Zn(II) in 2 F NH$_3$ and NH$_4$Cl buffer. [*Academic* (8).]

of scan is reversed, so that the voltage proceeds back toward zero with the same rate at which it had previously increased. The sudden drop (*D* to *E*) is caused by the reversal of the capacitive current. In the region *E* to *F*, the current, still cathodic, continues the process of depletion of zinc (II) from the vicinity of the electrode, which had been taking place in the *C*-to-*D* region. As the voltage reaches the point *F*, a diffusion steady state again becomes the controlling factor, and the current drops to the *G*-*H* area, where it is essentially the residual capacitive current. In the *H*-*I*-*J* region, metallic zinc, the product of prior reduction, is reoxidized, with a large anodic current, diffusion-controlled. At *J* the voltage scan is again reversed. The sudden rise in current, *J* to *K*, followed by a slow rise, *K* to *L*, represents the inverse of the *D*-to-*G* decrease. Finally the curve merges with the original *A*-*B* section, and from there the entire sequence is repeated. The fact that the current at *I* is not equal in magnitude to that at *C* may be due to incomplete removal of zinc metal from the mercury drop, and to a difference between diffusion coefficients for the aqueous species containing zinc (II), presumably $[Zn(NH_3)_4]^{2+}$ ion, and metallic zinc in the amalgam. The concentration gradient causing diffusion is, of course, very unequal for the two.

The electrode reaction used as an example here is highly irreversible, as shown by the large difference (about 0.6 V) between $E_{s,\,cathodic}$ and $E_{s,\,anodic}$. For a reversible process with $n = 2$, this difference (at 25°C) would be 0.28 V, predictable from Eq. (15-16). This is another convenient way to determine the degree of reversibility.

AC POLAROGRAPHY

A significant improvement in signal-to-noise ratio, and hence in the sensitivity of polarography, can be realized by superimposing a small alternating voltage (10 to 100 mV) on the dc ramp and then measuring the ac component of current. The optimum frequency is a few hundred hertz. The rapid, small variations in potential produce, via the Nernst equation, a corresponding rapid variation in the ratio of oxidized to reduced form of the electroactive species, which means that an alternating current must flow. The magnitude of the current is greatest where the two concentrations are equal, at the half-wave potential (for a reversible system).

The importance of this method lies in the possible discrimination against charging current. The ac component flowing through the double-layer capacitance is shifted in phase by 90°, whereas the faradaic current is only shifted by about 45°. Hence a phase-sensitive circuit can be employed to respond only to one of these components, rejecting the other. The method has been explored in detail by Bond and Canterford (9, 10).

Pulse polarography is another modification, wherein the voltage, instead of increasing steadily as in a ramp, is applied as a series of brief pulses of regularly increasing amplitude, one pulse per mercury drop. The charging current can be allowed to die out completely after the pulse is started and before the faradaic current is sampled.

All these polarographic methods (and a few others) have been reviewed and quantitatively compared by Bond and Canterford (11), whose results are summarized in Table 15-1.

QUALITATIVE ANALYSIS

Since the half-wave potential is characteristic of the substance undergoing reduction or oxidation at the microelectrode, this parameter can be utilized for its identification. The value of $E_{\frac{1}{2}}$ for a given substance depends on the nature of the supporting electrolyte, largely because of variation in the tendency to form complex ions. A few representative values are listed in Table 15-2. The importance of wise selection of electrolyte can be seen by comparing the data for lead and cadmium. These cations have identical half-wave potentials in NaOH, but are separated fairly well in KCl or H_3PO_4, and even further in KCN.

Many half-wave potential data are to be found in handbooks and monographs on polarography (5, 6, 12). However, it is frequently expeditious to plot polarograms of known substances for direct comparison with similar curves for unknowns. Published values for which information regarding the exact composition of the supporting electrolyte and nature of the reference electrode is lacking are of little value for purposes of identification.

The half-wave potential can be determined graphically from a conventional polarogram, as shown in Fig. 15-16. Portions AB and DF of the curve are extended as shown, and a tangent is drawn to the curve at its inflection point C. The line GH is bisected, and a line JK is drawn parallel to AB and DF. The abscissa of the point of intersection of JK

Table 15-1 Comparison of polarographic methods for determination of copper*

Method	Minimum concentration,† M	Comments
Conventional dc	6×10^{-6}	Standard for comparison
Rapid-scan dc	6×10^{-6}	Principal advantage is speed
Tast dc	2×10^{-6}	Long drop times most favorable
Phase-sensitive ac	1×10^{-6}	Recommended for trace analysis
Pulse	1×10^{-6}	Long drop times favorable

* Data from reference 11, with permission.
† Concentration limit recommended for quantitative analysis; the presence of copper ions can be detected at somewhat lower levels.

Table 15-2 Half-wave potentials of some common cations in various supporting electrolytes, V versus SCE*

	Supporting electrolyte				
Cation	*KCl* (0.1 *F*)	NH_3 (1 *F*) NH_4Cl (1 *F*)	*NaOH* (1 *F*)	H_3PO_4 (7.3 *F*)	*KCN* (1 *F*)
Cd^{2+}	−0.60	−0.81	−0.78	−0.77	−1.18†
Co^{2+}	−1.20†	−1.29†	−1.46†	−1.20†	−1.13† [to Co(I)]
Cr^{3+}	......‡	−1.43† [to Cr(II)] −1.71† [to Cr(0)]	....	−1.02† [to Cr(II)]	−1.38 [to Cr(II)]
Cu^{2+}	+0.04 [to Cu(I)] −0.22 [to Cu(0)]	−0.24 [to Cu(I)] −0.51 [to Cu(0)]	−0.41†	−0.09	NR§
Fe^{2+}	−1.3†	−1.49†	....	....	....
Fe^{3+}	....	....	−1.12¶ [to Fe(II)] −1.74¶ [to Fe(0)]	+0.06 [to Fe(II)]	....
Ni^{2+}	−1.1†	−1.10†	....	−1.18	−1.36
Pb^{2+}	−0.40	....	−0.76	−0.53	−0.72
Zn^{2+}	−1.00	−1.35†	−1.53	−1.13†	NR

* From data published by Meites (6).
† Irreversible reduction.
‡ . . . . indicates insufficient solubility or lacking information.
§ NR indicates that the ion is not reducible in this medium.
¶ 3 *F* KOH solution plus 3 percent mannitol.

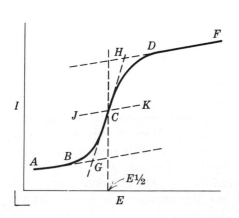

Fig. 15-16 Graphical method of locating the half-wave potential.

with the curve gives the value of $E_{\frac{1}{2}}$. This rather complicated procedure is necessary in the frequently encountered, nonideal case where DF is not so nearly parallel with AB as might be desired. According to the described procedure, slight errors of judgment in the location of the tangent line GH will have the least effect on the ultimate value of $E_{\frac{1}{2}}$.

In the case of rapid-scan polarograms, the $E_{\frac{1}{2}}$ value must be determined from a measured E_s with the aid of Eq. (15-16). If two or more species appear on the same curve (as in Fig. 15-14), exact half-wave potentials are difficult to determine, and qualitative identification should be made by comparison with known authentic samples.

QUANTITATIVE ANALYSIS

The magnitude of the diffusion current in dc polarography is related to the concentration of reducible species by the Ilkovič equation. Hence if the several factors of that equation are known or can be measured, the concentration of the species can theoretically be computed from the observed current. The only factor difficult to evaluate independently is the diffusion coefficient. It can in some instances be determined from measurements of the electric conductivity of the solution, or by chronopotentiometry. It can sometimes be estimated by comparison with known values for other ions of comparable size. It must be remembered that the diffusion coefficient is sensitive to the viscosity and temperature of the solution. Coefficients for many species can be found in the literature, but must be applied with great caution.

Many factors which are difficult to handle in a polarographic analysis may be avoided by making only comparative measurements. To do this it is necessary to prepare standard solutions of the desired substance, and determine their current-voltage curves under the same conditions as the unknown. Then the Ilkovič equation can be applied in the simplified form

$$i_d = KC \tag{15-18}$$

The proportionality constant can be evaluated graphically and need not be resolved into its theoretical equivalent.

An example of a typical analysis will make this clear. A sample of zinc is suspected of containing some cadmium, and a quantitative analysis is required. A 0.1-g sample is dissolved in hydrochloric acid, enough Triton X-100 added to eliminate maxima, and the solution diluted to a known volume with 1.0 F potassium chloride. A small portion is placed in the polarographic cell, and oxygen is swept out with a current of nitrogen. A preliminary current-voltage curve is plotted, with the recorder operating at about one-tenth its full sensitivity. The potential range from about -0.4 to -0.8 V is adequate for this analysis; it is not always necessary to run a complete curve. The trial curve will show qualitatively whether or not cadmium is present, according to whether a wave is found at about —0.64 V. The reduction potential of zinc is so great that it will not interfere with the wave for cadmium. If cadmium proves to be present, the preliminary curve will give an idea of the relative dilution needed to give the best analysis. If the solution is already too dilute, the sensitivity can be increased; if too concentrated, an aliquot can be diluted further with potassium chloride solution. Another current-voltage curve is now plotted for the final analysis. A standard solution of cadmium chloride in 1.0 F potassium chloride must also be prepared at about the same final dilution, and a polarogram plotted for it. The resulting curves will resemble Fig. 15-17. The values of i_d for both standard and unknown are measured from the graph. The concentration C_x of cadmium in the final dilution of the unknown can be calculated by a simple proportion. Alternatively, a calibration graph can be constructed, from which concentrations of this and any future cadmium unknowns can be read. It is, of course, wise to check such a calibration curve by analyzing two or more standard solutions of varying concentrations.

In a modification of this procedure, a known quantity of standard solution can be added to a measured portion of the unknown. Polarograms taken before and after the addition give all the information needed to calculate the concentration of the unknown. This procedure is known as the method of *standard addition*. It has the advantage that unknown and standard are certain to be measured under identical conditions (assuming constant temperature).

All quantitative polarographic measurements must take into account the appropriate residual current. If a considerable voltage is scanned before the start of the wave, the residual current can be approximated by extrapolation, as in Fig. 15-17. Otherwise it may be necessary to run

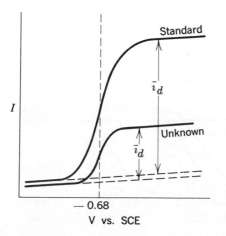

Fig. 15-17 Polarograms of an unknown and standard.

a blank; this would be the only way, for example, to measure the height of the first oxygen wave in Fig. 15-9. Error from this source will become larger as one goes to smaller concentrations, and some automatic method of compensation is called for. Such compensation is sometimes provided by a device which feeds a small current proportional to the scanning voltage in opposition to the cell current. This cannot give complete compensation, because the residual current is not strictly linear, but it can permit polarographic analysis down to 10^{-5} or even 10^{-6} F.

Another limitation to quantitative precision in polarography appears when the waves corresponding to two reducible species are relatively close together. If they are so close that no horizontal portion is evident between them, then there is no way in which the individual diffusion currents can be measured, though the sum of the two species can be determined. The only ways to improve the situation without resorting to chemical separation are to seek another supporting electrolyte or to adopt a rapid-scan or ac procedure, thus obtaining peaks instead of waves.

PILOT-ION PROCEDURE

Since the diffusion coefficients vary from ion to ion, there is no uniformity in the heights of the waves obtained with equivalent concentrations of different reducible ions. Figure 15-18 shows the waves of a number of ions at equal concentrations (13). The ordinates are given in terms of the quantity $i_d/Cm^{2/3}t^{1/6}$, which is sometimes defined as the *diffusion-current constant*. The Ilkovič equation predicts constancy for this ratio for any given temperature. Its value gives the relation between diffusion current and concentration, so that a knowledge of this quantity obviates the need for repeated calibrations. This possibility cannot be fully realized, since the values of m and t must be redetermined for each new capillary. How-

Fig. 15-18 Polarograms for a number of ions in 1 F HCl (idealized). The abscissas are volts; the ordinates are relative current values calculated on the basis of equal concentrations of ions, measured with a constant value of $m^{2/3}t^{1/6}$. Gelatin is present in each case. [*Industrial Engineering Chemistry, Analytical Edition* (13).]

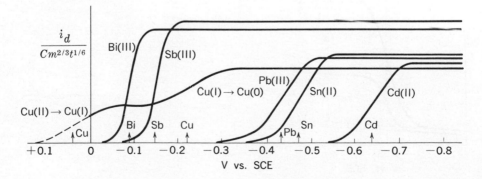

ever, the values for the various ions all change in the same proportion. Hence, once a series of these constants is determined for one capillary, it is necessary to repeat only the determination of the constant for *one* ion to establish those of the entire series for a new capillary. This makes it necessary to maintain only a single standard stock solution for each supporting electrolyte likely to be needed. This is called the *pilot-ion* method.

ORGANIC POLAROGRAPHY

Polarography presents an important tool for analysis and structure determination in organic chemistry. The principles are no different from those discussed above. The product of electrochemical action is, of course, insoluble in mercury, but is almost always soluble in whatever solvent or solvent mixture is suitable for the original material. Any solvent which will dissolve an electrolyte is potentially useful for polarography. Various alcohols and ketones, pure or mixed with water, have been used, as have molten urea, ammonium formate, dimethylformamide, ethylenediamine and others. Certain substituted ammonium salts, such as tetrabutylammonium iodide, are readily soluble in nonaqueous solvents and serve admirably as supporting electrolytes.

Many classes of organic compounds can be reduced at the DME: conjugated unsaturated compounds; certain carbonyl compounds; organic halogen compounds; quinones; hydroxylamines; nitro, nitroso, azo, and azoxy compounds; amine oxides; diazonium salts; certain sulfur compounds; certain heterocyclic compounds; peroxides; and reducing sugars. For details the student is referred to the literature (14). A typical example is the analysis of a mixture of aromatic carbonyl compounds. The following half-wave potentials (versus SCE) were observed in a supporting electrolyte of lithium hydroxide in aqueous ethanol: benzaldehyde, -1.51 V; *n*-propyl phenyl ketone, -1.75; isopropyl phenyl ketone, -1.82; *tert*-butyl phenyl ketone, -1.92. The waves were clearly defined, and the diffusion currents were linear with concentration over the range 0.2 to 2.5 mM. It is found that the pH and the ionic strength of the solution are more important in organic polarography than is usual in the inorganic field.

AMPEROMETRIC TITRATION

It is possible to carry out a titration in a voltammetric electrolysis cell and to follow its progress by observing the diffusion current after successive additions of reagent. This is analogous to potentiometric and conductometric titrations, and is known as *amperometric titration*. Since the diffusion current is generally proportional to concentration, the titration curve is found to consist of two straight-line segments, the intersection of which corresponds to the equivalence point. Three types of curves may be distinguished (Fig. 15-19). Curve a results from the titration of a reducible ion by a reagent which does not itself yield a polarographic wave. An example is the titration of lead ion by oxalate with the dropping-mercury cathode at a potential of —1.0 V versus SCE, in potassium nitrate. The initial diffusion current is relatively high, and it decreases regularly as the Pb^{2+} ion is removed by reaction with oxalate. After the equivalence point, further increments of reagent have no effect on the current.

The reverse titration, oxalate by lead ion, under similar conditions will give a curve such as b in Fig. 15-19. The Pb^{2+} ion cannot accumulate in the solution and give a diffusion current until all the oxalate is precipitated.

Curve c results from the titration of lead by dichromate at an applied potential of —1.0 V, where both lead and dichromate ions are capable of being reduced at the dropping cathode.

Amperometric titration is applicable not only to precipitation reactions but to many redox and complexometric titrations as well. It is inherently capable of greater accuracy than nontitrative polarographic methods, since each analysis involves a number of separate determinations so related that individual errors tend to cancel.

ROTATING PLATINUM ELECTRODE

The use of a rotating platinum electrode in place of the DME in titration increases sensitivity, because of the disruption of the diffusion layer by stirring. The electrode is usually fabricated with a 2- to 3-mm length of platinum wire extending horizontally from its vertical glass supporting tube. The tube is rotated at a few hundred revolutions per minute, which rate must be held quite constant to obtain consistent results. Another advantage of the platinum electrode is the greatly reduced residual current.

An excellent example is the titration of arsenite by potassium bromate in the presence of bromide (15). The arsenite solution is made 1 N in hydrochloric acid and 0.05 N in potassium bromide. The applied potential

Fig. 15-19 Typical amperometric titration curves.

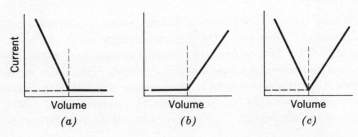

is $+0.2$ to 0.3 V versus SCE.   The reaction is

$$3AsO_2^- + BrO_3^- \rightarrow 3AsO_3^- + Br^-$$

The bromide serves as an indicator, since only after the arsenite is completely reacted can the bromide be oxidized:

$$BrO_3^- + 5Br^- + 6H^+ \rightarrow 3Br_2 + 3H_2O$$

Of all the substances present, free bromine is the only one which gives a polarographic wave.   Figure 15-20 shows the curve resulting from the titration of 100 ml of a 9.18×10^{-4} N solution of arsenious acid with 0.0100 N potassium bromate.   At this applied potential, oxygen is not reduced at the cathode; hence it is not necessary to remove oxygen before analysis.

In this titration with a rotating electrode, it is not necessary to make any readings prior to the equivalence point.   The operator adds reagent continuously at a moderate rate until some deflection is seen.   Then three or four readings in the presence of excess reagent will establish the sloping portion of the curve, which is extrapolated back to the zero axis to determine the equivalence point.

The number of titrations to which the amperometric method can be applied is much greater than that for potentiometric titration, because the electrodes are nonspecific.   There are so many ions and molecules which can yield polarographic waves at either a mercury or a platinum microelectrode that there is a good possibility of finding a suitable reagent for direct or indirect titration of nearly any substance.   The method is best suited for precise determination of low concentrations of the unknown.

BIAMPEROMETRIC TITRATIONS

A simplified procedure is made possible by impressing a small potential across two identical inert electrodes.   The apparatus required consists merely of a source of about 50 to 100 mV, and a galvanometer in series

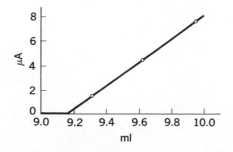

Fig. 15-20 Amperometric titration of arsenite by bromate. [*Journal of Physical Chemistry* (15).]

with the two platinum electrodes. No current can flow between the electrodes unless there are present both a substance which can be oxidized
at the anode and a substance which can be reduced at the cathode. Any
redox couple which is easily reversible will permit electrolysis to take place.
In the ferric-ferrous system, for example, Fe^{3+} ions can be reduced at the
cathode, and Fe^{2+} oxidized simultaneously at the anode. Some systems
other than reversible couples will also permit current to flow. H_2O_2, for
example, is oxidized at the anode to oxygen and reduced at the cathode
to hydroxide ions. The permanganate-manganous couple is not reversible
but, nevertheless, permits electrolysis, as Mn^{2+} is anodically oxidized to
MnO_2 and MnO_4^- undergoes cathodic reduction to MnO_4^{2-} or to MnO_2.
Some other electrolyzable systems are: I_2–I^-, Br_2–Br^-, Ce^{4+}–Ce^{3+}, $Fe(CN)_6^{3-}$–
$Fe(CN)_6^{4-}$, Ti^{4+}–Ti^{3+}, VO_3^-–VO^{2+}.

A few titrations performed in this manner are shown graphically in
Fig. 15-21 (16). The titration of iodine by thiosulfate (curve a) is one
of the earliest reactions to be followed by this method; the abrupt cessation
of current at the end point gave rise to the name *dead-stop* titration. Prior
to the equivalence point, both free iodine and iodide ions are present in
solution, and therefore current can flow, even with as little as 15 mV applied
potential. As the titration proceeds, iodine is reduced to iodide; at the
equivalence point no free iodine remains, and the solution cannot conduct.
Beyond the equivalence point, as more thiosulfate is added, no current
can flow, as thiosulfate-tetrathionate does not constitute a reversible couple.
This is widely used in the Karl Fischer moisture titration.

In the titration of ferrous iron by cerate (ceric sulfate) (curve b),
on the other hand, current can flow on both sides of equivalence, as both
ferrous-ferric and cerous-ceric pairs form reversible couples. There is a

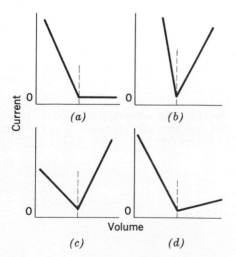

(a) *(b)* *(c)* *(d)*

Current — Volume

Fig. 15-21 Typical biamperometric titration
curves. (a) I_2 by $Na_2S_2O_3$; (b) Fe(II) by
Ce(IV); (c) V(V) by Fe(II); (d) Fe(CN)$_6^{-4}$
by Ce(IV). [*Analytical Chemistry* (16).]

point of (nearly) zero current corresponding to the complete removal of ferrous ions before excess cerate ions have been added.

The other curves can be interpreted along similar lines. The end points in all these titrations are sharply defined, as the current drops essentially to zero.

PROBLEMS

15-1 A 1.000-g sample of zinc metal is dissolved in 50 ml of 6 F hydrochloric acid, diluted to the mark in a 250-ml volumetric flask, and a drop of maximum suppressor added. A 25.00-ml portion is transferred to a polarographic cell, and oxygen is flushed out. A polarogram in the range 0 to 1 V (versus mercury-pool electrode) shows a wave at $E_{\frac{1}{2}} = -0.65$ V, $i_d = 32.0$ units of galvanometer deflection. A 5.00-ml portion of 5×10^{-4} F cadmium chloride is added directly to the polarograph cell which already contains the zinc solution, oxygen is again flushed out, and a second polarogram taken. The wave shows the same $E_{\frac{1}{2}}$, but the i_d is 77.5 units. Calculate the percent by weight of cadmium impurity in the zinc metal. Do not overlook the dilution effect. (Note that the supporting electrolyte is zinc chloride–hydrochloric acid.)

15-2 A 5×10^{-3} F solution of $CdCl_2$ in 0.1 F KCl shows a diffusion current at -0.8 V versus SCE of 50.0 μA. The mercury is dropping at a rate of 18.0 drops per minute. Ten drops are collected and found to weigh 3.82×10^{-2} g. (a) Calculate the diffusion coefficient D. (b) If the capillary were replaced by another, for which the drop-time is 3.0 s, and 10 drops weigh 4.20×10^{-2} g, what will be the new value of the diffusion current?

15-3 α-Benzoinoxime (cupron) is a precipitating agent for copper (17). In a medium consisting of 0.1 F NH_4Cl and 0.05 F NH_3 (pH = 9), cupron is reduced at the DME, giving a wave with $E_{1/2} = -1.63$ V versus SCE (curve a, Fig. 15-22). The double wave of copper in the same medium is shown as curve b. Sketch the curves which would be found in the amperometric titration of copper by cupron at applied potentials of -1.0 V and -1.8 V versus SCE. Which potential would be preferred for the titration of copper in the absence of interfering substances? Which would be more liable to interference by reducible impurities?

Fig. 15-22 Polarograms for (a) α-benzoinoxime, and (b) copper, in NH_3–NH_4Cl buffer. [*Industrial Engineering Chemistry, Analytical Edition* (17).]

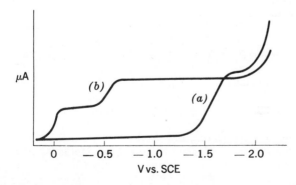

15-4 Small concentrations (0.1 to 10 ppm) of nitrate ion can be determined polarographically in 0.1 F zirconyl chloride, $ZrOCl_2$, as supporting electrolyte, by the difference in diffusion currents before and after reduction of nitrate by ferrous ammonium sulfate, with the DME 1.2 V negative to the SCE (18). The following data are recorded for two standards and an unknown:

Solution, NO_3^-	Diffusion current, μA	
	Before reduction	After reduction
10.0 ppm	87.0	22.0
5.0 ppm	48.5	15.2
Unknown	59.0	17.0

Calculate the concentration of nitrate in the unknown.

15-5 Calculate values of E_{DME} corresponding to currents of 1, 2, 3 μA, etc., up to 9 μA for three hypothetical reductions in each of which $E_{1/2} = -1.000$ V versus SCE, and $i_d = 10$ μA, but for which $n = 1, 2,$ and 3, respectively (temperature, 25°C). Plot these data on a single graph so that the three curves intersect at the half-wave point. For the same data, plot on another sheet the quantity $\log [i/(i_d - i)]$, as abscissas against E_{DME} as ordinates.

15-6 Tin(IV) in acidic pyrogallol solution gives a polarogram with two steps of equal height having $E_{1/2}$ values of -0.20 and -0.40 V versus SCE (19). Note that $E°$ for $Sn^{4+} + 2e^- \rightarrow Sn^{2+}$ is -0.10 V, and for $Sn^{2+} + 2e^- \rightarrow Sn^0$ is -0.38 V versus SCE. (a) Explain why two equal steps are observed. (b) Which form of tin, Sn(IV) or Sn(II), do you conclude is more strongly complexed by pyrogallol, and why?

15-7 Ag^+ ion in $NaClO_4$ as supporting electrolyte is reducible at the DME at the potential of the SCE. Cl^- ion in the same medium gives an anodic wave with $E_{1/2} = +0.25$ V versus SCE. It is possible to determine whether the complex $AgCl_2^-$ is reducible under these conditions by titrating Ag^+ by Cl^-, amperometrically, at the DME. Sketch and explain the titration curves which might result.

REFERENCES

1. Anderson, L. B., and C. N. Reilley: *J. Chem. Educ.*, **44**:9 (1967).
2. Delahay, P.: In I. M. Kolthoff and P. J. Elving (eds.), "Treatise on Analytical Chemistry," pt. I, vol. 4, chap. 44, Wiley-Interscience, New York, 1963.
3. Cooke, W. D., M. T. Kelley, and D. J. Fisher: *Anal. Chem.*, **33**:1209 (1961).
4. Reilley, C. N. and R. W. Murray: In I. M. Kolthoff and P. J. Elving, (eds.), "Treatise on Analytical Chemistry," pt. I, vol. 4, chap. 43, Wiley-Interscience, New York, 1963.
5. Heyrovský, J., and J. Kůta: "Principles of Polarography," Academic, New York, 1966.
6. Meites, L.: "Polarographic Techniques," 2d ed., Wiley-Interscience, New York, 1965.
7. Ewing, G. W., and J. E. Nelson: *J. Chem. Educ.*, **46**:292 (1969).
8. Schmidt, H., and M. von Stackelberg: "Modern Polarographic Methods," Academic, New York, 1963.

9. Bond, A. M.: *Anal Chem.,* **44:**315 (1972).
10. Bond, A. M., and J. H. Canterford: *Anal. Chem.,* **44:**732 (1972).
11. Bond, A. M., and D. R. Canterford: *Anal. Chem.,* **44:**721 (1972).
12. Meites, L.: In L. Meites (ed.), "Handbook of Analytical Chemistry," p. 5–38 et seq., McGraw-Hill, New York, 1963.
13. Lingane, J. J.: *Ind. Eng. Chem., Anal. Ed.,* **15:**583 (1943).
14. Zuman, P.: "Organic Polarographic Analysis," Macmillan, New York, 1964.
15. Laitinen, H. A., and I. M. Kolthoff: *J. Phys. Chem.,* **45:**1079 (1941).
16. Stone, K. G., and H. G. Scholten: *Anal. Chem.,* **24:**671 (1952).
17. Langer, A.: *Ind. Eng. Chem., Anal. Ed.,* **14:**283 (1942).
18. Rand, M. C., and H. Heukelekian: *Anal. Chem.,* **25:**878 (1953).
19. Bard, A. J.: *Anal. Chem.,* **34:**266 (1962).

Chapter 16

Electrodeposition and Coulometry

The cathodic deposition of transition metals for analytical purposes is probably the oldest of electroanalytical techniques. The currency metals in particular have been determined as major constituents of alloys and ores by exhaustive electrolysis on a preweighed cathode. Conventionally this has been carried out with current densities as high as a few tenths ampere per square centimetre, much too large to approximate thermodynamic equilibrium. Selective deposition of metals could be achieved in some cases by the addition of complexing agents. In other cases the use of mercury as cathode material allowed separations based on the high overvoltage for reduction of hydrogen ions (1).

The technique can be improved by the use of a *potentiostat*. This is an instrument that can control the potential of the cathode (or anode) of the electrolysis cell relative to a reference electrode. Many designs of potentiostats have been published (2), but only that using operational amplifiers will be described here (Fig. 16-1). Comparison with Fig. 15-11 will show that there is much in common between a potentiostat and a polarograph. For the present purpose a constant potential is desired, so the ramp generator is replaced with an adjustable voltage source. It is likely that more current will be needed than an ordinary operational ampli-

326

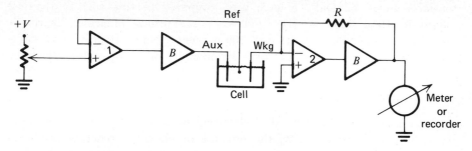

Fig. 16-1 A potentiostat using operational amplifier circuitry. The amplifiers marked *B* are boosters, which permit operation with greater current than the operational amplifiers alone can handle.

fier can handle, so a *booster* amplifier is added within the feedback loop; this can be considered as forming part of the control amplifier itself. The current-to-voltage converter, amplifier **2**, also must be provided with a booster.

  If this apparatus is to be used with gravimetric determination of the metal deposited, the working electrode can be grounded directly, and amplifier **2** and the recorder omitted. However, inclusion of these components permits electrical determination and possible saving of time.

  In controlled-cathode electrolysis, at a potential where only a single species is reducible, the *current* is limited by diffusion, hence is proportional to the concentration of the reducible species. It is therefore evident that both concentration and current will fall off exponentially with time. We can write:

$$\frac{C_t}{C_0} = \frac{i_t}{i_0} = 10^{-kt} \tag{16-1}$$

where C_t represents the concentration at time t, C_0 that at time $t = 0$, i_t and i_0 the corresponding currents, and k a constant. It can be shown that k is proportional to DAS/V, where D is the diffusion coefficient, A the area of the cathode, S the rate of stirring, and V the volume of solution. Thus a plot of log i against t will give a straight line with a negative slope equal to k. From such a plot one can determine the time required to deposit any given fraction of the desired species. In an experiment quoted by Lingane (2) for the deposition of copper, k was found to be 0.15 min^{-1}, from which it follows arithmetically that deposition was 99 percent complete in 13 min, and 99.9 percent complete in 20 min.

  Controlled-potential electrolysis has been found useful in some fields of chemistry other than analytical. In syntheses involving electrooxidation or electroreduction, improved operation has been found to result from potential control. Se and Te have been prepared in this way in the −2 oxidation state, also W(III) and W(V). Various pinacols, hydroxylamines, etc., have been reported in 100 percent yields. It provides a valuable method of separation of radioactive nuclides in submicrogram amounts.

COULOMETRY

An even more powerful method of electrical measurement is by integration. According to Faraday's law, the amount of chemical reaction produced by electrolysis is proportional to the quantity of electricity passed. This quantity, expressed in coulombs, is measured by the time integral of the current:

$$Q = \int i \, dt \tag{16-2}$$

Application of this relation permits the quantitative determination of any substance which can be made to undergo an electrochemical reaction with 100 percent current efficiency (i.e., no side reactions). This is the method known as *coulometric analysis*.

Two procedures are possible: operation at constant current, so that the amount of material deposited is proportional to the elapsed time, and operation at constant potential, in which case the current decreases from a relatively large value to practically zero. Both procedures have important areas of application.

The quantity Q can be measured directly by any of a number of kinds of integrators (3), chemical, electromechanical, or electronic in nature. One of these, easily assembled in the laboratory, uses operational amplifiers, as in Fig. 16-2. The current i_{in} flows to the input of the ampli-

Fig. 16-2 An integrator with digital readout, to be used as a coulometer.

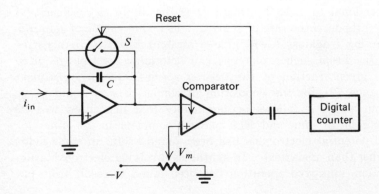

fier, but since it cannot enter the amplifier, goes to charging the capacitor C, just as in the ramp generator for polarography. If Q were limited to about 1 mC, then a single integrating amplifier would suffice, but in most applications many times this amount must be measured. This can be done by means of the added components in Fig. 16-2. When the voltage across the integrating capacitor builds up to a predetermined value V_m, the second amplifier, which has no feedback path (a *comparator*) suddenly changes its output from about -10 to $+10$ V (with the polarities shown). This activates a voltage-controlled gate (i.e., switch S, which can be a relay, or a FET, or other semiconductor device) to discharge the capacitor, which is then free to charge up again. This sequence continues as long as current flows into the integrator. Each time the comparator goes positive, a count registers on the counter, which thus gives a quantitative measure of Q (3, 4).

COULOMETRIC TITRATION

An indirect coulometric analysis normally consists in the electrolytic generation of a soluble species which is capable of reacting quantitatively with the substance sought. This falls within the broad definition of titration, as the reagent is added to the solution gradually (by electrolytic generation rather than from a buret), and some independent property must be observed to establish the equivalence point in the reaction. A process of this nature is commonly a hybrid between a direct and an indirect determination.

As an example, let us consider the coulometric determination of the concentration of ferric iron, in a solution containing HBr, by electrolytic reduction to the ferrous state. This is accomplished with the aid of a platinum cathode and a silver anode. The anodic half-reaction is $Ag + Br^- \rightarrow AgBr + e^-$; we are therefore only concerned with processes taking place at the cathode. Suppose that we first attempt a *direct* reduction. The situation will be clarified by reference to the current-voltage curves of Fig. 16-3. Curve 1 includes the reduction waves of the $FeBr_2^+$ ion and of the H^+ ion in the HBr solution. If we force a constant current of magnitude a through this cell, the cathode potential will assume the value $+0.40$ V (approximately) versus the SCE, where curves a and 1 intersect. As the electrolysis proceeds and iron is reduced, the plateau corresponding to the diffusion current of $FeBr_2^+$ is progressively lowered until the cathode potential suddenly jumps to approximately -0.3 V (curve 2), corresponding to reduction of H^+ ions. From this point on, ferric iron is reduced and hydrogen liberated simultaneously, so that the current efficiency with regard to iron reduction is less than 100 percent, and the analysis is no longer valid.

Let us now repeat the experiment with the addition of a considerable excess of titanium(IV), which is reducible in the presence of acid to tita-

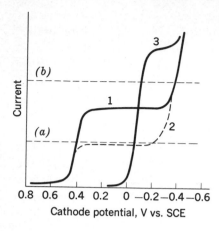

Fig. 16-3 Current-voltage curves as observed with a platinum cathode versus the SCE.

nium(III) at a platinum cathode with a potential slightly more negative than that of the SCE, curve 3. If again the current is established at level *a*, the iron will be reduced as before until its diffusion current is lowered to *a* but, at that point, the potential will jump, not to -0.3 V, but to -0.05 V versus SCE. From then on both iron and titanium will be reduced, but this does not result in any loss, because the titanium(III) reacts in the stirred solution to reduce ferric iron, so that the net result is the reduction of one iron atom per electron, no matter whether directly or indirectly. Thus the overall current efficiency is maintained at 100 percent, as required. (If the current had been set at level *b* in Fig. 16-3, then iron and titanium would have been reduced simultaneously from the very beginning, but the net result would have been the same.)

The equivalence point in this reaction is the point in time when the quantity of electricity is just equivalent to the total amount of ferric iron originally present in the sample. It can be identified (*1*) potentiometrically, (*2*) by an amperometric observation at about 0.25 V positive to the SCE, (*3*) by a biamperometric method, (*4*) by a photometric method involving addition of KSCN or other chromogenic reagent for iron(III), (*5*) by a photometric method with a redox indicator which will be reduced by titanium(III) only after all the iron is reduced, or possibly by other methods.

In the experiment just described, no interference resulted from the electrochemical reaction at the anode. In many titrations, however, a soluble product will be formed at the counterelectrode and, if no precaution is taken, it will react unfavorably either at the generator electrode or with the intermediate in solution. To avoid this kind of difficulty, the counterelectrode is often shielded by a glass tube with a fritted tip to discourage convection. Sometimes this is fully adequate, but in other cases it falls short of eliminating the error; it may be necessary to resort to an agar-gel salt bridge or other device.

A shield closed by an ion-exchange membrane instead of a glass frit will serve excellently in many situations (5, 6). Such membranes are available in two forms: cation exchangers which will not permit cations to pass through the membrane, and anion exchangers which exclude anions. For example, if we wish to titrate a base coulometrically by electrogeneration of H^+ ions, the generator electrode will be the anode, and its half-reaction the familiar $H_2O \rightarrow \frac{1}{2} O_2 + 2H^+ + 2e^-$. At the cathode the complementary reaction, $2e^- + 2H_2O \rightarrow H_2 + 2OH^-$, will take place. Obviously the hydroxide ion produced at the cathode must not be allowed to mix with the solution being titrated (the anolyte); an anion-exchange membrane separating the two compartments will prevent such mixing. This principle appears not to have received the study it deserves.

In Fig. 16-4 is shown schematically an apparatus suitable for the titration of base. The generator electrodes are connected to a constant-current source with an associated timer, while the indicator electrodes are the conventional glass–reference system of a pH meter. Constant-current supplies will be discussed in Chap. 26.

A large number of reagents have been prepared by electrolytic generation, including H^+, OH^-, Ag^+, and other metal ions, oxidants such as $Ce(IV)$, $Mn(III)$, $Ag(II)$, Br_2, Cl_2, I_2, and $Fe(CN)_6^{3-}$, reductants such as $Fe(II)$, $Fe(CN)_6^{4-}$, $Ti(III)$, $CuBr_2^-$, and $Sn(II)$, complexogens such as EDTA and CN^- ions.

By one or another of these reagents it has become possible to substitute for practically all the procedures of classical volumetric titrimetry their coulometric counterparts. This presents the great practical advantage that it is not necessary to prepare and store standardized solutions. The primary standard for coulometric titration is the combination of a constant-

Fig. 16-4 Apparatus for coulometric titration at constant current, with potentiometric end-point detection.

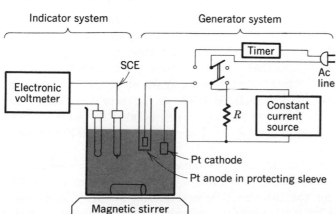

current source and an electric timer, which are applied to *all* titrations, no matter what their chemical nature. Coulometric titration also has the advantage that it is applicable to samples one or two orders of magnitude smaller than conventional procedures (samples of 0.1 down to 0.001 meq are usual, compared with 1 to 10 meq for volumetric titrations). Furthermore, as can be seen from the list above, some reagents can be employed which are unstable, or for other reasons not suitable for volumetic use, such as $Mn(III)$, $Ag(II)$, $CuBr_2^-$, and Cl_2.

The precision possible in coulometric titration can easily equal, and with precautions can exceed, that attainable by volumetric means. Eckfeld and Shaffer (7) have reported a careful study of precision in coulometric neutralization. They could measure coulombs (as microequivalents) to about ± 0.004 percent. One simple but effective precaution was to provide a slow flow of indifferent electrolyte in a salt bridge, so as to eliminate all possibility of contamination of solution or loss of sample through the fritted glass. This paper should be studied by anyone attempting precise work in coulometric titrations.

The coulometric method is less useful for larger concentrations, and this is its chief limitation. The reason is that it would be necessary to operate with much larger currents if unduly long times are to be avoided, and this tends to reduce the current efficiency from the required 100 percent, except in the unusual case in which only one electrode reaction is possible. A current of 5 to 10 mA is about as large as can safely be used.

Pretitration is often advisable. After the apparatus is assembled a small portion of the material to be analyzed is inserted and the electrolysis allowed to proceed until the desired end point is observed. Then the measured sample is added and titrated until the end point is again reached. This ensures that any impurities which can react with the generated titrant are removed in advance. It also obviates any uncertainty about the surface condition of the electrodes (formation of an oxide film, for example). The titration curve will have the appearance shown in Fig. 16-5. The material pretitrated reacts with generated reagent from time t_0 to t_1, following which reagent accumulates (at A). The sample is then added; it reacts immediately with the reagent which has just been generated, returning the curve to zero (at B), and then continues to react with generated titrant until it is all consumed, when the curve rises again at C. The two sloping portions of the curve are extrapolated back to the zero level at D and E, and the time between these two points, $t_2 - t_1$, is taken as the electrolysis time. The slopes at A and C will differ if appreciable dilution has occurred with addition of the sample.

Coulometric titration is easily automated, particularly because no burets or pumps are required. Passage of a constant current provides a linear time base for the titration. Several automatic or semiautomatic coulometric titrators are commercially available. There are also a number

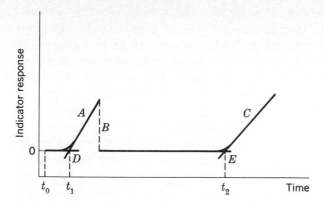

Fig. 16-5 Coulometric titration curve illustrating pretitration. The vertical axis refers to the indication of any detector (photometric, amperometric, etc.) the output of which is directly proportional to concentration.

of continuous coulometric analyzers for flow-stream monitoring, in which a recorder indicates the magnitude of current required to maintain constant the concentration of some component by causing it to react with an electrogenerated reagent. Coulometric generation can readily be adapted to an all-electronic version of the pH-stat discussed in Chap. 14.

ELECTROLYTIC PRECONCENTRATION

Cathodic electrodeposition is an extremely valuable tool in trace analysis for transition metals. Electrolysis can be continued for as long as necessary to accumulate sufficient metal for subsequent analysis. If the potential is made negative enough, all reducible metals present will be deposited together, but selective deposition can be effected by potentiostatic control. Electrolysis may be carried to exhaustion, but this may require inordinately long times, so more commonly deposition is continued for a specific period of time (8). In some situations exhaustive electrolysis would be impossible, for example, in situ analysis of a flowing stream.

The choice of the electrode on which deposition is to occur will depend on the method selected for analysis. For electrochemical analysis (the most common, see next section), mercury is usually preferred. Pyrolytic graphite, a form of carbon that can be cleaved into thin sheets, is particularly useful as a deposition electrode if subsequent analysis by x-ray fluorescence is contemplated (8).

STRIPPING ANALYSIS

This term refers to analysis of a mixed deposit of metals by anodic voltammetry or chronopotentiometry. In a typical procedure, a portion of solution is electrolyzed for perhaps 15 or 30 min with a small mercury cathode, at a potential sufficiently negative to reduce all metals up to the limit imposed by the hydrogen overvoltage. The mercury electrode is then made the anode, as the potential is swept in the positive-going direction, stripping off the metals deposited in the previous step. The stripping process is carried out under polarographic conditions in a time of 2 or 3 min.

One type of electrode consists of a hanging drop of mercury in a cell such as that of Fig. 16-6. A Teflon scoop is positioned between a polarographic capillary and a mercury-plated platinum contact wire, so that the operator can catch one or more drops of mercury and transfer them to the platinum, where they will adhere. In another modification, an electrode is fabricated by filling a glass or Teflon tube with a conductive mixture of powdered graphite and a heavy oil or wax. Mercury is plated onto the end surface as a thin film.

A solid electrode cannot be used satisfactorily with voltammetric stripping, because atoms of a minor constituent of the plated mixture may be covered or trapped by a major constituent and hence not be oxidized completely. This difficulty does not arise with metals dissolved in mercury.

Analytical stripping is usually done by anodic voltammetry, scanning the voltage in the positive direction starting from the potential employed

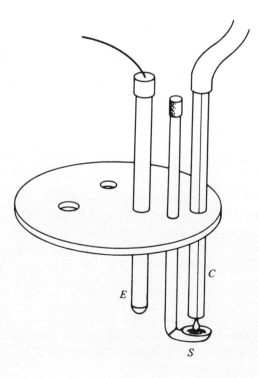

Fig. 16-6 Hanging mercury drop electrode assembled in a plastic beaker cover. S is a scoop with which a drop of mercury can be transferred from the capillary C to the platinum-tipped electrode E.

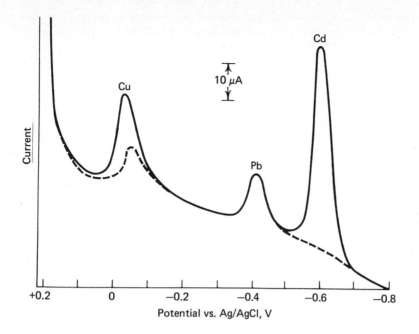

Fig. 16-7 Anodic stripping voltammogram. The dashed line is a blank, the solid line corresponds to a solution 2×10^{-7} M each of Cd^{2+} and Cu^{2+} in 0.05 M KCl; sweep rate, 2.5 V/min. Note that Pb shows up equally in blank and analytical determination. [*Journal of Chemical Education* (9).]

in the deposition. As each metal is removed in turn from the mercury, a peak appears on the current-voltage curve. The quantity of each is determined from the height of the peak, measured from a similar curve produced from a blank. Figure 16-7 shows the results of an experiment, reported by Ellis (9), on a known solution 2×10^{-7} M in both Cu^{2+} and Cd^{2+}. The electrode was of the graphite-wax type with a plated mercury film. Deposition was for 5 min at -1.0 V (versus Ag/AgCl). Note that the reagent blank showed appreciable amounts of both copper and lead. This is not uncommon in stripping analysis; the method is so extremely sensitive that supposedly pure solvents and reagents are often found to contain unsuspected trace metals.

There have been many modifications of stripping analysis devised, each with its own special advantages and drawbacks. The field has been reviewed by Barendrecht (10), by Shain (11), and by Ellis (9).

PROBLEMS

16-1 A constant current of 1.500 A is allowed to flow for a period of 1 h (to the nearest second) through a number of electrolytic cells connected in series. All electrodes are platinum. The cells contain an excess of the electrolytes listed. For each, tell what substance is deposited at the cathode, and calculate the quantities in grams (if solid or liquid) or in milliliters at STP (if gaseous).

(a) $Cu(NO_3)_2$		(d) HgI_2	
(b) $NaOH$		(e) $Pb(NO_3)_2$	
(c) $K_4Fe(CN)_6$		(f) $Ag(NH_3)_2Cl$	

16-2 A cadmium amalgam is to be prepared for use in a Weston standard cell by electrolyzing cadmium chloride with a mercury cathode and silver anode. The desired concentration is 12 percent cadmium by weight. Starting with 20 g of mercury, how long should electrolysis continue at 5 A to attain this concentration?

16-3 Ceric ion is to be determined by coulometric titration with electrolytically reduced ferrous ion. The end point is observed potentiometrically with platinum and SCE connected to a pH meter. Preliminary studies show that the potential of the platinum

Time, s	Coulometer reading	Potential, Pt-SCE
	Preliminary titration	
. . . . .	20	+0.830
436.0	120	0.801
472.8	170	0.790
	Analysis titration	
472.8	170	0.893
. . . . .	470	0.861
. . . . .	720	0.820
. . . . .	750	0.814
. . . . .	770	0.810
925.0	790	0.803
939.5	810	0.794

wire at the end point should be +0.800 V versus SCE. The acidified solution contained 0.0005 mol of ferric iron in a volume of 250 ml. Interfering substances were removed by a preliminary titration of the same type. For this purpose about 0.2 μeq of ceric ion was added, and current was passed through the generating electrodes ($Fe^{3+} + e^- \rightarrow Fe^{2+}$) until the indicator showed 0.800 V. Then a 1.00-ml portion of the unknown ceric solution was added, and generation continued until the end point was reached again. The tabulated data were taken. The coulometer readings are in arbitrary units such that a change of one unit corresponds to 4.79×10^{-4} μF.

Calculate the concentration of the unknown in micrograms of cerium per milliliter.

16-4 The copper from a 0.400-g sample of a brass was deposited electrolytically with a suitably controlled cathode potential. The cell was placed in series with a recording ammeter. The record obtained is reproduced in Fig. 16–8. The quantity of electricity involved can be determined by integration, approximately, by counting the squares beneath the curve, or by the method of trapezoids. Perform this integration and compute the percentage of copper in the alloy. Why does this curve not follow the exponential law?

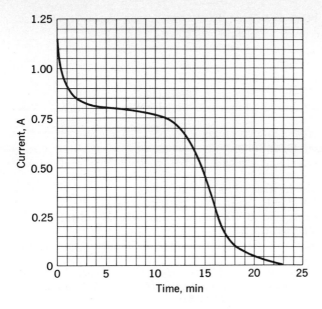

Fig. 16-8 Time-current curve for deposition of copper from a sample.

16-5 What value of current (in milliamperes) would be required for coulometry at constant current so that a seconds timer would read directly in microequivalents of electrode reaction?

16-6 Anthranilic acid (*o*-aminobenzoic acid) can be brominated by electrolytically generated bromine at pH 4 to produce tribromoaniline. Hargis and Boltz (12) have shown that small amounts of copper can be determined by precipitating Cu(II) anthranilate, $Cu(C_6H_4NH_2CO_2)_2$, dissolving the precipitate, and coulometrically titrating the liberated anthranilic acid.

   The copper in a 1.000-g sample of a biological material is converted to ionic form, and precipitated with excess anthranilic acid. The precipitate is filtered, washed, and redissolved, and the acid brominated with a constant current of 6.43 mA. The time required is 22.40 min. Calculate the amount of copper in the original sample, in parts per million.

16-7 It was desired to analyze a factory effluent water for Cu, Pb, and Cd by means of stripping analysis at an electrode consisting of a thin film of Hg plated onto a supporting metal. In one experiment, the potential of the electrode was held at −1.0 V versus SCE for 90 min in a 200-ml sample, by which time the current had dropped to a negligible value. It was then scanned anodically, and three peaks were obtained corresponding to the dissolution of the three metals. Integration indicated 12.5 μC at the potential relating to Cd, 41.0 μC for Pb, and 38.2 μC for Cu. Calculate the concentration of each metal in the effluent, in terms of parts per billion (parts per 10^9).

16-8 A solution 10^{-8} F in Cu^{2+} is to be analyzed by anodic stripping from a hanging mercury drop. If the volume of the drop is 0.0010 cm^3 and the electrolysis current is 0.500 μA, how long must the plating process be continued to give a 10^{-3} F amalgam?

16-9 In Fig. 16-2, if $C = 10$ μF and $V_m = 10$ V, what count would be obtained for 1 mmol of electrons (i.e., 0.001 F) entering the integrator?

REFERENCES

1. Maxwell, J. A., and R. P. Graham: *Chem. Rev.*, **46**:471 (1950).
2. Lingane, J. J.: "Electroanalytical Chemistry," 2d ed., Wiley-Interscience, New York, 1958.
3. Ewing, G. W.: *J. Chem. Educ.*, **49**:A333 (1972).
4. Pareles, S. R.: *Anal. Chem.*, **45**:998 (1973).
5. Eisner, U., J. M. Rottschafer, F. J. Berlandi, and H. B. Mark, Jr.: *Anal. Chem.*, **39**:1466 (1967).
6. Ho, P. P. L., and M. M. Marsh: *Anal. Chem.*, **35**:618 (1963).
7. Eckfeld, E. L., and E. W. Shaffer, Jr.: *Anal. Chem.*, **37**:1534 (1965).
8. Vassos, B. H., R. F. Hirsch, and H. Letterman: *Anal. Chem.*, **45**:792 (1973).
9. Ellis, W. D.: *J. Chem. Educ.*, **50**:A131 (1973).
10. Barendrecht, E.: In A. J. Bard (ed.), "Electroanalytical Chemistry," vol. 2, pp. 53–109, Dekker, New York, 1967.
11. Shain, I.: In I. M. Kolthoff and P. J. Elving (eds.), "Treatise on Analytical Chemistry," pt. I, vol. 4, chap. 50, Wiley-Interscience, New York, 1963.
12. Hargis, L. G., and D. F. Boltz: *Talanta,* **11**:57 (1964).

Chapter 17

Conductimetry

The preceding chapters have dealt with specific electrochemical reactions at electrode surfaces. Nonfaradaic currents have been treated as *noise*, undesirable phenomena tending to obscure the desired information. One nonfaradaic quantity that can carry useful chemical information is electrolytic conductance.

A cell consisting of two platinum electrodes in an ionic solution may be represented by an equivalent electrical circuit made up of resistances and capacitances, as in Fig. 17-1. In this diagram, R_{L1} and R_{L2} represent the resistances of the lead wires, usually negligible, C_{DL1} and C_{DL2} are the double-layer capacitances at the two electrodes, C_P is the interelectrode capacitance (in parallel with the cell), and R_{sol} is the resistance of the solution between electrodes. The components marked Z_1 and Z_2 represent the *faradaic impedances* at the two electrodes, i.e., the electrical equivalent of any possible electrode reactions. Direct current cannot flow through a capacitor, so if a direct current is impressed on this network, except for a brief transient, nothing will happen unless the voltage is high enough to cause reactions, when current will flow through the Z components and R_{sol}.

If an alternating potential is applied, alternating current will flow through the C_{DL}'s and R_{sol}, and at the same time through C_P. Each C_{DL}

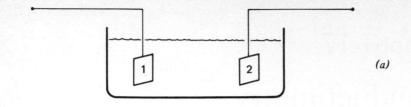

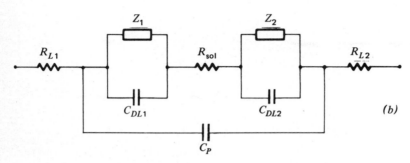

Fig. 17-1   A conductance cell (*a*) and its equivalent circuit (*b*).

provides such an easy path for alternating current that voltage cannot build up across the corresponding Z to the point where faradaic current can flow.   Hence if C_P can be kept negligibly small and the C_{DL}'s large, the effect of R_{sol} can be studied by itself.   In practice the double-layer capacitance can be increased manyfold by coating the platinum electrodes with spongy platinum black (*platinizing*) which greatly increases the surface area. C_P becomes significant only with high-resistance solutions, when large electrodes close together must be selected in order to keep resistance measurements within the range of the instrument.

THEORY

It is convenient to derive a relation between the ionic content of the solution and its *conductance* L, defined as the reciprocal of the resistance.   The conductance is related to the ratio of the area a of the electrodes to the distance d between them (the reciprocal of which is designated the *cell constant* θ), and to the total ionic concentration by the expression

$$L = 10^{-3}\theta^{-1} \sum_i z_i C_i \lambda_i \tag{17-1}$$

where θ is measured in reciprocal centimetres, C_i is the molar concentration of the ith ion with charge z_i, λ_i is the corresponding equivalent ionic conductivity, and the summation covers all ions of both signs.

Table 17-1 Equivalent ionic conductivity at infinite dilution (in units of S·cm²·mol⁻¹ at 25°C)*

Cations	λ^0	Anions	λ^0
H^+	349.8	OH^-	198.6
K^+	73.5	$\frac{1}{4}Fe(CN)_6^{4-}$	110.5
NH_4^+	73.5	$\frac{1}{3}Fe(CN)_6^{3-}$	101.0
$\frac{1}{2}Pb^{2+}$	69.5	$\frac{1}{2}SO_4^{2-}$	80.0
$\frac{1}{3}La^{3+}$	69.5	Br^-	78.1
$\frac{1}{3}Fe^{3+}$	68.0	I^-	76.8
$\frac{1}{2}Ba^{2+}$	63.6	Cl^-	76.4
Ag^+	61.9	NO_3^-	71.4
$\frac{1}{2}Ca^{2+}$	59.5	$\frac{1}{2}CO_3^{2-}$	69.3
$\frac{1}{2}Sr^{2+}$	59.5	$\frac{1}{2}C_2O_4^{2-}$	74.2
$\frac{1}{2}Cu^{2+}$	53.6	ClO_4^-	67.3
$\frac{1}{2}Fe^{2+}$	54.0	HCO_3^-	44.5
$\frac{1}{2}Mg^{2+}$	53.1	$CH_3CO_2^-$	40.9
$\frac{1}{2}Zn^{2+}$	52.8	$HC_2O_4^-$	40.2
Na^+	50.1	$C_6H_5CO_2^-$	32.4
Li^+	38.7		
$(n\text{-}Bu)_4N^+$	19.5		

* Data mainly from Frankenthal (1).

The quantity λ is a property of ions that gives quantitative information concerning their relative contributions to the conductance of a solution. Its value is to some extent dependent on the total ionic concentration of the solution, increasing with increasing dilution. It is convenient to tabulate numerical values of λ^0, the limiting magnitude of λ as the concentration approaches 0 (infinite dilution). This is essentially the same as the *ionic mobility*, sometimes mentioned. Representative values are given in Table 17-1.

The cell constant θ must be known for some applications. As direct measurement is inconvenient except in cells specially designed for the purpose, it is determined by measuring the conductance of solutions of known *specific conductance* ($\kappa = L\theta$). Some values for calibration purposes are given in Table 17-2. The unit of conductance is the *siemens* (symbol S), equal to one reciprocal ohm, sometimes called *mho*.

INSTRUMENTATION

The traditional instrument for electrolytic conductance is the Wheatstone bridge modified for ac operation (Fig. 17-2). The bridge is energized from the source E, at either 1 kHz or the power-line frequency. The arms R_1 and R_2 can be selected by the switch to give precise ratios of 0.1, 1.0, or 10. R_x represents the resistance of the conductance cell, shunted

Table 17-2   Specific conductances of KCl solutions at 25°C (2)

Concentration, $g \cdot kg^{-1}$ of solution	$\kappa,$ $S \cdot cm^{-1}$
71.1352	0.11134
7.41913	0.012856
0.74526	0.0014088

by C_x. R_3 is a precision variable resistor with a calibrated dial, shunted by a small variable capacitor C_3 with which to balance out the capacitance of the cell.

The alternating potential appearing across the diagonal of the bridge is amplified, rectified, and metered. In use, the selector switch and the dial of R_3 are adjusted until the meter shows zero deflection, at which point the resistance of the cell is given by

$$R_x = \frac{R_1}{R_2} R_3 \qquad\qquad (17\text{-}2)$$

Another approach to the measurement of conductance is based on an operational amplifier control circuit (Fig. 17-3). Because of the ability of the amplifier to keep the potentials of its two inputs equal, the ac voltage E_{in} appears directly across the cell, but the current that this produces through it is balanced by an equal current from the amplifier's output

Fig. 17-2 Ac Wheatstone bridge for conductance measurements.   The resistance values are illustrative only.

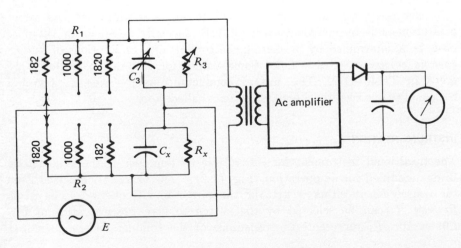

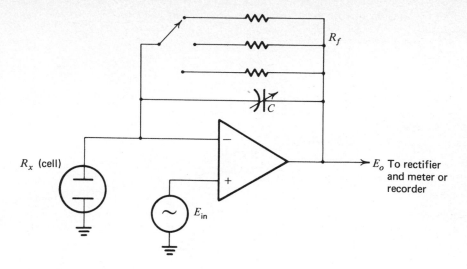

Fig. 17-3 An operational amplifier circuit for electrolytic conductivity measurements.

through a selected feedback resistor R_f. The output voltage E_{out} is easily shown to be

$$E_{out} = E_{in}\left(\frac{R_f}{R_x} + 1\right) = E_{in}(R_f L_x + 1) \qquad (17\text{-}3)$$

where $L_x \equiv 1/R_x$. If R_f is large enough, the 1 can be neglected and the output (ac) voltage will be proportional to the solution conductance. A more elaborate version is described by Mueller et al. (3). Note that this apparatus does not require balancing like a bridge, but gives a continuous reading.

Daum and Nelson (4) have described an interesting apparatus for rapid and precise continuous recording of conductance, suitable for kinetic studies over a wide range of concentrations. They apply in succession two current pulses, equal in magnitude but opposite in sign. Integration of the resulting potential is shown to give a measure of conductance without regard to cell capacitance.

APPLICATIONS

Most applications of conductance measurements have been concerned with aqueous solutions. Water itself is a very poor conductor. Its specific conductance due to dissociation into H_3O^+ and OH^- ions is approximately 5×10^{-8} S $\cdot$ cm^{-1} at 25°C, but ordinary distilled or deionized water falls far short of this. Water stills and demineralizers are often provided with conductance monitors.

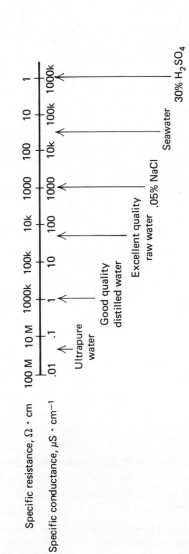

Fig. 17-4 Specific resistance and conductance ranges for some typical solutions. (*Beckman Instruments.*)

Figure 17-4 shows specific resistances and conductances for some typical samples.

Solutions of strong electrolytes show a nearly linear increase in conductance with concentration up to the vicinity of 10 to 20 percent by weight. At higher concentrations the conductance decreases again, as interionic attraction hinders free movement of ions through the solution. Many special-purpose conductance meters are available, calibrated for particular solutes, with scales such as "0 to 30 ppm NaCl," "2 to 25 percent H_2SO_4," or "0 to 10 gr · gal^{-1} seawater."

CONDUCTOMETRIC TITRATIONS

The conductance method can be employed to follow the course of a titration, provided that there is a significant difference in conductance between the original solution and the reagent or the products of reaction. It is not necessary to know the cell constant, since relative values are sufficient to permit locating the equivalence point. It is essential, however, that the spacing of the electrodes does not vary during a titration.

The conductance produced by any ion is proportional to its concentration (at constant temperature), but the conductance of a particular solution will in general not vary linearly with added reagent, because of the dilution effect of the water being added along with reagent. Hydrolysis of reactants or products, or partial solubility of a precipitated product, will also cause departures from linearity.

The shape of a titration curve can be predicted easily. The concentration of each ion at any point in the titration is calculated by the usual methods based on stoichiometry, equilibria, and dilution. The concentration, multiplied by the λ^0 value from Table 17-1, gives the relative contribution of the ion, and summation of all ions present will give a point on the curve.

Conductometric titrations are not used as much as formerly, at least partly because of the lack of specificity of the method.

RADIOFREQUENCY CONDUCTIMETRY

If the frequency of the oscillator powering the conductance meter is raised from the usual 1 kHz into the megahertz region, the necessary instrumentation changes considerably, as does the mathematical theory, but the resulting information is not greatly different. For further details the reader is referred to reviews by Reilley (5) and by Pungor (6).

PROBLEMS

17-1 A conductance cell containing a 0.0100 M solution of potassium chloride is connected in series with a 1000-Ω standard resistor and a battery which has an output of 6.00 V. The current which flows immediately after closing the circuit is shown by an ammeter to be 0.00571 A. By means of Ohm's law, calculate the conductance of the sample. From this and the known specific conductance, compute the cell constant. (Assume a temperature of 25°C.)

17-2 It has been established (7) that the specific conductance κ of seawater at 25°C depends on the salinity S according to the relation

$$\kappa = 1.82 \times 10^{-3}S - 1.28 \times 10^{-5}S^2 + 1.18 \times 10^{-7}S^3$$

where S is expressed in grams of salts per kilogram of seawater. The average value of S for undiluted seawater is 35.

A sample of water taken near the mouth of a river showed a specific conductance of 1.47×10^{-2} S·cm^{-1}. What is its salinity? How much fresh water has been mixed with each kilogram of seawater at this location? (Assume that the conductance of the river water is negligible and the differences in density have no effect.)

The cubic equation can be solved by the method of successive approximations. Calculate a value S' by neglecting the terms in S^2 and S^3. Calculate a second value S'' by substituting S' in the higher terms, and a third value S''' by using S'' in the higher terms. S''' should represent a sufficiently good approximation to the true value for S.

17-3 A 25.00-ml portion of 0.1000 F sodium hydroxide was placed in a conductance cell. It was observed that the electrodes were covered, so no water was added. The sample was titrated with 0.100 F hydrochloric acid. The following data were recorded:

Buret reading, ml	Resistance, Ω	Buret reading, ml	Resistance, Ω
0.00	45.3	30.00	149.9
5.00	60.3	35.00	128.4
10.00	79.6	40.00	114.4
15.00	103.8	45.00	104.9
20.00	137.5	50.00	97.7
25.00	186.9		

For each observation, calculate the conductance, and apply the appropriate correction for dilution. Plot both the corrected and uncorrected values against added volume of acid. Discuss.

17-4 Sketch on one graph the curves you would predict for the conductometric titration of 0.1 N AgNO$_3$ solution by 1 N solutions of each of the following: HCl, KCl, NH$_4$Cl, CaCl$_2$, NaCl, and LiCl. Which would result in the greatest precision? Explain.

17-5 Henry, Hazel, and McNabb (8) have reported the conductometric titration of organic bases by boron tribromide, or vice versa, in such aprotic solvents as nitrobenzene. For example, an inverted V-shaped conductance curve is obtained when boron tribromide in nitrobenzene is titrated by the addition of quinoline. Outline the reactions involved and sketch a titration curve showing graphically the contributions of the various species present to the total conductance. (It will be necessary to refer to the original paper for details.)

REFERENCES

1. Frankenthal, R. P.: In L. Meites (ed.), "Handbook of Analytical Chemistry," p. 5-29 ff., McGraw-Hill, New York, 1963.
2. Jones, G., and B. C. Bradshaw: *J. Am. Chem. Soc.,* **55:**1780 (1933).
3. Mueller, T. R., R. W. Stelzner, D. J. Fisher, and H. C. Jones: *Anal. Chem.,* **37:**13(1965).
4. Daum, P. H., and D. F. Nelson: *Anal. Chem.,* **45:**463 (1973).
5. Reilley, C. N.: In P. Delahay (ed.), "New Instrumental Methods in Electrochemistry," chap. 15, Wiley-Interscience, New York, 1954.
6. Pungor, E.: "Oscillometry and Conductometry," Pergamon, New York, 1965.
7. Ruppin, E.: *Z. anorg. Chem.,* **49:**190 (1908).
8. Henry, M. C., J. F. Hazel, and W. M. McNabb: *Anal. Chim. Acta,* **15:**187 (1956).

Chapter 18

Introduction to Interphase Separations

Separation is not per se an analytical technique, but it is so commonly required prior to analyses that it is a subject of major concern to analytical chemists. It has a valid place in this book, because many of the procedures involve instrumentation which may be highly sophisticated. We will not be concerned with those separation techniques which are primarily intended for preparation or purification of materials.

A large and important class of separation schemes involves the transfer of one or more substances from one phase to another. Hence we can classify separation methods according to the type of phases between which equilibrium takes place (or is approached). There are four such classes: gas-liquid, gas-solid, liquid-liquid, and liquid-solid.

It will be advantageous to establish a general notation to systematize the several analytical situations which may arise.* We will let A and B represent substances to be separated. The two phases involved will be designated by subscripts 1 and 2. Consider the case of a solute A distributed between the two phases; let the equilibrium concentrations be

---

* Different authors use variations on the notation here presented, so that one must be alert to avoid confusion.

designated $C_{1(A)}$ and $C_{2(A)}$. Then we can define a *partition coefficient*,

$$K_{(A)} = \frac{C_{1(A)}}{C_{2(A)}} \tag{18-1}$$

It is convenient to express the concentrations in terms of moles per liter (M/V):

$$K_{(A)} = \frac{M_{1(A)}}{M_{2(A)}} \frac{V_2}{V_1} \tag{18-2}$$

The ratio of numbers of moles in the two solvents can also be expressed as the ratio of $p_{(A)}$ to $q_{(A)}$, where $p_{(A)}$ is the fraction of the total amount of A in solvent 1, and $q_{(A)}$ the corresponding fraction in solvent 2. Thus $p_{(A)} + q_{(A)} = 1$.

By combining these relations it is easily shown that

$$p_{(A)} = \frac{K_{(A)}}{K_{(A)} + V_2/V_1} \tag{18-3}$$

and

$$q_{(A)} = \frac{1}{K_{(A)}(V_1/V_2) + 1} \tag{18-4}$$

It should be noted that the concentrations in the two phases do not necessarily refer to identical chemical species. For example, in the distribution of acetic acid between water (phase 1) and benzene (phase 2), C_1 will refer to the total of ionized and un-ionized aqueous species, and C_2 will include both monomeric and dimeric forms.

The distribution ratio for a given system, even at constant temperature, may not be strictly constant. This is partly because the distribution equilibrium may be affected by competing equilibria within one or both phases, as suggested above for acetic acid, so that K may well change with total concentration. Also, K cannot be taken as a true thermodynamic constant unless activity coefficients are included, and these are usually not readily determinable.

To separate A and B from each other, it is desirable that the partition coefficients be as widely different as possible. We will define α, the *separation factor*, as

$$\alpha = \frac{K_{(A)}}{K_{(B)}} \tag{18-5}$$

Combining Eqs. (18-1) and (18-5) gives

$$\alpha = \frac{C_{1(A)} C_{2(B)}}{C_{2(A)} C_{1(B)}} \tag{18-6}$$

which is independent of volumes.   This ratio should be either much larger or much smaller than unity for best separation.   The separation is most effective if $K_{(A)} \approx 1/K_{(B)}$; this is often not attainable, but can be approached by manipulating the volume ratio.

In practice we are almost invariably concerned with repetitive separations; only rarely will a single equilibrium suffice.   The repetition can be continuous, as in chromatography, or stepwise, as in solvent extraction with separatory funnels, but actually the difference between these approaches is a matter of degree rather than kind.   If the stages in a stepwise procedure are made sufficiently numerous, the net effect is indistinguishable in theory from a truly continuous method.   On the other hand, it is sometimes convenient to treat a continuous separation mathematically as though it were stepwise, using the theoretical-plate concept originated in fractional distillation theory.

Experimentally we can distinguish between *countercurrent* and *crosscurrent* systems.   In the former, both phases are continuously replenished with fresh solvent; an important example is elution chromatography.   In the latter (crosscurrent) only one of the phases is replenished; an example is the repetitive batchwise extraction of a solute by an immiscible solvent.

Complete separation by either the crosscurrent or the countercurrent method is theoretically impossible, but separation can be carried to any required degree, so that high purity can be attained at the expense of yield.

It must be emphasized that the present discussion is general; the mechanism of transfer of the sample between phases may involve ion exchange, surface adsorption, solubility, volatility, or other phenomena.

CROSSCURRENT EXTRACTION

Let us assume that a substance A exists initially solely in phase 2.   A portion of phase 1 is added and the system equilibrated, following which the fractions of the total quantity of A present in the two phases will be $p_{(A)}$ and $q_{(A)}$.   The phases are then separated mechanically, and a new portion of phase 1 added and equilibrated.   The fraction remaining in phase 2 is now $q_{(A)}q_{(A)} = q_{(A)}^2$; the fraction in the combined portions of phase 1 is $1 - q_{(A)}^2$.   The process can be repeated as many times as necessary, say n times.   The fraction remaining in phase 2 and the total fraction extracted become

$$q_{(A)}^n = \frac{1}{(K_{(A)} + 1)^n} \tag{18-7}$$

$$p_{(A),\text{total}} = 1 - q_{(A)}^n \tag{18-8}$$

These relations can be used to determine the number of equilibrations necessary to effect a desired degree of separation, or to determine the separation attainable by a given number of equilibrations, provided that K is known.

If two substances, A and B, both initially in phase 2, are to be separated, it follows that the ratio of the fraction of A remaining in phase 2 to the fraction of B also remaining in phase 2, after n stages, is

$$\left(\frac{q_{(A)}}{q_{(B)}}\right)^n = \frac{K_{(B)} + 1}{K_{(A)} + 1} \tag{18-9}$$

and the ratio of the total amounts passing into phase 1 is

$$\frac{p_{(A),\text{total}}}{p_{(B),\text{total}}} = \frac{1 - q_{(A)}^n}{1 - q_{(B)}^n} \tag{18-10}$$

Consider as an example the stereoisomeric pair, maleic and fumaric acids. The partition coefficients for these compounds between equal volumes of water and ether (each solvent saturated with the other) are:

$$K_{(\text{mal})} = 9.65 \quad \text{and} \quad K_{(\text{fum})} = 0.90$$

giving a separation factor,

$$\alpha = \frac{K_{(\text{mal})}}{K_{(\text{fum})}} = \frac{9.65}{0.90} = 10.7$$

If one starts with an equimolar mixture of the two acids in water and extracts successively with ether, the aqueous concentration of fumaric acid will decrease about 10 times faster than that of maleic acid. Application of Eq. (18-9) predicts that the ratio of concentrations in the aqueous phase after n extractions will be:

n	$C_{\text{mal}}/C_{\text{fum}}$
0	1.00
1	5.60
2	31.6
3	178.
4	1000.

COUNTERCURRENT EXTRACTION

This is a powerful method of separating in a reasonable time substances with more nearly identical distribution ratios. Whereas in crosscurrent methods only one of the phases is continuously renewed, countercurrent techniques call for both to be renewed. One can visualize the sample standing still while phase 1 moves by it in one direction, and phase 2 in the opposite. Components of the sample, if they differ at all in their K values, will have different tendencies to be pulled one way or the other.

It is seldom convenient, at least on a laboratory scale, to have both phases actually mobile; one is usually stationary while the other moves past it. The sample components move also, but at a slower rate than the mobile phase, as their partial affinity for the fixed phase tends to hold them back.

We will first consider stepwise equilibrations. Let us assume that we are dealing with an indefinitely long row of separatory funnels, which we will refer to as *tubes*. We can represent these schematically in Fig. 18-1 as rectangles composed of two segments corresponding to equal volumes of the two phases. The tubes are designated by r *numbers*. Substance A is present initially only in tube $r = 0$, distributed in the ratio of p to q in accordance with Eq. (18-3). (For the moment we will simplify by omitting the subscript A). Now let each segment of phase 1 be moved

Fig. 18-1 Stepwise countercurrent distribution. Equal volumes of two solvents are equilibrated in each tube at each step.

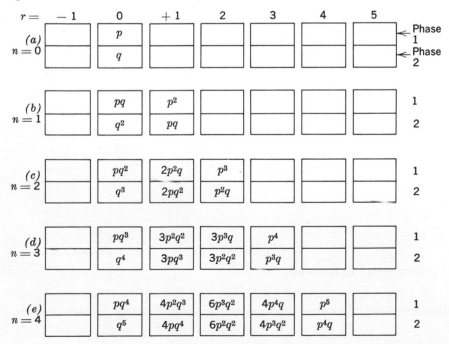

one space to the right with respect to phase 2, as at (b) in the figure. After the transfers, each segment is reequilibrated, with the results shown in terms of p and q; the total amount of A in position $r = 1$ must be p and, when this equilibrates with a fresh portion of phase 2, the ratio in the two phases must be $K = p/q = p^2/qp$. Analogously, in position $r = 0$, the quantity q will distribute itself in the ratio $p/q = pq/q^2$.

If we now repeat the relative motion, we obtain the sequence shown in (c). In this case, the total amount in tube $r = 1$ is the sum of pq (coming from $r = 0$ in phase 1) plus another pq (from $r = 1$ in phase 2), giving $2pq$, which reequilibrates in the ratio of $K = p/q$, or $p(2pq) = 2p^2q$ in phase 1 and $q(2pq) = 2pq^2$ in phase 2.

By similar reasoning, the quantities after further transfers and equilibrations are as shown in (d) and (e) of the same figure. It will be noted that the numerical coefficients are identical to those of the binomial expansion.* The quantities in successive positions of the upper phase (phase 1) are $p(p + q)^n$, and in the lower phase $q(p + q)^n$, where n is the number of transfers, r is the number of the tube, and

$$(p + q)^n = \frac{n!}{r!(n - r)!}\, p^r q^{n-r} \qquad (18\text{-}11)$$

This equation will give directly the total fractional amount $f_{n,r}$ of the corresponding substance in tube r after n equilibrations. It will be more conveniently utilized if we substitute the value of K in terms of p and q, to obtain

$$f_{n,r} = \frac{n!}{r!(n - r)!} \frac{K^r}{(K + 1)^n} \qquad (18\text{-}12)$$

We are now in a position to calculate the relative fractions of two substances A and B present in any tube r after n equilibrations. Let us take, for example, substances for which $K_{(A)} = 0.10$ and $K_{(B)} = 12.0$, and calculate the total fraction of each present in tube $r = 2$ after $n = 10$ transfers. The factorial portion of Eq. (18-12) becomes $10!/2!8!$ which equals 45. The equation then tells us that the total fractional amount of A is $(45)(0.10)^2(1.10)^{-10} = 4.71 \times 10^{-8}$. Thus, given equal quantities of A and B to start with, there will be about 30 million times more A than B in tube 2 after 10 stages.

The figures for the same substances at $r = 8$ after 10 stages show that here there will be about 1 million times more B than A. The fractional amounts of A and B (with the K values specified above) are plotted in Fig. 18-2, as calculated for each value of r, for both $n = 10$ and $n = 5$. Note that, for $n = 10$, positions corresponding to $r = 4$, 5, and 6 contain

---

* For example, $(a + b)^4 = a^4 + 4a^3b + 6a^2b^2 + 4ab^3 + b^4$.

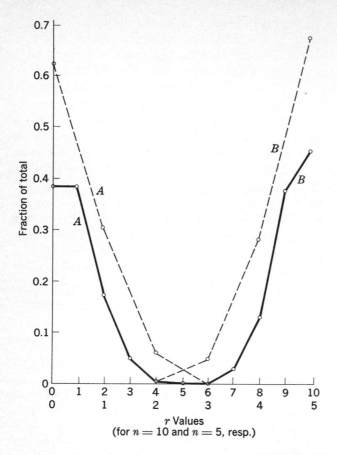

Fig. 18-2 Separation of two substances, A and B. Dashed lines: $n = 5$, separation is nearly complete. Solid lines: $n = 10$, separation is complete. $K_{(A)} = 0.1$; $K_{(B)} = 12$.

no significant quantity of either A or B, so that 10 transfers and equilibra-tions represent a waste of time and effort. Five stages ($n = 5$) would achieve essentially complete separation, since at $r = 2$ there is negligible B, and at $r = 3$, negligible A. The two curves are not quite symmetric, because $K_{(A)}$ was taken (intentionally) not exactly the reciprocal of $K_{(B)}$, though not far from it.

A more complex case is illustrated in Fig. 18-3, the attempted sepa-ration of three substances with $K_{(A)} = 0.90$, $K_{(B)} = 1.15$, and $K_{(C)} = 12.0$, calculated only for $n = 10$. Separation here is not adequate, particularly for A and B, which are not resolved to any useful extent. All three could be separated with a much larger number of stages, but binomial calcu-lations become unwieldy beyond about $n = 20$ or 25.

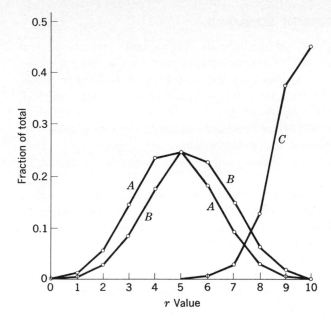

Fig. 18-3 Partial separation of three substances, A, B, and C, for $n = 10$. $K_{(A)} = 0.90$; $K_{(B)} = 1.15$; $K_{(C)} = 12$.

When the number of equilibrations must be greater than this, calculation can be facilitated by use of the *gaussian* or *normal* distribution, which can be considered the limiting form of the binomial distribution as n becomes large. The gaussian distribution can be stated as

$$f_{n,r} = \frac{1}{\sqrt{2\pi pq}} \exp\left(-\frac{(r_{max} - r)^2}{2npq}\right)$$

$$= \frac{K + 1}{\sqrt{2\pi nK}} \exp\left(-\frac{(r_{max} - r)^2 (K + 1)^2}{2nK}\right) \qquad (18\text{-}13)$$

where r_{max}, the value of r corresponding to the maximum in the distribution curve, is given by $r_{max} = np$. The standard deviation σ is $\sqrt{npq}$, wherein it must be noted that n may not always be an integer, since the gaussian equation is derived on the basis of a continuous function. The advantage of this equation is that tabular data are readily available whereby one can determine the fraction of the enclosed area, hence the fraction of the desired substance, between any selected value of r and r_{max}. Note that Eq. (18-13) predicts that for $r = r_{max}$, $f_{n,r} = \sqrt{2\pi npq}$.

CONTINUOUS COUNTERCURRENT SEPARATION

The most significant analytical methods in this category are the several varieties of *chromatography*. Chromatography involves a mobile phase, either liquid or gas, which passes over the surface of a stationary phase, which may be a solid or may be a liquid immobilized by some method, as by absorption on the surface of a solid. The sample is inserted at or near the point where contact is first made between the two phases. Its components are then carried along at various rates depending on their relative affinities for the two phases and, if the experiment is successful, are cleanly separated.*

Chromatographic methods are usually classified according to their most obvious features, such as paper chromatography, ion-exchange chromatography, and gas-liquid chromatography. Sometimes it is not clear exactly what separation mechanism is operative, and we find noncommittal terms such as *column chromatography*.

Whatever the mechanism, the same general mathematical approach can be taken. In our discussion it will be convenient to refer to the stationary phase as the *column*, even though sometimes the "column" will be a sheet of paper, more often called a chromatographic bed.

The sample may be introduced in either phase, as convenient, but always in as small and compact a form as possible, to provide a sharp starting point. As the experiment proceeds, two major effects will be noted: (*1*) the movement of the *zone* or slug of solute with reference to the column, and (*2*) the broadening of the zone. Other effects may also be observed, including decreased symmetry of the zone, which may take the form of *tailing*.

The position of the zone is best observed at its maximum; the width is less easily defined. If the shape of the zone is gaussian (often a good approximation), then its width can be specified in terms of the *standard deviation* σ, defined as half the width of the gaussian curve measured at 0.607 of the maximum height.† The value of 4σ is often taken for the width, since better than 95 percent of the area beneath a gaussian lies within a width of 4σ, and because this distance is easily measured graphically. The relation between the various quantities can be seen in Fig. 18-4. Note that each peak can be approximated by drawing a triangle with sides tangent at the points of inflection of the gaussian, which is just the height where σ is defined. This shows that the distance w is equal to 4σ.

---

* We are referring to elution chromatography only, and will not discuss the less useful frontal and displacement techniques.

† 0.607 = $\sqrt{e}$, where e is the base of natural logarithms; this is consistent with the use of the term standard deviation in the previous section.

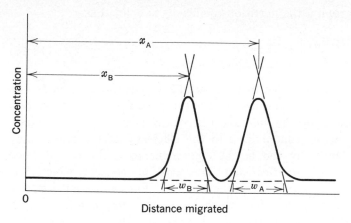

Fig. 18-4   Peak resolution in continuous countercurrent separation.

It can be shown, by both theory and experiment, that the separation between peaks, $x_A - x_B$ in Fig. 18-4, increases in proportion to the distance traveled, while the width of the peak increases as the square root of the distance:

$$\frac{w_A}{w_B} = \left(\frac{x_A}{x_B}\right)^{\frac{1}{2}}$$

(18-14)

From this relation it seems that, by simply increasing the length of the column, any degree of separation could be achieved, since the peaks will draw apart faster than they broaden.   This is true, but of limited usefulness for practical reasons.   The longer the column, the longer the total analysis time, so that a compromise must be reached.   In addition, the peaks may broaden so much that they cannot be measured precisely with respect to their location on the distance or time coordinate; integration beneath the curve also becomes less accurate.

THEORY OF CHROMATOGRAPHIC MIGRATION

The theory of migration is based on the repeated transfer of solute molecules (or ions) back and forth between phases.   Any one molecule (on the average) will spend time t_s in the stationary phase and time t_m in the mobile phase, as it passes through the column.   During the time t_m, it moves forward with the velocity of the carrier v; during time t_s it does not move forward at all.   Its motion, then, is stepwise, as it transfers into and out of the mobile phase.   The relative magnitudes of t_s and t_m will determine how quickly the solute moves along the column.

In chromatography, the partition coefficient is defined as

$$K = \frac{C_s}{C_m} = \frac{M_s}{M_m}\frac{V_m}{V_s} = k\,\frac{V_m}{V_s} \tag{18-15}$$

where subscripts s and m refer to the stationary and mobile phases [cf. Eqs (18-1) and (18-2)]. The constant k is the *partition ratio* or *capacity ratio*, and is dependent on the volumes, whereas K is volume-independent.

The various factors contributing to the efficiency of separation may be treated conveniently by the height equivalent to a theoretical plate (HETP) approach. A *theoretical plate* is a fictitious concept, which does not correspond to any actual entity in the column. For evaluation purposes it is a very convenient parameter, which is its only *raison d'être*. It is defined as that length of column which will yield an effluent in equilibrium with the mean concentration over that length in the stationary phase. For high efficiency, a large number N of theoretical plates is desirable and, to avoid very long columns, the HETP must be as short as possible. Thus the lower the HETP, the more efficient the column.

The number N_x of theoretical plates in a distance x along the column is shown from statistical considerations to be

$$N_x = \frac{x}{H} = \left(\frac{x}{\sigma}\right)^2 = 16\left(\frac{x}{w}\right)^2 \tag{18-16}$$

in which H is the average HETP over the distance x. The value of H relates directly to the width of a peak as recorded on the chromatogram; it is equal to σ^2/x. It is more convenient to measure w than σ on a chromatogram, so we make use of the expression

$$H = \frac{L}{N} = \frac{L\sigma^2}{x^2} = \frac{Lw^2}{16x^2} \tag{18-17}$$

where L is the length of the column.

It is clearly desirable for w, and hence H, to be as small as possible so that recorded peaks will be narrow and sharp. The peak width is adversely affected by many different phenomena occurring in the column itself. For the simpler case of gas-liquid chromatography it has been shown by van Deemter and others that the HETP can be expressed by a relation, the *van Deemter equation*, of the form

$$H = A + \frac{B}{v} + Cv \tag{18-18}$$

where A, B, and C are constants for a given system, and v is the velocity of the carrier gas in centimetres per second.

The A term arises from the fact that not all solute molecules in the mobile phase travel exactly the same distance in passing through the column; this effect is enhanced by nonuniformity of particle size. A is given by the relation

$$A = 2\lambda d_p \qquad (18\text{-}19)$$

where d_p is the average diameter of the solid particles, and λ expresses the irregularity of packing. The B term, which becomes less important as the velocity v increases, has to do with the longitudinal diffusion of the solute within the mobile phase. If the diffusion coefficient is designated by D_g, then

$$B = 2\gamma D_g \qquad (18\text{-}20)$$

where γ is the ratio of the actual velocity of solute molecules down the column to the velocity of the carrier gas v. The B/v term can be lowered by decreasing the temperature or increasing the flow rate. The last term, Cv, which predominates at higher flow rates, is contributed by transverse diffusion in the mobile phase, as from one channel to another, and by the kinetic lag in attaining equilibrium between phases. The pertinent relation is

$$C = \frac{8}{\pi^2} \frac{k}{(1 + k)^2} \frac{d_f{}^2}{D_s} \qquad (18\text{-}21)$$

where k is the partition ratio [see Eq. (18-15)], and d_f is the average thickness of the film of stationary liquid phase.

A fourth term must be added if more precision is required; this involves second-order interactions between the previously mentioned factors.

The van Deemter equation is plotted in Fig. 18-5, to show the qualitative relations between the three terms. There is an optimum flow velocity v_{opt} for the system at which H will be a minimum. In practice this is usually determined by trial and error. The optimum velocity will not be the same for different substances in a mixture, and should be selected for the component most difficult to separate. Chromatographers sometimes choose to use a velocity greater than v_{opt} in order to cut the time of analysis, even though sacrificing some resolution.

In liquid-liquid chromatography, the relation corresponding to the van Deemter equation is somewhat more complex than Eq. (18-18) in that the quantity C is itself made up of several terms.

RETENTION TIME AND VOLUME

In the discussion concerning Fig. 18-4 distance x measured on the recording was treated as an independent variable. This is valid if the recorder paper

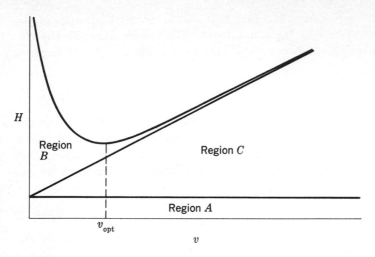

H

Region
B

Region C

Region A

v_{opt}

v

Fig. 18-5   Plot of van Deemter equation.

is driven at constant speed, but clearly time is a more fundamental variable. The time elapsed between the injection of a sample onto the column and the elution of a component (measured at its peak) is defined as the *retention time t_R*.   Often more convenient is an adjusted retention time t'_R, corrected for the time of passage of an unretained component through the column. Air is often satisfactory as such an unretained component: $t'_R = t_R - t_{\text{air}}$. It can be shown that an alternative expression for the partition ratio is

$$k = \frac{t'_R}{t_{\text{air}}} \tag{18-22}$$

Measurements can also be expressed in terms of the *retention volume V_R*, related to the retention time through the flow rate F:*

$$V_R = t_R F \tag{18-23}$$

The retention volume can also be adjusted for measurement from an air peak:

$$V'_R = t_R F - t_{\text{air}} F \tag{18-24}$$

In gas chromatography, correction must be made for the difference in pressure at the inlet and outlet of the column.   (This correction is not necessary

---

* The flow rate F has units of cubic centimetres per second, contrasted with the linear velocity v in centimetres per second, as used in the van Deemter equation.

if the mobile phase is a liquid, since liquids are not compressible.)   A temperature correction is also needed.   These factors can be combined to give

$$V_g = t'_R Fj \frac{273}{T_c} \frac{1}{W_s} \tag{18-25}$$

where j is the pressure correction,* T_c is the temperature of the column (in kelvins), and W_s is the weight of stationary phase in the column. V_g represents the volume of carrier fluid required to move one-half of the solute through a hypothetical column containing 1.00 g of stationary liquid, operating at 0°C, and causing no pressure drop.   One can show that

$$V_g = \frac{273K}{T_c \rho_s} \tag{18-26}$$

where ρ_s is the density of the stationary phase. V_g provides a convenient way in which to report retention data.

RESOLUTION

The resolution is a measure of the degree of separation of adjacent peaks. It is defined as

$$R = \frac{\Delta t_R}{4\sigma} = \frac{x_{(A)} - x_{(B)}}{0.5(w_{(A)} + w_{(B)})} \tag{18-27}$$

where $\Delta t_R = t_{R(A)} - t_{R(B)}$ (see Fig. 18-4).   Figure 18-6 shows the appearance of peaks for several values of R.   Resolution is generally considered complete for $R = 1.5$.

It can be shown that R and N can be related by the equation

$$N = 16R^2 \left(\frac{\alpha}{\alpha - 1}\right)^2 \left(\frac{k + 1}{k}\right)^2 \tag{18-28}$$

This useful relation enables one to determine the number of theoretical plates needed for a desired separation, or the resolution obtainable for a given column.   Note that the resolution, as would be expected, is a function of the separability ratio α, and also of the volume ratio, since $k = K(V_s/V_m)$ [Eq. (18-15)].

* The factor j is given by

$$j = \frac{3[(P_i/P_0)^2 - 1]}{2[(P_i/P_0)^3 - 1]}$$

for gas chromatography.   For liquid chromatography, $j = 1$.

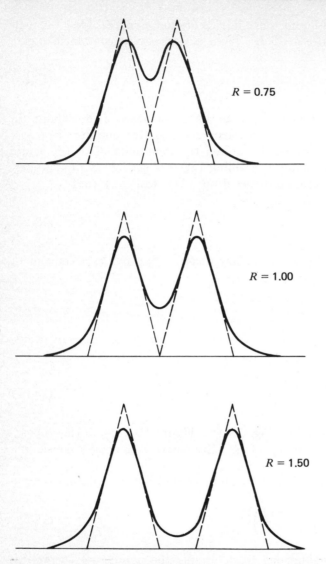

Fig. 18-6 Two peaks of equal height, for which the resolution R is 0.75, 1.0, and 1.5, as marked. The dashed lines are the triangular approximations to the individual peaks.

PROBLEMS

18-1 Copper is extracted by acetylacetone at pH 2 to 5, to the extent of 87.3 percent. How many successive unit extractions are necessary to remove 99.99 percent of the copper from a copper sulfate solution?

18-2 A volume of 100 ml of acetylacetone containing 70.8 mg of beryllium as the acetylacetonate was shaken for 15 min with 100 ml of acidulated water (pH 3) saturated with acetylacetone. The aqueous layer was then separated and analyzed gravimetrically. It was found to contain 1.7 mg of beryllium. Calculate the distribution coefficient for beryllium.

18-3 Apply Eq. (18-11) to the countercurrent extractive separation of maleic and fumaric acids. Calculate the amount of each in each of 10 tubes ($r = 10$) after 10 transfers ($n = 10$). Plot the results in the manner of Figs. 18-2 and 18-3.

18-4 Suppose that, for a particular gas chromatographic system, the van Deemter coefficients are:

$$A = 0.001 \text{ cm}$$
$$B = 5 \text{ cm}^2 \cdot \text{s}^{-1}$$
$$C = 5 \times 10^{-4} \text{ s}$$

Calculate the optimum flow velocity v_{opt} and the corresponding minimum HETP.

GENERAL REFERENCES

Dal Nogare, S., and R. S. Juvet, Jr.: "Gas-Liquid Chromatography: Theory and Practice," Wiley-Interscience, New York, 1962.

Giddings, J. C.: "Dynamics of Chromatography," pt. I, vol. I, Dekker, New York, 1965.

Morris, C. J. O. R., and P. Morris: "Separation Methods in Biochemistry," Wiley-Interscience, New York, 1963.

Rogers, L. B.: In I. M. Kolthoff and P. J. Elving (eds.), "Treatise on Analytical Chemistry," pt. I, vol. 2, chap. 22, Wiley-Interscience, New York, 1961.

Chapter 19

Gas Chromatography

This technique is beyond doubt the most extensively employed for analytical purposes of all the instrumental separation methods, and so merits detailed consideration. It provides a quick and easy way of determining the number of components in a mixture, the presence of impurities in a substance and, in many instances, prima facie evidence of the identity of a compound. The only requirement is some degree of stability at the temperature necessary to maintain the substance in the gas state. Thus a gas chromatograph (GC) is an essential tool for the chemist concerned with the synthesis or characterization of covalent compounds of moderate molecular weight (1, 2).

Figure 19-1 shows schematically the essential parts of a GC. There is a great deal of latitude between a basic unit, which will serve for many identifications, and a highly sophisticated instrument suitable for the varied and stringent requirements of research. We will consider first the physico-chemical systems—the choice of materials for the fixed and mobile phases appropriate for various kinds of samples—and then the detailed instrumental features.

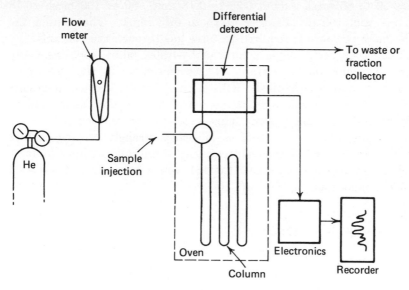

Fig. 19-1 Schematic of a typical GC.

THE STATIONARY PHASE

GC is divided into two subclasses, according to the nature of the stationary phase. In one (called GSC, for *gas-solid chromatography*) the fixed phase consists of a solid material such as granular silica, alumina, or carbon. The separation process involves adsorption on the solid surface. It is quite limited in applicability, largely because of the tailing caused by nonlinear adsorption isotherms, partly because of the difficulty in reproducing surface conditions, and partly because of excessive retention of reactive gases which reduces the available area. Surface catalysis may also play a restricting role. GSC is of chief value in the separation of permanent gases and low-boiling hydrocarbons.

By far the more important class is GLC, *gas-liquid chromatography*, in which the fixed phase is a nonvolatile liquid held as a thin layer on a solid support. The *solid support* ideally should have no effect on the chromatographic process, but only serve as a mechanical matrix for the liquid phase. The most common support is *diatomaceous earth*, also known as *kieselguhr*, which is available in many forms. This highly porous siliceous material is prepared in either of two general ways for chromatographic work. It may be treated with alkali and calcined, giving a white product with some residual alkalinity, or it may be calcined with a binder but without the alkaline flux. In the latter case, the product is red or pink, is somewhat acid, and is generally known as crushed *firebrick*. The particle size should be fairly uniform and not too fine. Typical diameter ranges are 60 to 80 mesh (about 0.25 to 0.18 mm), 80 to 100 mesh (0.18 to 0.15 mm), and 100 to 120 mesh (0.15 to 0.13 mm). The smaller the grains, the more pressure is required to force gas through the columns.

Diatomaceous earth, being a form of hydrated silica, contains many free hydroxyl groups on its surface. These can serve as sites at which solute molecules can be adsorbed. This is undesirable, because it results in a sluggish release of solute from the liquid film to the carrier gas as evidenced by *tailing* of the peaks. This effect can be reduced by the use of a polar liquid phase that itself is adsorbed strongly on the surface of the solid. Often more effective is treatment of the solid with a *silanizing agent* such as hexamethyldisilazane (HMDS), which converts the Si—O—H groups to the innocuous Si—O—Si(CH₃)₃:

$$-Si-O-Si- \;+\; [(CH_3)_3Si]_2NH \rightarrow$$

with —OH groups on both Si, labeled HMDS below the reagent, giving product:

$$-Si----O----Si-$$
with each Si bonded down to O, then to Si, each terminal Si bearing H_3C, CH_3, and CH_3 groups $\quad + NH_3$

Figure 19-2 illustrates these points with chromatograms of mixed polar solvents under several conditions (2). In (1) is shown the result of using the nonpolar liquid Nujol coated on a firebrick support (Chromosorb P); the asymmetry (tailing) is so marked as to make the chromatogram worthless. Changing to the more polar dinonyl phthalate (DNP) at (2) improves the situation somewhat, as does silanizing the support (3). An acceptable recording is only obtained. (4) when both of these changes are introduced.

CAPILLARY COLUMNS

It is possible to eliminate the granular support by using a long capillary of metal, glass, or organic polymer, in which the walls act as support for the stationary liquid phase. Typically dimensions are 2.5 mm in diameter and 100 m in length, with an HETP of less than 1 mm. The advantages are the ability to handle extremely small samples ($<1\ \mu g$), and high efficiency in terms of the large number of theoretical plates that can be contained in a given oven size.

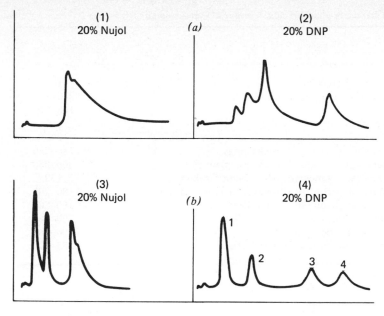

Fig. 19-2 (a) Tailing due to adsorption on Chromosorb-P, and (b) its reduction following silanization. The solutes are: 1, ethanol; 2, methyl ethyl ketone; 3, methyl propyl ketone; and 4, n-butanol. [*Wiley-Interscience* (2).]

THE STATIONARY (LIQUID) PHASE

Hundreds of liquids have been reported as particularly suited to specific separations. These differ for the most part with respect to their degree of polarity and the temperature range over which they are useful. For the majority of applications a limited number of liquids will suffice. Table 19-1 lists the 13 materials that the Perkin-Elmer Corporation has found to provide a versatile selection. The temperatures quoted are the upper useful limits; they also depend on other factors, hence are not to be considered hard restrictions. Thus some detectors will tolerate larger partial pressures of volatile stationary liquid than will others. Also, it may be permissible to heat a column to the upper limit cited or even beyond, if it is held there for only a very short time. The lower limit (not listed) depends on such factors as freezing or greatly increased viscosity.

The *polarity* of the liquid phase is not usually specified in terms of dielectric constant, but rather empirically by its ability to separate appropriate compounds under chromatographic conditions. Nonpolar solutes, such as pentane, butane, and propane, can be resolved easily on a nonpolar liquid such as squalane, whereas their peaks fall much closer together on a column of similar dimensions but containing a polar liquid

Table 19-1   Some stationary-phase liquids, in order of increasing polarity*

Material	Max. temp., °C
1. Squalane ($C_{30}H_{62}$, branched)	150
2. Apiezon-L grease (A.E.I., England)	250–300
3. Didecyl phthalate	165–170
4. Di-(2-ethylhexyl) sebacate	150
5. Methyl silicone oil, low viscosity (DC-200, Dow-Corning)	200
6. Phenyl silicone oil (DC-550, Dow-Corning)	180–220
7. Methyl silicone gum (SE-30, General Electric)	300–350
8. Polyethyleneglycol (Carbowax 1540, Union Carbide)	150
9. Polyalkyleneglycol (Ucon Oil LB-550-X, Union Carbide)	180–200
10. Polyalkyleneglycol (Ucon Oil 50-HB-2000, Union Carbide)	180–200
11. Polyphenylether (OS-138)	200–225
12. Butanediol succinate polyester "BDS"	200–205
13. Diethyleneglycol succinate polyester "DEGS"	205–210

* Perkin-Elmer Corporation.

such as one of the succinates. The converse applies to the separation of polar solutes such as alcohols.

The amount of liquid carried by the solid supporting material is specified in terms of percent *loading* by weight. The usual coating procedure is to dissolve the required amount of the liquid in a volatile solvent, mix it thoroughly with the dried solid support in an open container, and then remove the solvent by evaporation. The solid with its liquid coating has the appearance of a free-flowing sand which can be poured into a long straight metal tube (with tapping or vibrating to promote even packing). The tube, fitted with pressure-type connectors, is coiled loosely *after* filling.

Capillary columns are usually coated by forcing a small amount of a 10 percent solution of the coating material in a volatile solvent through the column.

Another column packing is available which can be considered intermediate between the bare solids of GSC and the coated supports of GLC (3). This is a packing consisting of porous beads of a copolymer of styrene and divinylbenzene. It appears that the components of the sample exchange directly between the gas phase and the porous amorphous beads, the latter acting more like a solvent than an adsorbent. This material gives remarkably clean separations. The maximum permissible temperature is about 250°C.

Any of these columns will usually need further conditioning prior to use. This is accomplished by flushing with nitrogen gas for a few hours at the highest permissible temperature.

CARRIER GAS

By far the most common carrier gas is helium, in spite of its cost. There are two principal reasons for this choice. One of the most useful detectors depends on the thermal conductivity of the gas, a property that is much greater for hydrogen and helium than for any other gases. The other advantage of helium, also shared by hydrogen, is that because of its low density greater flow rates can be employed, thus reducing the time required for a separation. Hydrogen has two drawbacks, its fire and explosion hazard and, more fundamentally, its reactivity toward reducible or unsaturated sample components.* Other gases, such as argon or nitrogen, are required by certain detectors, as will be detailed later.

SAMPLE INJECTION

An outstanding feature of GC is its ability to utilize small samples—from 0.1 to 50 μl of a liquid is usual. There are three methods of inserting samples: by valve, by ampul, and by syringe.

The syringe technique is the most widely used. The device employed is essentially the same as the medical hypodermic syringe, and is available in many calibrated sizes to deliver from 0.1 μl up. The chromatograph is provided with an inlet port closed with a replaceable septum of rubber, neoprene or, especially for high-temperature work, silicone rubber, through which the needle can be inserted. Syringes can be used with gases or with liquids of low viscosity.

The valve method is especially convenient for sampling gas streams. Figure 19-3 shows an example consisting of a pair of identical dual-path stopcocks. In the position shown, the carrier gas is passed through to the column, which is in a standby condition. To take a sample, stopcock no. 1 is turned 90°, so that the reservoir (of calibrated volume) is filled with sample gas. Then no. 1 is returned to its original position, and no. 2 turned through 90°, so that the measured quantity of gas is flushed onto the column. Many ingenious modifications of this device have been constructed, using a single stopcock with multiple openings, or the equivalent valve with a linear sliding motion.

The introduction of samples by ampul is probably the most precise, but least convenient. The sample, cooled if necessary, is sealed in a fragile glass ampul and weighed. The ampul is then inserted into a special heated chamber at the head of the column, where it is crushed mechanically while

---

* Hydrogen has a specific advantage, however, in that it can be generated electrolytically. An ingenious application to a self-contained GC for use in space vehicles has been described (4).

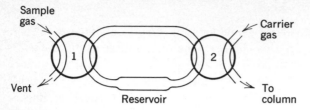

Fig. 19-3   Gas sampling valves.

surrounded by flowing carrier gas.   The temperature is such that the sample is vaporized almost instantaneously and swept onto the column.

In order to obtain narrow peaks it is important to minimize dead spaces in all parts of the apparatus.   Also, it is essential that the entire sample be swept into the gas stream as nearly as possible instantaneously. To this end the inlet chamber is made small in volume and is heated (for liquid samples) well above the boiling point.   Overheating of rubber septa should be avoided, as they can evolve surprisingly large amounts of volatile components to contaminate the sample.

Solids can only be analyzed directly by GC if they have high enough vapor pressure to ensure immediate volatilization in the heated inlet space. Many nonvolatile solids can be decomposed by heating to produce characteristic gaseous products that can be chromatographed.   This is an extensively used procedure, known as *pyrolysis GC* (PGC).   Most commonly the sample is placed directly into a small coil of platinum wire so that it can be heated to several hundred degrees Celsius in a few seconds while the carrier gas is flowing over it.   The pyrolysis products are passed directly onto the column.   Better temperature control can be achieved by the use of a heater of nickel or other magnetic material operated at its Curie point.   This transition point stabilizes the temperature, but operation is limited to a few specific temperatures.   Heating by a pulse of light from a ruby laser has been reported, and shows much promise (5).   Figure 19-4 shows some typical results (6).

Another path by which nonvolatile organic materials can be studied is the formation of volatile chemical derivatives.   An example of this method is the separation of amino acids as *N*-acetyl and *n*-amyl esters on a Carbowax column (7).   Another useful series of derivatives can be prepared by *silylation*, insertion of the TMS group, $-Si(CH_3)_3$, or DMS, $-SiH(CH_3)_2$, in place of reactive hydrogen in compounds containing such functional goups as $-OH$, $-COOH$, $-SH$, $-NH_2$, and $=NH$ (8).   This was first reported for making volatile derivatives of sugars, but is applicable to many other classes of compounds.   Several reagents are available for the formation of these derivatives, including trimethylchlorosil-

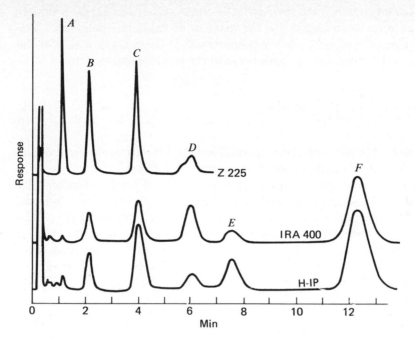

Fig. 19-4 Pyrolysis chromatograms of three ion-exchange resins at 100°C. The column was a polyphenyl ether supported on Celite. The fragments are: *A*, benzene; *B*, toluene; *C*, xylene and ethylbenzene (unresolved); *D*, styrene; *E*, ethyltoluene; and *F*, vinyltoluene. [*Analytical Chemistry* (6).]

ane $(CH_3)_3SiCl$, hexamethyldisilazane, N,O-bis(trimethylsilyl)acetamide, $(CH_3)_3Si$—O—C(CH_3)=N—Si$(CH_3)_3$, and their —SiH$(CH_3)_2$, analogs.

DETECTORS

In principle the measurement of any property that has different values for different gases can be incorporated into a detector for GC, and several dozen have been described (9). As pointed out by Halász, detectors can be classified in two major families (10). In the first family are those that respond to the *concentration* (in mole fraction) of solute in the carrier, whereas those in the second family respond to the *mass flow rate* of the solute (in moles per unit time).

Members of the second family characteristically destroy the sample in the process of detecting it, while those in the first family do not. This can be important, as sometimes it is desirable to collect successive fractions of the solute for further characterization. On the other hand, precise quantitative analysis is more readily implemented with mass flow-rate detectors. Due to the fact that the entire sample component is consumed, the integrated area beneath the signal-time (or signal-volume) curve must correspond exactly to the mass m of substance detected. The height of the curve at any point is proportional to the mass flow rate of the sample $v_s = dm/dt$; hence the area beneath the recorded curve is

$$A = \int v_s \, dt = \int \frac{dm}{dt} \, dt = m \qquad (19\text{-}1)$$

and thus m is obtained directly. Examples of this family are several detectors that depend on combustion of the sample in a flame.

First-family detectors can also give quantitative results, but only by careful control of variables so that unknowns are duplicated by standards, especially with respect to total gas flow rate v (sample plus carrier). The signal from the detector is a measure of the mole fraction x_s of solute, a dimensionless quantity, so the area beneath a peak on the recorded chromatogram is $\int x_s \, dt$. But $x_s = v_s/(v_s + v_c)$, where $v_s + v_c = v$, the sum of the flow rates of sample and carrier. So we can write

$$A = \int x_s \, dt = \int \frac{v_s}{v_s + v_c} \, dt \qquad (19\text{-}2)$$

This is proportional to m only if v is held constant, which is not a simple matter experimentally (flow regulators exert control over v_c only). Very small samples will have negligible effect on v, so that

$$A = \frac{1}{v} \int v_s \, dt = \frac{1}{v} \int \frac{dm}{dt} \, dt = \frac{m}{v} \qquad (19\text{-}3)$$

Hence, with detectors of this type, if the measurement is to be absolute, the measured area must be multiplied by the flow rate. For determinations relative to a standard, this factor can be included in the overall calibration. Concentration detectors include several very useful examples, the thermal conductivity, argon, helium, and electron-capture detectors, as well as those based on the absorption of radiant energy.

FIRST-FAMILY DETECTORS: THE THERMAL CONDUCTIVITY (TC) DETECTOR

The TC detector usually consists of a block of metal with two cylindrical cavities machined into it, each cavity provided with a centrally positioned thin-wire filament or thermistor (11) (Fig. 19-5). These resistive elements form two arms of a Wheatstone bridge (R_1 and R_2 in Fig. 19-6). If the bridge is balanced and the total current indicated by the ammeter A is varied by changing R_6, the galvanometer G will continue to show no deflec-

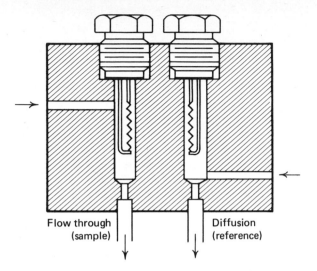

Fig. 19-5 One form of TC detector. The gas on the reference side is allowed to enter the chamber only by diffusion, giving greater stability, whereas on the sample side the gas flows through, giving greater speed of response. (*Gow-Mac Instrument Company.*)

tion.   An increase in current causes a rise in temperature of both R_1 and R_2, and hence a change in their resistance, upward for a metallic wire, downward for a thermistor; the change is equal for both arms, and the bridge remains balanced.   If, however, the gas surrounding one of the resistors is replaced by a different one, the heat developed in the two arms will in general be conducted away through the gases at different rates,

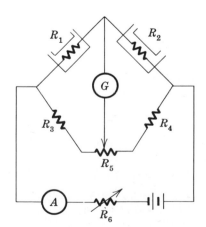

Fig. 19-6 Electric circuit of TC detector. R_1 and R_2 are temperature-sensitive elements, R_3 and R_4 are ratio arms, R_5 is a zero adjustment, and R_6 is a current limiter; in practice the galvanometer G is replaced by an electronic voltmeter.

Table 19-2 Thermal conductivity of a few gases (cal · s^{-1} · cm^{-1} · deg^{-1})

Gas	Thermal conductivity
H$_2$	44.5
He	36.0
Ne	11.6
CH$_4$	8.18
O$_2$	6.35
N$_2$	6.24
CO$_2$	3.96
CH$_3$OH	3.68

the two arms will then be at different temperatures, and the bridge will no longer be balanced. It is assumed of course that R_3 and R_4 are identical, and preferably of low or zero temperature coefficient. Thus the galvanometer can be made to respond to changes in the thermal conductivity of the gas surrounding R_1 or R_2.

Alternatively, the bridge can consist of four filaments or thermistors, all identical, arranged so that the sample surrounds opposed resistors simultaneously (R_1 and R_4, for example) while the reference gas passes over the other two. This doubles the sensitivity.

The TC bridge is operated in a differential manner. One resistor (or opposed pair) is exposed to the carrier gas prior to the introduction of the sample, the other to the column effluent. The detector must be held at a temperature at least as high as that of the column to prevent condensation. Helium and hydrogen conduct heat considerably better than any other gases (see Table 19-2). Therefore one of these is the best choice when the detection is by TC.

TC finds considerable application in industry apart from GC, because the equipment is simple, having no moving parts, and the precision is good. One of the most extensive applications is in the determination of CO$_2$ in flue gases, which provides a direct indication of furnace efficiency.*

THE ELECTRON-CAPTURE (EC) DETECTOR

For many years GC made use of TC detectors almost exclusively, because of their inherent simplicity and wide applicability. With the advent of capillary columns, which are limited to smaller samples, and of trace analy-

---

*The TC unit is sometimes called a *katharometer*. A review of its applications outside of GC is given by Cherry (12), and a detailed analysis of its chromatographic application has been presented by Lawson and Miller (11).

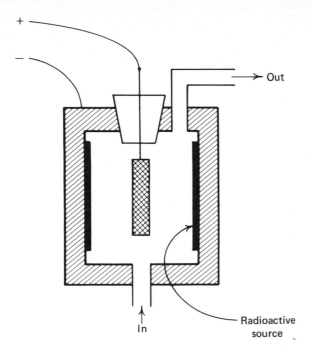

+

−

Out

In

Radioactive
source

Fig. 19-7 EC detector, schematic.

sis generally, greater sensitivity was required than the TC unit could provide.

One method of detection that can give greater sensitivity is a modification of the ionization chamber long used for radiation detection (13). The effluent from the chromatographic column is allowed to flow through such a chamber, which is subjected to a constant flux of beta-ray electrons from a permanently installed radioisotope (Fig. 19-7). A titanium foil containing adsorbed tritium makes the most satisfactory source, though ^{63}Ni can also be used. Both are pure beta sources, which makes for easy shielding against radiation hazard. The ion current through the chamber will be in the nanoampere region; hence a high-impedance, high-gain amplifier (an electrometer) will be required.

The carrier gas used with the electron-capture detector is either argon or nitrogen. Argon tends to be promoted to a metastable state by beta radiation, so a quenching gas such as methane must be added. The sensitivity to organic solutes is dependent on their ionization cross section. The detector is particularly sensitive to halogenated compounds. It was largely because of this detector that the ubiquitous distribution of pesticides in the world environment became evident.

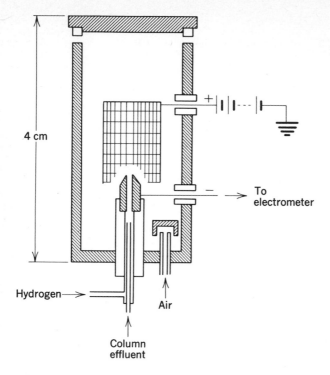

4 cm

+|·|·|⊢ --- |⊢

To electrometer

−

Hydrogen ⟶

Air

Column effluent

Fig. 19-8 Hydrogen flame detector. (*Barber-Colman Company.*)

SECOND-FAMILY DETECTORS: THE FLAME IONIZATION DETECTOR (FID)

Most organic compounds are readily pyrolyzed when introduced into a hydrogen-oxygen flame, and produce ions in the process. The ions can be collected at a charged electrode and the resulting current measured by an electrometer amplifier (14). Figure 19-8 is a diagrammatic sketch of a FID. The carrier gas (usually helium or nitrogen) emerging from the column is mixed with about an equal amount of hydrogen and burned at a metal jet in an atmosphere of air. The jet (or a surrounding ring) is made the negative electrode, and a loop or cylinder of inert metal surrounding the flame is made positive. The sensitivity to organic solutes varies roughly in proportion to the number of carbon atoms; it is perhaps a thousandfold more sensitive than the TC detector. The FID has become one of the most popular detectors, superior for quantitative analysis, since there is no flow dependence.

The FID can be made almost solely responsive to organophosphorus compounds, even in submicrogram amounts, by the incorporation of a block of CsBr or Rb_2SO_4 as part of the structure immediately surrounding the flame (9). The flame is operated in a hydrogen-rich mode, rather than in an oxygen-rich mode as in the unmodified detector. The physical basis on which this detector works has not been entirely elucidated. It is the detector of choice in measuring trace levels of phosphate pesticides.

THE FLAME-PHOTOMETRIC DETECTOR (FPD)

Another selective detector is based on the luminous emission from a hydrogen-rich flame in the presence of compounds containing either sulfur (394 nm) or phosphorus (526 nm) (15). The detector (Fig. 19-9) consists of a hydrogen-air burner with a photomultiplier so located as to see only the upper portion of the flame, which does not emit appreciably in the absence of sulfur and phosphorus. Interchangeable optical filters permit selection of one or the other of these two elements.

Aue and Hill (16) have shown that this device is able to detect heavy metals (Fe, Sn, Pb) in organometallic compounds, with nanogram sensitivity. They monitored the light output and the ionic conductivity simultaneously, with two recording channels.

An important feature in all analytical GC is the dead volume of the system, including especially the detector. With a 6- to 8-mm (inner diameter) packed column, the rate of flow of gas passing through is adequate to keep any ordinary detector flushed out. But with a capillary column, the rate may be too slow, so that the contents of the detector become nearly stagnant, thus broadening the peaks that the column has just cleanly separated. It is essential to design the detector to have as

Fig. 19-9   Flame photometric detector, schematic.

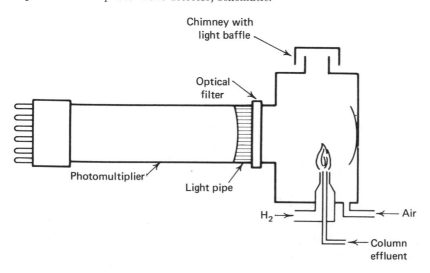

**Table 19-3 Some detectors for gas chromatography:
Comparative data (typical) (9)**

Detector*	D†	Linear range
TC	2×10^{-12} g · ml^{-1}	10^5
FID	2×10^{-12} g · s^{-1}	10^7
ECD (T)	2×10^{-14} g · ml^{-1}	10^4
FID (P)	2×10^{-15} g · s^{-1}	10^3
FPD (P)	2×10^{-12} g · s^{-1}	None
IR	"High μg"	None
MS	"High ng"	10^6

* TC, thermal conductivity; FID, flame ionization detector; ECD (T), electron capture (tritium); FID (P), flame ionization, for phosphorus; FPD (P), flame photometric, for phosphorus; IR, infrared detector; MS, mass spectral detector.
† Detectivity, $D = 2N/S$, where N is an equivalent noise voltage, and S is a sensitivity parameter.

little holdup volume as possible, but there are practical limits to this approach. The problem can usually be eliminated by adding an extra supply of pure carrier gas directly to the detector, to keep it flushed out. This is called *scavenging*.

Table 19-3 gives some comparative data regarding the sensitivity and linear range of a number of detectors (9). The sensitivity figures are the minimum detectable amounts for a signal-to-noise ratio of 2. Those detectors for which no linear range exists must be calibrated against known samples to establish a working curve.

DUAL DETECTION

Since different detectors have different sensitivities for various classes of compounds, added information about samples can often be obtained by the use of two (or even more) detectors simultaneously at the output of the same column. The two detectors may be placed in series, so that the column effluent passes first through one then the other, provided that the first detector is nondestructive. Otherwise, the two detectors may be connected in parallel, with a flow-splitting device to direct part of the gas to one and part to the other.

Figure 19-10 shows an example taken from the work of Hartmann et al. (17), in which dual traces were taken from EC detectors and FIDs. Several peaks can be seen in each trace which are absent or much less

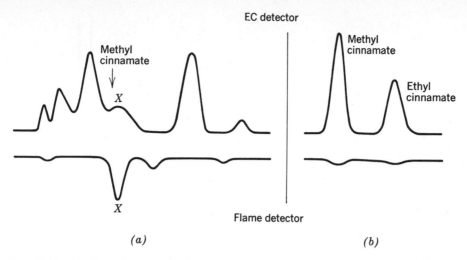

Fig. 19-10 (a) Dual-detection GC of the high-boiling constituents of peppermint oil. Peak X was initially thought to be methyl cinnamate on the basis of the known retention time, indicated by the arrow. The results in (b) on authentic methyl cinnamate show much greater response on the EC detector than on the FID, which rules it out as a source of peak X. (*Varian Aerograph.*)

pronounced in the other. The relative responses of a number of compounds to these two detectors are given in Table 19-4. It is evident that a dual record like this can be of great help in the identification of substances. The fact that a peak occurs in one trace and not in the other can rule some compounds out of consideration, but can never alone prove an identity.

TEMPERATURE PROGRAMMING (18)

In the separation of a number of compounds of similar type but widely varying volatility, a difficulty arises if the experiment is carried out at constant temperature: The low-boiling components are eluted quickly and bunch together on the record chart, while the less volatile species take

**Table 19-4 Approximate ratio of sensitivities:
Electron capture to flame ionization (17)**

Hexane	10^{-6}	Methyl salicylate	1.2
Carvone	0.01	Diethyl maleate	53
Pulegone	0.01	Diacetyl	53
Menthol	0.1	Ethyl cinnamate	65
Ethyl crotonate	0.9	Benzylideneacetone	65
Benzaldehyde	1.0	Cinnamaldehyde	200
Anisaldehyde	1.0	Carbon tetrachloride	10^6

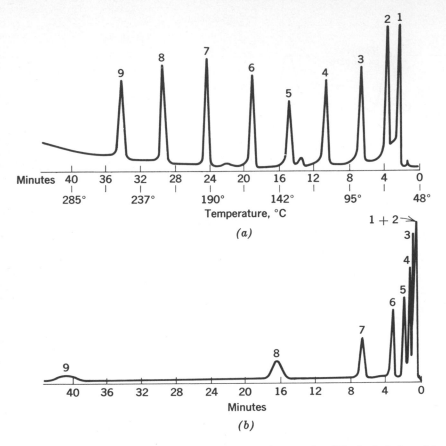

Minutes

Fig. 19-11 The effect of temperature programming on the GC of alcohols. (*a*) Programmed temperature chromatogram; (*b*) isothermal chromatogram. The components are: (1) methanol; (2) ethanol; (3) 1-propanol; (4) 1-butanol; (5) 1-pentanol; (6) cyclohexanol; (7) 1-octanol; (8) 1-decanol; and (9) 1-dodecanol. [*Analytical Chemistry* (19).]

much longer and their peaks are much broader and shallower. This can be overcome by increasing the temperature of the entire column at a uniform rate. When this is done, the partition coefficients of the compounds (temperature-dependent equilibrium constants) change; the result is that the peaks are much more evenly distributed along the chart paper, and are more nearly equal in sharpness. Figure 19-11, taken from the original paper on the subject, illustrates this concept admirably (19).

This figure shows an increase in the base line on going to higher temperatures (to the left in the figure), which is characteristic of programmed-temperature GC. It is due to increased *bleeding* or volatilization of the stationary liquid as the temperature is raised, which would not be evident if the column were held at one temperature. In Fig. 19-11, this effect is not very pronounced, but sometimes it can reduce excessively both sensitivity and accuracy. The effect can be minimized or even eliminated by the use of two parallel columns with identical packing placed together in the same oven. The two columns have identical detectors or, in the case of the TC detector, the effluent from the two columns pass through opposite sides of the same bridge. The carrier gas is split into two streams *before* entering the columns, and the sample is inserted into one side only. The detectors are electrically connected so as to balance out the effect of the bleeding liquid, which is the same in the two columns.

COMMERCIAL GAS CHROMATOGRAPHS

Flexible GC instruments suitable for exacting research are mostly modular in construction. The user can maintain a large inventory of interchangeable columns, and quickly insert any selected one (or pair) as appropriate for a particular sample. One can also select one of several detectors. Temperature-programming controls are built in. The recorder is generally a separate unit, rather than integrated with the main instrument chassis, as is customary with spectrophotometers. This is because there is no scanning feature in GC, which needs to be tied to the independent variable of a recording.

If the chromatograph is intended for a single purpose, as it might well be in a quality-control laboratory, only a single detector and column type need be purchased. There are also several stripped-down chromatographs available, usually equipped only with a TC detector and manual temperature control. These are fully adequate for many routine analyses.

QUALITATIVE ANALYSIS

As suggested above, some degree of qualitative information can be had from observation of the relative sensitivity of various column liquids and various detectors. But this approach will seldom go further than to identify the class to which a compound belongs. For further information one must turn to observation of *retention times* (or *volumes*). This refers to the time elapsed between the injection of the sample and its appearance at the detector. For a given column, flow rate, and temperature, the retention time of a particular gas will be constant, but it is not practicable to transfer such data from one set of conditions to another except by some application of the internal standard principle. Various suggestions have been made as to the best way of doing this.

Relative retention times are often specified. In the determination of these values, a standard substance is added to the mixture prior to running through the column, and retention times are taken relative to this

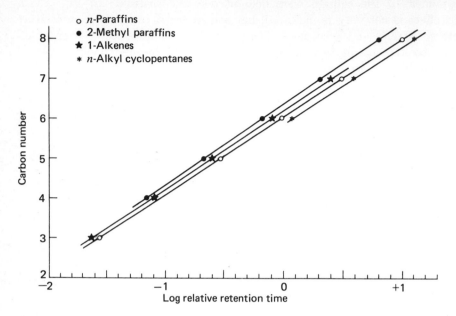

Fig. 19-12   Relative retention time as a function of carbon number for several series of hydrocarbons.

internal standard. n-Pentane is widely used for this purpose but, for work on a polar column at elevated temperature, some other substance, such as methyl palmitate, may be more appropriate.

There are a number of semiempirical relations between relative retention figures and other parameters.   A plot of the logarithm of retention times against the number of carbon atoms in a homologous series of compounds gives a family of parallel lines, as shown in Fig. 19-12 [taken from the data of Ladon and Sandler (20)].   The slope indicates that retention time approximately doubles for the addition of one carbon atom to an aliphatic chain.   Since boiling points also relate to chain length, it is to be expected that they will likewise give linear plots against the log of retention time.   Figure 19-13 shows such a plot for the same compounds as in Fig. 19-12.   Note that certain series that give distinct lines against carbon number merge into a single line when plotted against boiling points.

A decided improvement is the *Kováts retention index* system, which presents a uniform scale rather than a single fixed point for comparison (21).   The retention index I was originally defined in terms of volumes, but we will use the equivalent time expression:

$$I = 100 \frac{\log t_x - \log t_z}{\log t_{z+1} - \log t_z} + 100 \tag{19-4}$$

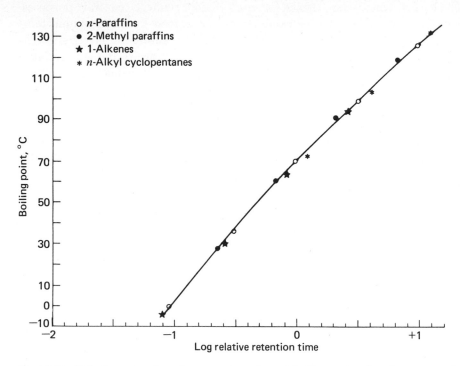

Fig. 19-13 Relative retention time as a function of boiling point for the same series shown in Fig. 19-12.

where t_x is the corrected retention time for substance x (i.e., measured from the air peak), t_z and t_{z+1} are the corresponding times for normal hydrocarbons with z and $z + 1$ carbon atoms, respectively. The logs are taken because this function produces a linear scale for successive hydrocarbons (see Fig. 19-12). This definition requires that t_x lie between t_z and t_{z+1}. The index is nearly linear with temperature, at least over short ranges.

Even more information can be obtained by comparing the Kováts indices for the same series of compounds as separated on polar and nonpolar stationary phases. Index differences (ΔI) from one column to the other can yield supporting data concerning molecular structure. The details of the method can be found in a report by Schomburg (**22**), who has applied it to a series of 156 hydrocarbons containing the cyclopropane ring.

SIMULATED DISTILLATION

GC has found a valuable field of application as a replacement for analytical distillation, particularly in the petroleum field. A fractional distillation, as conventionally carried out with high precision, takes something like

100 h for completion, hence is useless for refinery control purposes. Equally good or even better results can be obtained by dual-column, temperature-programmed GC in only 1 h (23). An automatic electronic system continuously integrates the detector signal and prints the accumulated total at intervals of a few seconds. These data, plotted against column temperature, give a curve identical in shape to that produced by the 100-h distillation. A correction must be made to the temperature scale, if true boiling points are required, because the partial pressure of a component in the carrier gas is not 1 atm, as required in the definition of boiling point. Several manufacturers have distillation simulators in their product lines.

QUANTITATIVE ANALYSIS

It is possible to collect sample components as they are eluted from the column or following passage through any nondestructive detector. The sample collector may take many forms, from a test tube standing in a paper cup full of ice, to elaborate automated and refrigerated sample collectors. The collected substances can be weighed directly or analyzed by any appropriate method. This is not usually done simply for quantitation, but for further study and identification of unknown components.

For known substances, quantitative determinations are generally performed on the recorded chromatogram. If the peaks are sharp and narrow, little error results from simple height measurement. For broader peaks the included area must be determined. The simplest way is to make use of the triangular approximation of the gaussian curve, mentioned in the preceding chapter, whereby the area is equated to the height divided by half the base width 2σ (Fig. 18-4).

The recorders provided with many GCs have built-in integrators that print directly along the edge of the chart a series of spikes proportional in frequency to the area beneath the curve. Counting the spikes associated with a peak gives a measure of the corresponding material.

GC AS A MEMBER OF A TEAM

GC can play a valuable role in combination with any other instrumental technique that can accept gaseous or volatile liquid samples and that is compatible in speed. The most important are mass spectrometry and infrared spectrophotometry.

In the case of infrared, the effluent gas stream can be led directly through a gas cell in a rapid-scan (24) or interferometric (25) spectrophotometer. The optical system of a scanning instrument must be of special design to ensure that sufficient radiant flux can pass through a long, narrow

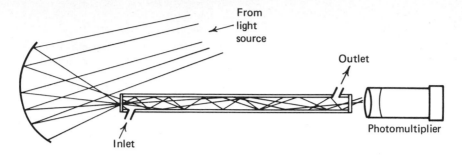

Fig. 19-14 Absorption cell, using multiple reflections, as a GC detector.

sample space with minimum dead volume. (One commercial model has 30-cm cells only 5 mm in inside diameter.) Collecting adequate radiation into a narrow beam requires expensive compound mirrors or their equivalent. An alternative possibility is to focus the radiation on the end of the cell, as in Fig. 19-14. The cell acts as a light pipe, leading the radiation through the sample to the detector, with multiple reflections at the walls.

The spectra obtained as just described show rotational fine structure not discernible in the liquid phase. Hence they are not directly comparable to most library spectra. This defect has been overcome (26) by using a liquid film as sample medium. The film consists of a nonvolatile liquid, which may or may not be the same as that used for the stationary phase of the column. The cell then becomes effectively an extension of the column (Fig. 19-15). Each component in turn is captured momentarily by the portion of the liquid film in the path of the infrared, and then eluted as more of the carrier gas flows by. The resulting spectra show the usual liquid-phase characteristics.

The combination of GC with mass spectral techniques is especially fruitful, since the mass spectrometer is at its best when presented with small gas-phase samples, and the effluent from the GC consists of a sequence of just such samples. The major problem in interfacing the two is the need for stripping away the carrier gas, which would swamp the pumping system in conventional mass spectrometers. Further discussion will be found in Chap. 21.

Fig. 19-15 Infrared GC detector forming an extension of the column.

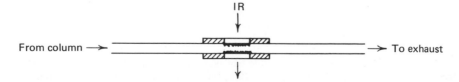

PROBLEMS

19-1 It is sometimes advantageous to analyze a hydrocarbon mixture quantitatively by oxidizing the column effluent prior to detection. The water vapor formed can be removed by a trap, and the CO_2 admitted to the TC detector. In a particular experiment, seven components gave peaks with integrated areas as follows:

Peak no.	Compound	Relative area
1	n-Pentane	2.00
2	n-Hexane	5.72
3	3-Methylhexane	2.21
4	n-Heptane	8.15
5	2,2,4-Trimethylpentane	1.92
6	Toluene	3.16
7	n-Octane	5.05

(*a*) Point out some advantages and disadvantages of this preoxidation procedure. (*b*) Should the oxidation step precede or follow passage through the column, and why? (*c*) Compute the composition of the sample giving rise to the data cited, in terms of mole percent of total hydrocarbons.

19-2 A chromatogram shows peaks as follows, in terms of distance from the injection point, measured on the recording paper:

Air	2.2 cm
n-Hexane	8.5 cm
Cyclohexane	14.6 cm
n-Heptane	15.9 cm
Toluene	18.7 cm
n-Octane	31.5 cm

Calculate the Kováts indices for toluene and cyclohexane.

19-3 Two components, A and B, were found, in a particular gas chromatogram, to give peaks at 5.0 and 7.0 cm on the chart paper, measured from the point of injection. An air peak appeared at 1.0 cm. Other pertinent data were:

Chart speed	6 cm · min^{-1}
Flow rate	$F = 10$ cm^3 · s^{-1}
Pressure at top of column	$P_i = 2$ atm
Pressure at bottom	$P_0 = 1$ atm
Column temperature	$T_c = 100°C$
Weight of stationary phase	$W_s = 60$ g
Density of stationary phase	$\rho_s = 2$ g · cm^{-3}

For each component, calculate K and V_g.

19-4 Submicrogram quantities of cyanide can be determined by GC analysis following conversion to cyanogen chloride, CNCl (27). The reagent is aqueous chloramine-T (sodium p-toluenesulfonchloramide). The product CNCl is extracted with hexane and injected onto a GC column. The carrier gas is argon with 5 percent CH_4. The EC detector gives a retention time of 2.4 min for CNCl and 4.8 min for hexane, under the particular conditions reported. The hexane peak serves as an internal standard. Calibration showed an average slope of 2.53 ml $\cdot$ μg^{-1} for a plot of the peak area ratio (CNCl/hexane) against concentration of cyanide ion. A 1-ml sample of blood (from the coroner's office) was processed as described; a CNCl peak with an area of 31.61 units was recorded, along with a hexane peak of 0.2333 unit. What was the cyanide content of the blood, in micrograms per milliliter?

REFERENCES

1. Bennett, C. E., S. Dal Nogare, and L. W. Safranski: Chromatography: Gas, in I. M. Kolthoff and P. J. Elving (eds.), "Treatise on Analytical Chemistry," pt. I, vol. 3, chap. 37, Wiley-Interscience, New York, 1961.
2. Dal Nogare, S., and R. S. Juvet, Jr.: "Gas-Liquid Chromatography," Wiley-Interscience, New York, 1962.
3. Hollis, O. L.: *Anal. Chem.*, **38**:309 (1966).
4. Stevens, M. R., C. E. Giffen, G. R. Shoemake, and P. G. Simmonds: *Rev. Sci. Instrum.*, **43**:1530 (1972).
5. Vanderborgh, N. E., and W. T. Ristau: *Am. Lab.*, **5**(5):41 (1973).
6. Parrish, J. R.: *Anal. Chem.*, **45**:1659 (1973).
7. Johnson, D. E., S. J. Scott, and A. Meister: *Anal. Chem.*, **33**:669 (1961).
8. Sweeley, C. C., R. Bentley, M. Makita, and W. W. Wells: *J. Am. Chem. Soc.*, **85**:2497 (1963).
9. Hartmann, C. H.: *Anal. Chem.*, **43**(2):113A (1971).
10. Halász, I.: *Anal. Chem.*, **36**:1428 (1964).
11. Lawson, A. E., Jr., and J. M. Miller: *J. Gas Chromatogr.*, **4**:273 (1966).
12. Cherry, R. H.: Thermal Conductivity Gas Analysis, in D. M. Considine (ed.), "Process Instruments and Controls Handbook," p. 6-186, McGraw-Hill, New York, 1957.
13. Lovelock, J. E., G. R. Shoemake, and A. Zlatkis: *Anal. Chem.*, **36**:1410 (1964).
14. Sternberg, J. C., W. S. Gallaway, and D. T. L. Jones: In N. Brenner (ed.), "Gas Chromatography, 3rd International Symposium, 1961," p. 231, Academic, New York, 1962.
15. Brody, S. S., and J. E. Chaney: *J. Gas Chromatogr.*, **4**:42 (1966).
16. Aue, W. A., and H. H. Hill, Jr.: *Anal. Chem.*, **45**:729 (1973).
17. Hartmann, C. H., et al.: *Aerograph Res. Notes,* Varian Aerograph, Walnut Creek, Calif., fall, 1963; spring, 1966.
18. Harris, W. E., and H. Habgood: "Programmed Temperature Gas Chromatography," Wiley, New York, 1966.
19. Dal Nogare, S., and C. E. Bennett: *Anal. Chem.*, **30**:1155 (1958).
20. Ladon, A. W., and S. Sandler: *Anal. Chem.*, **45**:921 (1973).
21. Ettre, L. S.: *Anal. Chem.*, **36**(8):31A (1964).
22. Schomburg, G., and G. Dielmann, *Anal. Chem.*, **45**:1647 (1973).
23. Green, L. E., L. J. Schmauch, and J. C. Worman, *Anal. Chem.*, **36**:1512 (1964).
24. Penzias, G. J.: *Anal. Chem.*, **45**:890 (1973).
25. Kizer, K. L.: *Am. Lab.*, **5**(6):40 (1973).
26. Lephardt, J. O., and B. J. Bulkin: *Anal. Chem.*, **45**:706 (1973).
27. Valentour, J. C., V. Aggarwal, and I. Sunshine: *Anal. Chem.*, **46**:924 (1974).

Chapter 20

Liquid Chromatography

In this chapter we shall consider those forms of chromatography employing a liquid rather than a gas as the moving phase (1–3). The stationary phase may be a solid or a liquid on a solid support. The mechanisms responsible for distribution between phases include surface adsorption, ion exchange, relative solubilities, and steric effects.

The earliest form of chromatography to be seriously studied as an analytical tool was the separation of natural materials, such as plant pigments, from solution. Liquid chromatography (LC) was eclipsed for many years because of the tremendous success of its gas analog. Recently, however, it has met with renewed interest as a result of the application of modern design principles.

In the older technique a glass column, perhaps 1 by 30 cm, was filled with a suitable granular solid, with or without an immobilized liquid film, and the carrier liquid with the sample was poured through. The chief difficulty was slow speed. If the granules were small enough to give good separation, then the delivery under gravity alone might decrease to a few drops per minute. The obvious way to increase the throughput is to force the liquid by a positive-displacement pump or by gas pressure. Older

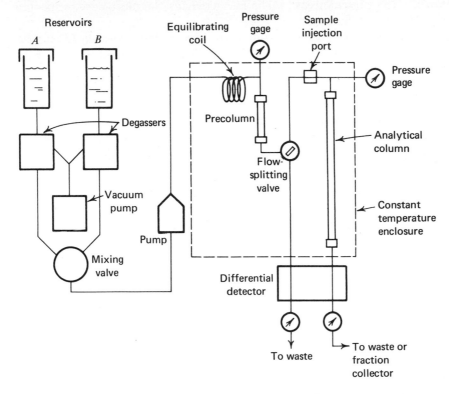

Fig. 20-1   Flow diagram for a typical liquid chromatograph.

apparatus could not take high pressure safely, so major developments had to await a complete redesign.

LC is apt to be somewhat more cumbersome than GC, but it has the major advantage of operating at a lower temperature, the range being obviously limited by the freezing and boiling points of the solvent. This means that LC is a viable technique for the separation of such substances as proteins and nucleosides that cannot survive the elevated temperature that would be necessary to volatilize them.

As an example of the kind of progress that has been made in LC, consider the separation of nucleosides (1). Uridine, guanosine, adenosine, and cytidine could be cleanly separated in 1967 in a 6-mm diameter column, at a pressure of 10 to 20 psig* in 60 min. Two years later the same separation in a 1-mm column at 400 psig took 24 min and, in 1970, in the same column, at 5000 psig, only 1.25 min was required. Clearly, apparatus to work at such high pressure must involve highly specialized design

---

* psig = pounds per square inch, gage, i.e., above atmospheric pressure. 1 psig = 51.7 torr = 6.90×10^3 pascals.

features. The present treatment is restricted for the most part to modern high-pressure systems.

In Fig. 20-1 is shown a flow diagram of a typical versatile LC instrument. Two reservoirs (A and B) of solvents* are provided, each with a degassing chamber to remove dissolved air. The two liquids are mixed (or either of them taken singly) by proper positioning of a mixing valve. This valve can be programmed, as with a geared-down motor drive, so that the proportions of the two solvents can be changed during the progress of a chromatographic separation if desired. The mixed liquid is pumped at high pressure into the chromatograph proper, enclosed within a controlled-temperature chamber. The liquid first passes through a coil of tubing to bring it to the working temperature, and then through a *precolumn* loaded with the same solid support and stationary liquid phase as that of the main column. The purpose of the precolumn is to ensure that the mobile liquid is saturated with fixed phase liquid, thus preventing any removal of the latter from the analytical column. The sample is introduced between the pre- and main columns, either by syringe or by a valve, as in GC. The column effluent passes through one side of a differential detector to a fraction collector or to waste. The reference material for the detector consists of a portion of the solvent taken from the main stream just prior to sample injection. The more critical of these components will be treated separately in the following sections.

PUMPS

The high pressure that must be applied to the liquid to force it through the column at a satisfactory rate can be obtained either by a motor-driven pump or by pressure transfer from a cylinder of compressed nitrogen. The latter requires a diaphragm or piston to keep the nitrogen from direct contact with the liquid; otherwise considerable amounts would dissolve, to reappear as bubbles in the column and detector as the pressure becomes less. (The production of bubbles is still likely, because of the normal dissolution of air in the liquid, and necessitates the use of a degasser before the pump, where dissolved gas is removed by heat and partial vacuum.)

More widely used is a motor-driven pump, either reciprocating or single-stroke, as with a syringe. Most LC detectors are of the *concentration* or *first-family* type, and are therefore flow-sensitive. Hence the pump must be able to deliver liquid at a constant rate, without pulsations. Manufacturers use various ingenious methods of achieving this condition (3b). One way is to provide two cylinders and pistons, so timed that one fills quickly while the other delivers slowly, and then takes over the delivery function as its mate becomes empty, after which the cycle repeats.

* The reason for providing two solvent reservoirs will be made evident in a later paragraph.

In this way flow can be held constant for longer periods than a single cylinder of reasonable size could accommodate.

COLUMNS

For LC, columns are typically of 1- to 6-mm inside diameter, 30 to 100 cm in length, and of stainless steel or glass. Short columns are straight, but longer ones up to the maximum feasible length (about 600 cm) are coiled. The solid support for an immobile liquid phase (LLC) can be diatomaceous earth (e.g., firebrick) which is porous, or material such as Du Pont's Zipax or Waters' Corasil, which consists of particles with hard cores and a porous surface layer. The former give large sample capacity, but diffusion of solutes into the liquid-filled pores slows down the separation; the latter type provides high-speed capability at the expense of the size of sample that can be handled. For adsorption chromatography (LSC) the solids most commonly used are silica gel or alumina, or such proprietary materials as Zipax and Corasil (uncoated).

LC columns are frequently operated at room temperature. The rate of diffusion and hence speed of separation can be increased somewhat by higher-temperature operation, so a heating jacket or oven is sometimes required. Temperature control is not nearly as significant as in GC.

GRADIENT ELUTION

The role played in GC by temperature programming is taken over in LC by a continuous alteration in the composition of the liquid phase, known as *gradient elution* or *solvent programming*. This causes a change in the partition coefficients of the components, and can lead to improved resolution. The composition is gradually changed during the progress of a separation by means of the programmed mixing valve of Fig. 20-1. The two liquids to be combined can be miscible solvents of differing polarity (4) (but neither can be a solvent for the fixed-phase liquid), or they can be solutions of varying ionic strength. Other composition variables, such as pH, may be appropriate in special situations. Gradient elution is used to advantage in many LC applications; it is essential in high-pressure, ion-exchange chromatography.

An example of a separation with and without gradient elution is reproduced in Fig. 20-2. Elution without this gradient feature (with a moving phase of constant composition) is called *isocratic*, and is analogous to *isothermal* elution in GC.

DETECTORS

Many more modes of detection are used in LC than in GC, because there is no universal detector comparable to GC's TC or FID detectors.

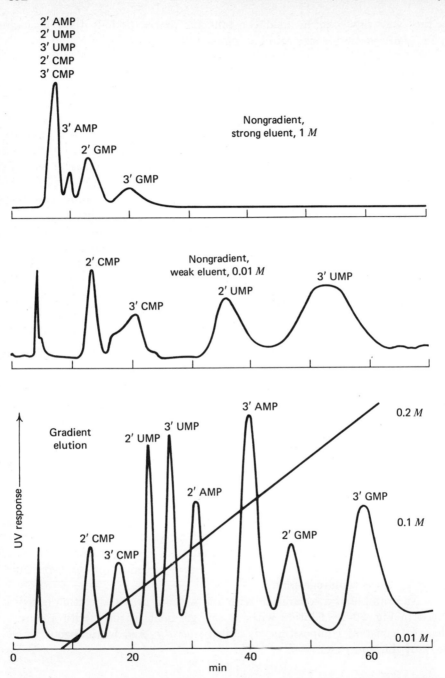

Fig. 20-2 Comparison of gradient and isocratic elution in the separation of nucleotides. The sloping line in the lower chart represents the concentration of salts in the eluting liquid, against the molarity scale on the right. The abbreviations are the standard designations of various nucleotides. [*Varian Aerograph* (1).]

PHOTOMETRIC DETECTORS

The photometric detector depends on the measurement of ultraviolet or visible absorption, and is one of the most popular modes of detection presently used in LC. The principal difficulty is in designing the optical system so that enough radiant energy can be made to pass through the narrow absorption cells. The cells are limited to a diameter of about 1 mm to avoid dead space; otherwise passage through the cell would degrade the separation. A typical photometer is diagrammed in Fig. 20-3. The sample and reference cells consist of cylindrical channels cut in a block of stainless steel or Teflon. The channels are closed by fused silica windows. Ultraviolet radiation from the lamp, condensed by a lens, passes through the two cells and a filter to dual photocells. The source is commonly a low-pressure mercury arc lamp, which gives most of its output at 254 nm. Some manufacturers provide for alternative illumination at 280 nm by fluorescence; a silica plate or rod is coated with a suitable phosphor and inserted between lamp and cell, to convert 254 nm to the longer wavelength. Many compounds of biochemical interest absorb at 254 or 280 nm, hence these detectors are appropriate.

Several manufacturers combine the principles of dual-wavelength photometry and LC detection, with simultaneous measurements at 254 and 280 nm. The electronics permits plotting absorbance at both wavelengths or the ratio of the two, as desired, against the time axis.

For more general usefulness, of course, the continuously selectable wavelength available with a spectrophotometer is preferable. This provides an additional degree of selectivity not possessed by other LC detec-

Fig. 20-3 Typical photometric LC detector. L, lens; W_1 and W_2, ultraviolet-transmitting windows; F, optical filter; D_1 and D_2, photocells.

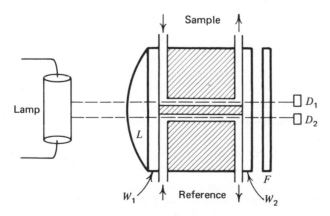

tors, in that one can discriminate against unwanted components by varying the wavelength. A number of spectrophotometers are adaptable to LC service. Those instruments designed specifically for this use have the necessary intense light source and condensing optics to give adequate sensitivity, but this is not always true for standard instruments with LC accessories. The spectrophotometer can be set to monitor any selected wavelength, or the flow can be stopped temporarily while a spectrum is scanned.

REFRACTIVE INDEX DETECTORS

Another possible detector for LC is the *differential refractometer*. Differences in the index of refraction can be measured to 1 part in 10^6 or better, and this corresponds to a few parts per million of an organic solute in water (this estimate is based on handbook values for maltose at 25°C). This provides a detection method for LC that is nonspecific and depends for sensitivity solely on the considerable difference in the measured quantity between solutes and solvent. The advantage of refractive index detection is that it can be employed with solvents whose physical properties (such as ultraviolet absorption) might interfere with other modes of detection. The disadvantage is that it cannot be used in gradient elution, because the change in index due to solvent programming completely overwhelms any signals produced by the eluted components.

Figure 20-4 shows the principles of operation of two refractometric detectors. The form in (*a*) is based on the angular displacement of a light beam on passing through two liquid-filled prisms. If the two liquids are identical, the displacement will be nil, but even a change of a few parts per million in the refractive index will make a detectable difference in the angle of the beam. The beam is split laterally between two photocells, and the difference between their signals is recorded.

The detection system in (*b*) depends on the relative intensity of light reflected from and transmitted through an interface between two transparent media. The sample and reference streams are led through a pair of narrow passages between one face of a 45° glass prism and a slightly roughened stainless-steel plate. A collimated beam of light enters the prism as shown; it is partly reflected at the interface with the liquid, and partly transmitted to produce a lighted spot on the steel back plate. The illumination of the steel plate is observed through a telescope fitted with a photocell. It can be shown mathematically that, provided the angle of incidence is chosen correctly (a little less than the critical angle of total reflection), then the power of the observed radiation will depend linearly on the index of refraction of the liquid. The light source and focusing optics must be mounted on an arm that can be rotated through a small angle around the prism to make the initial adjustment. It should be pointed out that, in this detector, the designer had the option of using

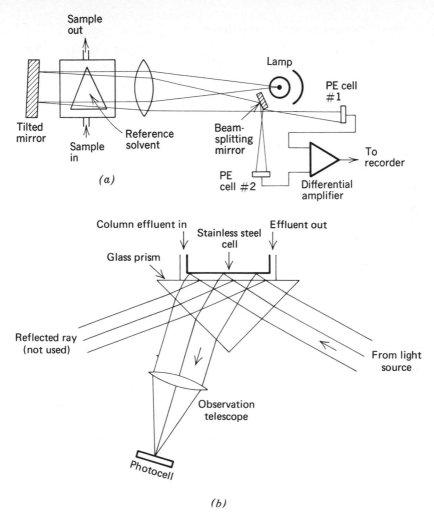

Fig. 20-4 Refractometric LC detectors. (*a*) Waters Associates; (*b*) Laboratory Data Control.

either the transmitted light (as he actually chose) or the light reflected from the interface, as both of these quantities are related to the refractive index. The other choice would give an inverse rather than a direct relationship, and in addition the observing telescope would have to be movable so as always to intercept the maximum reflected light.

These two refractometric detectors are about equal in sensitivity, being able to detect a change in index of 10^{-7}. The deflection type (Fig. 20-4*a*) is more convenient to operate, but has a smaller linear range than that of Fig. 20-4*b*.

MOVING-WIRE (FLAME-IONIZATION) DETECTORS

Several attempts have been made to adapt GC detectors for use in LC. The problem, of course, is to eliminate the solvent. The Pye chromatograph* makes use of the mechanism shown in Fig. 20-5, as a detector for organic compounds (5). A stainless-steel wire, perhaps 0.1 mm in diameter, serves to transport the column effluent through a series of manipulations. The wire is first cleaned and superficially oxidized by heating in a stream of air. It is then coated with liquid from the column, as a thin film (for example, 10 μl may be spread over 4.5 m of wire). The wet wire passes through a small oven to vaporize the solvent, and then into an oxidizing oven where the sample is burned to CO_2 and H_2O. The product CO_2 is entrained with mixed H_2 and N_2 and passed over a Ni catalyst which reduces it to CH_4. The CH_4 is led directly into a FID. The linear range of this detector system is better than 1000 to 1, and the limit of detection is of the order of 1 $\mu g \cdot ml^{-1}$.

The advantage of the moving-wire detector is its unique adaptability to gradient elution; the disadvantage is that the detector is mechanically complex and thus prone to malfunction.

OTHER DETECTORS

There have been many other detector types reported (3). Related to ultraviolet absorption detectors are those based on fluorescence measurements.

* Pye Unicam, Ltd., via Philips Electronic Instruments.

Fig. 20-5 Moving-wire, flame-ionization LC detector. [*Journal of Chromatographic Science* (5).]

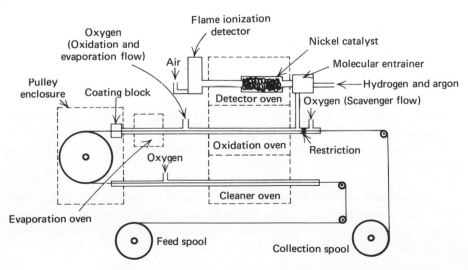

This is especially important for the many biochemical applications of LC. Excitation is usually with the 254-nm line of the mercury arc. This detector is highly selective and sensitive. The user must be alert to the possibility of quenching by other solution components or by dissolved oxygen.

Still another technique for identifying and measuring eluates is by a *reaction detector* (3b). The liquid stream from the column is mixed continuously with a stream containing a reagent that will produce a chromophoric or fluorescent product with expected samples. The combined flow is then passed through an appropriate optical detector. An application of this principle is the use of the reagent fluorescamine (Fluram) in the analysis of proteins and amino acids (6). This polynuclear aromatic compound reacts with primary amines to give products that fluoresce strongly at 475 nm upon excitation at 390 nm.

A detector based on the continuous measurement of electrolytic conductivity plays a unique role in LC. It is used not primarily for detecting eluted ionic compounds but rather for monitoring the ionic-strength gradient frequently used in high-pressure, ion-exchange chromatography.

Radioactivity measurements are appropriate for studies of naturally or artificially active materials (7, 8).

LIQUID-SOLID CHROMATOGRAPHY (LSC)

This form of LC depends on the relative strengths of adsorption of solute molecules on hydroxyl group sites on silica or alumina substrates without the presence of a fixed liquid phase. The adsorption is generally characteristic of functional groups in organic substances, hence is particularly useful in separating classes of compounds. There may also be steric effects, so that positional isomers, such as cis-trans pairs, can sometimes be resolved.

A major difficulty in quantitative separations is the nonlinearity of the adsorption isotherms encountered as the concentration of solute is raised. It is desirable that the linear region (i.e., where the amount adsorbed is proportional to concentration) extend as far as possible. In practice it is found that this linear region is longer for weakly adsorbed species than for those that are strongly adsorbed. This is interpreted as indicative of discrete adsorption sites active with respect to dipole and hydrogen-bond interactions, whereas the weaker van der Waals forces can be effective over a larger area.

It is possible to reduce the activity of an adsorbent by adding a small amount of water along with the mobile solvent (9). The water molecules become adsorbed selectively on the most active sites, leaving available the somewhat less active ones. This has the effect of increasing markedly the linear region for organic solutes. The adsorption of water

**Table 20-1   Solvents for
gradient elution in LSC***

n-Heptane	
Carbon tetrachloride	
Chloroform	
Ethylene dichloride	
2-Nitropropane	Polarity
Nitromethane	increases
Propyl acetate	
Methyl acetate	
Acetone	
Methanol	
Water	

* *Analytical Chemistry* (4).

is reversible, and so the activity of the solid can be altered by changing the water content of the solvent.

It is useful to tabulate solvents in the order of their polarity, the sequence in which they must be employed in gradient elution. Table 20-1 follows the listing given by Scott and Kucera (4). (Actually these authors found it desirable in some instances to premix adjacent solvents in their list, so that the effect of changing from any given liquid to the next more polar would be nearly the same at all points in the sequence. The reference should be consulted for details.)

LIQUID-LIQUID CHROMATOGRAPHY (LLC)

In this category, also known as *partition liquid chromatography*, we consider systems in which the granular solid serves as a mechanical support for a stationary phase, just as in GLC, while a second liquid forms the mobile phase. Because of the requirement that the two liquids be immiscible, it follows that they will differ markedly in degree of polarity. Either the more polar or the less polar liquid may be immobilized. Most commonly a polar solvent, such as alcohol or water, is held on a porous support of silica (diatomaceous earth), alumina, or magnesium silicate. A nonpolar solvent can be held on the same materials following silanization, which renders them hydrophobic; this is sometimes called *reverse-phase* partition chromatography.

Figure 20-6 is an excellent example of a separation by LLC (10). The chromatogram in (*a*) was run at a flow rate of 32 ml·h⁻¹, and required 14 min for elution of the most tightly held component. The time could be shortened (*b*) without loss of resolution by programming the flow rate to increase linearly by 10 percent per minute; the elapsed time was cut in half.

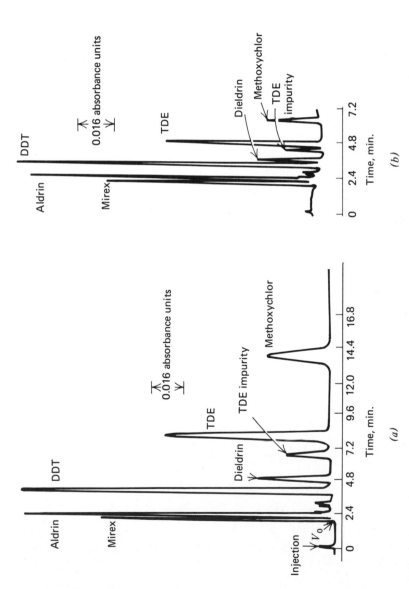

Fig. 20-6 LLC separation of insecticides. (*a*) Flow rate, 32 ml · h⁻¹; (*b*) flow rate varied linearly from 30 to 100 ml · h⁻¹. Stationary phase, β,β'-oxydipropionitrile; mobile phase, isooctane at 375 psig pressure; sample, 25 μg of each compound except 5 μg of methoxychlor; detector, ultraviolet absorption. [*Analytical Chemistry* (10).]

It is possible to calculate a value for K [Eq. (18-2)] from the observed LLC separations. In the majority of systems the values so obtained are identical with partition coefficients determined by conventional batch methods, which of course is an excellent confirmation of theory.

This can be done through application of some of the relations presented previously:

$$K = k \frac{V_m}{V_s} = \frac{(t_r - t_0)F}{W_s \rho_s} \tag{20-1}$$

where t_r is the retention time of the component of interest, t_0 is that of an unretained component, F is the flow rate, and W_s is the weight of stationary liquid phase of density ρ_s. All these are measurable quantities.

COMBINED LSC AND LLC

A column material that can be considered intermediate between a solid and an immobilized liquid is formed by bonding chemically a liquidlike material to the surface of silica or other solid support. This is conveniently done through silylation reactions, as previously described, to give a structure typified by

The side chains may be hydrophobic, as shown, hence applicable to reverse-phase separations, or they may contain polar functional groups, rendering them hydrophilic, suitable for normal-phase separations. An example of the latter is β,β'-oxydioprionitrile (OPN) bonded to silica.

These materials have the great advantage that the coating cannot be washed off, and a precolumn is no longer required. Several examples of this and other LC techniques are given in Chap. 9 of reference 1.

MOLECULAR SIEVES AND GEL PERMEATION CHROMATOGRAPHY

The term *molecular sieve* refers to a class of natural and synthetic crystalline materials, including *zeolites*, characterized by a high degree of porosity and having all pores of the same size (11). The synthetic varieties are available in granular form with pore diameters from about 0.4 to 1.5 nm, the same order of magnitude as the dimensions of low-molecular-weight organic molecules. Molecules appreciably smaller than the pore size readily diffuse into the interior of the grains and are adsorbed there very strongly, while molecules larger than the pores obviously cannot enter at all.

These materials can be used in columns, and provide excellent removal of small molecules from a flowing stream, allowing larger ones to pass through unhindered. Table 20-2 lists some materials that are adsorbed by the several sieve types from one manufacturer. It is clear that this kind of separation can be of great value, but as it is not based on any distribution ratio, the mathematical relations of Chap. 18 do not apply, and the method is not strictly chromatographic.

Table 20-2   Separation of components on Linde molecular sieves*

Adsorbed on both 4A and 5A	Adsorbed on 5A but not on 4A	Adsorbed on 13X but not on 4A or 5A
Ethane, propane	Propane and higher n-paraffins	Branched paraffins
Ethylene, acetylene	Butene and higher n-olefins	Benzene and other aromatics
Methanol, ethanol, n-propanol	n-Butanol and higher n-alcohols	Branched, secondary, and tertiary alcohols
	Cyclopropane	Cyclohexane
Water, ammonia, carbon dioxide, hydrogen sulfide	Freon-12	Carbon tetrachloride, sulfur hexafluoride, boron trifluoride

* From publications of the Linde Division of Union Carbide Corporation, via H. C. Mattraw and F. D. Leipziger, in I. M. Kolthoff and P. J. Elving (eds.), "Treatise on Analytical Chemistry," pt. I, vol. 2, p. 1102, Wiley-Interscience, New York, 1961. The pore sizes of the sieves are as follows: 4A, 0.4 nm (4 Å); 5A, 0.5 nm; 13X, 1.0 nm. At temperatures below about $-30°C$, appreciable quantities of carbon monoxide, nitrogen, oxygen, and methane are adsorbed on both 4A and 5A.

Table 20-3   Exclusion limits for Sephadex*

Type	Limiting molecular weight excluded
G-25	3500–4500
G-50	8000–10,000
G-75	40,000–50,000
G-100	100,000
G-200	200,000

* Data from Pharmacia Fine Chemicals, Inc.

An extension of the molecular sieve concept that does obey the laws of chromatography is known as *gel filtration* or *gel permeation* chromatography (12). The stationary phase consists of beads of porous polymeric material. One of the most widely used is a cross-linked dextran (a carbohydrate derivative) which is marketed under the name Sephadex.* Others are the copolymer of styrene and divinylbenzene mentioned in the preceding chapter, and a variety of polyacrylamide gels. These materials have much larger pores than zeolites, and are saturated with solvent (usually, but not always, water) prior to use. The solvent causes the particles to swell considerably, which is one of the attributes of a gel. The sieve effect is now of such a magnitude that it is more convenient to specify it in terms of the molecular weight of a solute which is just excluded. Table 20-3 gives such data for several varieties of Sephadex; the other polymers mentioned have comparable grades. The exclusion limits are rather wide, because the shape of the molecules (globular, linear, folded, etc.) will have considerable effect.

Figure 20-7 shows the results of a separation of oligosaccharides on a 3 by 120 cm column of Sephadex G-25, 0.15- to 0.074-mm-diameter particles, eluted with distilled water (13). The flow rate was 20 ml·h⁻¹, and the experiment lasted for 40 h. The molecular weight of the heaviest fraction in this separation, a tetraose, is about 700, much smaller than the exclusion limit for G-25 Sephadex. If a substance of, say, 6000 molecular weight were present, it would have passed through the column without retention and produced a peak at the point corresponding to the efflux of one column of water (at V_0 on the graph).

ION-EXCHANGE CHROMATOGRAPHY

Ion-exchange resins consist of beads of highly polymerized, cross-linked, organic materials containing large numbers of acidic or basic groups.

* AB Pharmacia, Uppsala, Sweden, and Pharmacia Fine Chemicals, Inc., Piscataway, N.J.

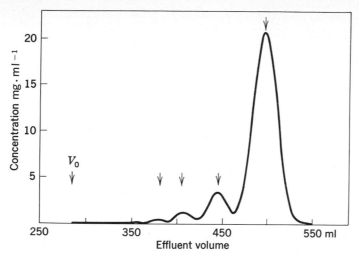

Fig. 20-7 Separation of oligosaccharides in a column of Sephadex
G-25. Arrows indicate, from left to right, the void volume V_0,
isomaltotetraose, isomaltotriose, isomaltose, and glucose. The
sugars were identified by paper chromatography. (*Pharmacia
Fine Chemicals, Inc.*)

Although the resins are insoluble in water, the active groups are hydrophilic
and have varying degrees of affinity for ionic solutes. There are four types
of resins, which are listed with some illustrative applications in Table
20-4. The useful pH ranges are significant. Below pH 5 the weak acid

Table 20-4 Ion-exchange resins for chromatography*

Resin class	Nature of resin	Effective pH range	Chromatographic applications
1. Strongly acidic cation exchange	Sulfonated polystyrene	1–14	Fractionation of cations; inorganic separations; lanthanides; B vitamins; peptides; amino acids
2. Weakly acidic cation exchange	Carboxylic polymethacrylate	5–14	Fractionation of cations; biochemical separations; transition elements; amino acids; organic bases; antibiotics
3. Strongly basic anion exchange	Quaternary ammonium polystyrene	0–12	Fractionation of anions; halogens; alkaloids; vitamin B complexes; fatty acids
4. Weakly basic anion exchange	Polyamine polystyrene or phenolformaldehyde	0–9	Fractionation of anionic complexes of metals; anions of differing valence; amino acids; vitamins

* From data on Amberlite resins, Rohm & Haas Company, Philadelphia, Pa., via Mallinckrodt Chemical Works, St. Louis, Mo.

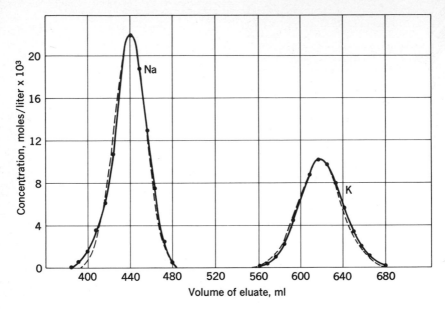

Fig. 20-8 Ion-exchange separation of sodium and potassium on a cation-exchange resin, Dowex-50, eluted with 0.7 F HCl. [*Analytical Chemistry* (14).]

resins are so slightly dissociated that cation exchange becomes negligible; the converse is true for weakly basic types above pH 9.

Three examples will be described to show the versatility of ion-exchange chromatography. The first is the separation of simple cations on a strongly acid exchanger. For monovalent ions, the relative affinities for water are $Li^+ < H^+ < Na^+ < NH_4^+ < K^+ < Rb^+ < Cs^+ < Ag^+ < Tl^+$ (that is, Li^+ is held least strongly on the resin). A comparable scale for divalent ions is $UO_2^{2+} < Mg^{2+} < Zn^{2+} < Co^{2+} < Cu^{2+} < Cd^{2+} < Ni^{2+} < Ca^{2+} < Sr^{2+} < Pb^{2+} < Ba^{2+}$. Figure 20-8 shows the complete separation of Na^+ and K^+ ions (14). The mixed sample was placed on the top of a column and eluted with 0.7 F HCl. The collected samples were evaporated to dryness to eliminate excess HCl, and then redissolved and analyzed by titration of the Cl^- ion by the Mohr procedure. The error averaged 0.25 mg of alkali halide in samples up to about 350 mg. The dashed curves represent theoretical predictions based on gaussian distributions. The average HETP was determined to be 0.5 mm.

The next example suggests the possibilities of gradient elution in ion-exchange chromatography (Fig. 20-9), although the figure was the result of stepwise rather than continuous change in eluant (15). The sample consisted of a number of transition metal salts; the column contained a strongly basic anion-exchange resin in the chloride form. The column was initially filled with 12 M HCl, and the sample inserted at the top. Elution

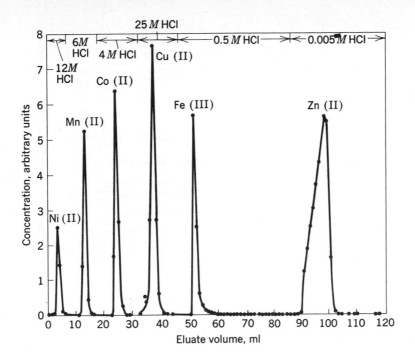

Fig. 20-9 Ion-exchange separation of several transition metals on an anion-exchange resin, Dowex-1, eluted with successively more dilute HCl. [*Journal of the American Chemical Society* (15).]

was carried out by successively more dilute solutions of HCl. The Ni(II) was not retained at all, even in the presence of concentrated HCl, though none of the other metals present moved appreciably. When diluted to $6\ M$, the acid caused the elution of Mn(II); Co(II) came out at $4\ M$, Cu(II) at $2.5\ M$, Fe(III) at $0.5\ M$, and Zn(II) only at $0.005\ M$. This sequence reflects the relative stabilities of the complex chloride anions, as well as the varying affinity of the resin and of water for these ions (and for chloride ion). The chromatography of lanthanides on a cation-exchange column in the presence of citrate buffers is another outstanding example of a difficult separation involving the interplay of two sets of equilibrium constants (7).

Ion-exchange chromatography is an extremely valuable tool in the separation of complex mixtures of compounds of biochemical interest. Scott and coworkers (16) have resolved peaks corresponding to over 100 constituents of urine, using parallel columns of anion and cation exchangers, with gradient elution.

Figure 20-10 shows 46 identified peaks in an ion-exchange chromatogram of mixed amino acids, observed with a photometric detector. A num-

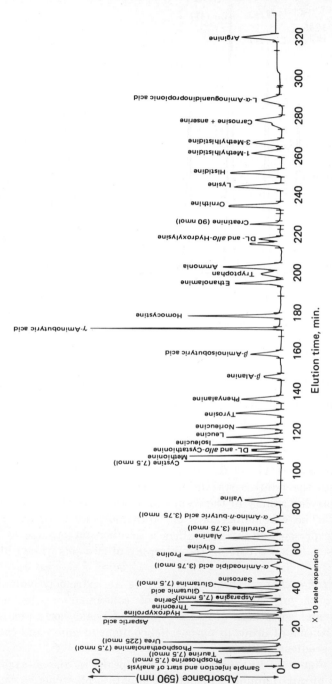

Fig. 20-10 Ion-exchange separation of amino acids from a calibration mixture. Note that the elapsed time was 5.5 h. (*Durram Instrument Corporation.*)

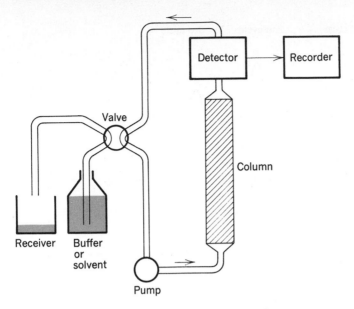

Fig. 20-11 Apparatus for recycling chromatography.

ber of manufacturers offer complete special-purpose chromatographs for the analysis of amino acid mixtures.

It should be mentioned that ion exchange has important analytical applications other than chromatographic. These include removal of interfering ions of opposite charge, and preliminary increase in concentration of trace quantities of ionic material. Refer to the literature for further discussion (17, 18).

RECYCLING CHROMATOGRAPHY

In the separation of substances with close values of the distribution coefficient by LC methods as previously described, a very long column may be needed. This is inconvenient to pack and operate; also, the weight of the packing material is apt to crush the solid beads near the bottom, markedly reducing the efficiency. Better results can be obtained by pumping the effluent back onto the column repeatedly. It is possible to do this because the zones occupied by the two similar solutes cover at any moment only a small part of the column, the rest of which is not being utilized. A column, a detector, a sampling valve, and a pump are assembled in a closed loop (Fig. 20-11). The solution passes upward through the column, opposing gravitation, thus tending to prevent unduly dense consolidation of the packing material.

Figure 20-12 shows results obtained by recycling a biochemical preparation eight times on a Sephadex column (19). The detector was an ultra-

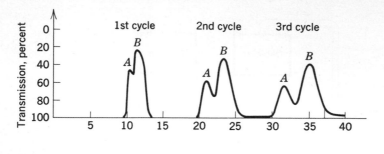

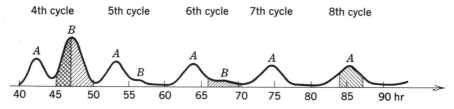

Fig. 20-12   Recycling separation of ceruloplasmin on Sephadex G-100; absorption at 254 nm in a 3-mm cuvet. Hatched areas indicate bleedings of the column. [*Academic* (19).]

violet absorptiometer operated at 254 nm. Three cycles could be completed before the advancing first component nearly caught up with the tail of the second component; at this point the valve was opened to bleed off component B and simultaneously admit an equal volume of fresh buffer. A small amount of B remained and was bled off after the sixth cycle. Pure A was pumped out in the eighth cycle. The column height in this experiment was a little less than 1 m; without the recycling feature, a column 8 m high would have been required. This procedure does not shorten the operating time. Apparatus for this service is available commercially.

NONCOLUMN LC TECHNIQUES: THIN-LAYER CHROMATOGRAPHY (TLC)

TLC is the designation given to LSC or LLC carried out on a powdered solid phase coated onto a smooth supporting plate of glass or plastic (20, 21). The coating can be of any of the materials suitable for column LC. A binder such as starch or calcium sulfate must be added so that the adsorbent will adhere to the plate, but care must be taken lest the binder mask the properties of the adsorbent.

   The sample is applied in a small drop a short distance from one end of the plate, and the solvent evaporated off. The plate is then placed vertically in a closed jar, with its lower edge dipping into a pool of eluting solvent. The solvent ascends the plate by capillary action. The components of the sample are carried along by the moving solvent, as in any form of LC. Separation is stopped by removal of the plate when the solvent front approaches the top edge. Thus the sample components are not eluted out through a detector as in column procedures, but remain on the plate as a series of distinct spots.

If the components are colored, they can be located readily, but more often they are invisible and must be located by other means. Illumination with ultraviolet will excite fluorescence in many compounds. Another possibility is to impregnate the plate in advance with a fluorescent dye; the presence of an ultraviolet-absorbing compound will result in a dark spot on exposure to ultraviolet light, as the compound quenches the fluorescence of the dye. If this approach is not applicable, a color- or fluorescence-producing reagent can be sprayed onto the dried plate to render the spots visible.

Once the spots are located they can be removed, if desired for further study, by scraping the powder from the plate with a razor blade and then eluting the specimen with solvent. Alternatively, the spot can be measured in situ, with a specially designed absorptiometer or fluorometer; the plate is illuminated with ultraviolet or visible light, and the radiant energy transmitted or emitted as fluorescence is measured. Scattering by the coated plate prevents the accuracy of these instruments from being as great as one might wish, but they serve well for many purposes, including qualitative identification.

APPARATUS FOR TLC (20)

Plates for TLC can be purchased with fluorescent or other coatings already applied, but the application of coating material by the user is often preferred. For this purpose a number of manufacturers supply special devices to spread the adsorbent and binder evenly over the plate. This is a rather critical operation, especially if quantitative comparison of plates is required. Except with the smallest plates (e.g., microscope slides), it is usual to run several chromatograms side by side on the same plate. This is facilitated by scoring the plate with parallel grooves where the coating is removed. The scores, about 1 cm apart, prevent the spots from spreading and interfering with their neighbors.

The samples for multiple chromatograms must be applied to the plate reproducibly with respect to position and volume. A spotting guide or template can be used to advantage, together with a calibrated syringe.

PAPER CHROMATOGRAPHY

A closely related method uses a strip of filter paper as support for chromatographic separations. The technique is similar to that for TLC, but without the need for special coatings. Cellulose tends to hold a film of water firmly adsorbed on its surface. Therefore paper chromatography is usually a liquid-liquid partition.

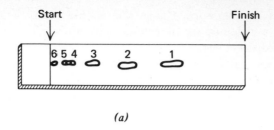

(a)

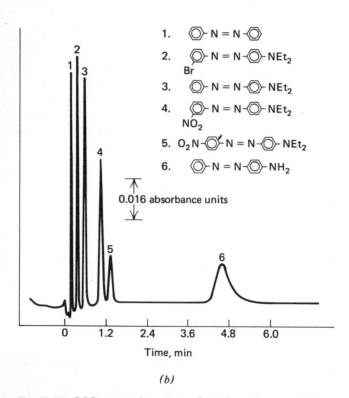

(b)

Fig. 20-13 LSC separation of azo dyes (lower) compared to the TLC pilot separation (upper). In both, the solid phase was silica and the mobile phase 10 percent CH_2Cl_2 in hexane. [*Analytical Chemistry* (10).]

Either TLC or paper chromatography can serve as a pilot analysis to help establish the optimum conditions for further separation on a column. It is often quicker and more convenient to test various solvents and adsorbents on a plate or paper than with a full-scale LSC or LLC instrument. Figure 20-13 shows the separation of a series of diazobenzenes both by TLC and on a column (10).

PROBLEMS

20-1 In a separation of nucleosides on an LC column with an ultraviolet detector, peaks appeared that were identified as follows:

Air	4.0 min
Uridine	30 min
Inosine	43 min
Guanosine	57 min
Adenosine	71 min
Cytidine	96 min

On another column of different dimensions but packed with the same stationary phase, the air peak came at 5.0 min, and uridine at 53 min. Another component was eluted at 100 min. Identify it.

REFERENCES

1. Hadden, N., et al.: "Basic Liquid Chromatography," Varian Aerograph, Walnut Creek, Calif., 1971.
2. Kirkland, J. J. (ed.): "Modern Practices of Liquid Chromatography," Wiley-Interscience, New York, 1971.
3. Veening, H.: *J. Chem. Educ.* (a) **47**:A549, A675, A749 (1970); (b) **50**:A429, A481, A529 (1973).
4. Scott, R. P. W., and P. Kucera: *Anal. Chem.*, **45**:749 (1973).
5. Scott, R. P. W., and J. G. Lawrence: *J. Chromatogr. Sci.*, **8**:65 (1970).
6. Udenfriend, S., S. Stein, P. Böhlen, W. Dairman, W. Leimgruber, and M. Weigele: *Science*, **178**:871 (1973).
7. Ketelle, B. H., and G. E. Boyd: *J. Am. Chem. Soc.*, **69**:2800 (1947).
8. McGuinness, E. T., and M. C. Cullen: *J. Chem. Educ.*, **47**:A9 (1970).
9. Engelhardt, H., and H. Wiedemann: *Anal. Chem.*, **45**:1641 (1973).
10. Majors, R. E.: *Anal. Chem.*, **45**:755 (1973).
11. Meier, W. M., and J. B. Uytterhoeven (eds.), "Molecular Sieves," Advances in Chemistry, No. 121, American Chemical Society, Washington, 1973.
12. Cazes, J.: *J. Chem. Educ.*, (a) **43**:A567, A625 (1966); (b) **47**:A461, A505 (1970).
13. Flodin, P.: "Dextran Gels and Their Applications in Gel Filtration," p. 57, AB Pharmacia, Uppsala, Sweden, 1962.
14. Beukenkamp, J., and W. Rieman, III: *Anal. Chem.*, **22**:582 (1950).
15. Kraus, K. A., and G. E. Moore: *J. Am. Chem. Soc.*, **75**:1460 (1953).
16. Scott, C. D., D. C. Chilcote, and N. E. Lee: *Anal. Chem.*, **44**:85 (1972).
17. Samuelson, O.: "Ion-Exchange Separations in Analytical Chemistry," Wiley, New York, 1963.
18. Rieman, W., III, and H. Walton: "Ion Exchange in Analytical Chemistry," Pergamon, New York, 1970.
19. Porath, J., and H. Bennich: *Arch. Biochem. Biophys.*, Suppl., **1**:152 (1962).
20. Lott, P. F., and R. J. Hurtubise: *J. Chem. Educ.*, **48**:A437, A481 (1971).
21. de Zeeuw, R. A.: *CRC Crit. Rev. Anal. Chem.*, **1**:119 (1970).

Chapter 21

Mass Spectrometry

The mass spectrometer is an instrument that will sort out charged gas molecules (ions) according to their masses. It has no real connection with optical spectroscopy, but the names *mass spectrometer* and *mass spectrograph* were chosen by analogy because the early instruments produced a photographic record resembling an optical line spectrum.

INSTRUMENTATION

There are several distinct types of mass spectrometers, but all possess components to perform the following functions: (*1*) ionization of the sample, (*2*) acceleration of the ions by an electric field, (*3*) dispersion of the ions according to their mass-to-charge ratio, and (*4*) detection of the ions and production of a corresponding electric signal. These will be taken up in the paragraphs that follow.

Since the mass spectrometer depends on a stream of gaseous ions following well-defined trajectories through combinations of electric and magnetic fields, it is essential that all parts from ion source to detector be evacuated. The pressure must generally be no greater than about 10^{-5} torr (approximately 10^{-3} Pa). This requires efficient and continuous pump-

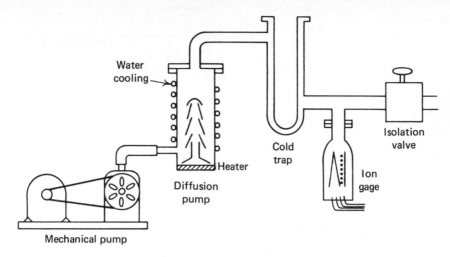

Fig. 21-1 Typical vacuum system for a mass spectrometer.

ing, for which oil-diffusion pumps trapped with liquid nitrogen are most commonly employed.   An accurate and sensitive pressure gage is essential. A typical vacuum system is shown in Fig. 21-1.   An extensive discussion of the various components has been given by Roboz (1).

THE ION SOURCE

Only gaseous ions can be utilized by the mass analyzer (the usual name for the dispersion unit); hence analysis is effectively limited to samples that either are normally gases or have appreciable vapor pressure.   Non-volatile solids can be studied only if they can be converted to gases by chemical reaction (pyrolysis) induced by intense thermal or electrical energy.   The ions are typically produced in a small boxlike enclosure called the *ion source* (2).

The most widely used type of ion source utilizes bombardment by a beam of electrons to form ions.*   Figure 21-2 shows the structure of such a source.   The gaseous sample is introduced into the source by a variety of techniques to be described later.   The electron beam traverses

---

* Other types of ion sources (2) are: (1) *field ionization,* in which an electrostatic gradient of approximately 10^8 V/cm is produced at the surface of a sharp metallic point or blade, strong enough to remove an electron from a sample molecule; (2) *surface ionization,* in which the nonvolatile sample is coated onto a tungsten ribbon and then heated to produce ions directly; (3) *spark ionization,* in which a high-voltage radiofrequency (approximately 800-kHz) spark is passed between electrodes, one of which contains or is made of the sample material; (4) *chemical ionization,* which will be discussed in a later paragraph.

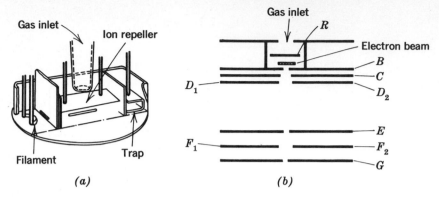

Fig. 21-2 Electron-impact source of ions for a mass spectrometer.   (*a*) Ionization chamber; (*b*) schematic diagram of source, showing the ion repeller *R*, first slit *B*, draw-out plate *C*, focusing plates *D*, collimating plates *E* and *G*, and beam-centering plates *F*.   [*McGraw-Hill* (16).]

a space where ionizing collisions occur, and the resulting ions are acted on by an electric field.

The ionizing electrons are liberated by thermionic emission from a heated tungsten or rhenium filament, and accelerated by an adjustable potential.   After crossing the ionization region, they are collected on an anode called an *electron trap*.   The adjustable potential controls the kinetic energy available for the ionization of gas molecules.

Bombardment of molecules by energetic electrons usually produces more positive ions than negative, and so the majority of mass spectral applications involve positive ions only.   The ions are extracted from the region in which they are formed by a field applied between a positive repeller electrode and a negative accelerator.   They pass through a narrow slit, or its equivalent, and several focusing electrodes into the mass analyzer.   For proper subsequent focusing, it is essential that the ions emerging from the source be as nearly uniform as possible in kinetic energy.   During acceleration they acquire energy $E = eV$, where e represents the charge on the ion, in units of electronic charge, and V is the applied potential.   The kinetic energy of the ions as they pass through the exit slit is given by

$$E_{\text{kin}} = eV + E_0 = \tfrac{1}{2}mv^2 \tag{21-1}$$

where m is the mass of the ion, v is its velocity, and E_0 is the kinetic energy the ion may have had as a result of the ionizing impact.   The energy of the emergent ions will always have a finite spread, because of variations in E_0 and because of the fact that ions are formed at different distances from the slit.   Careful design of the source structure can minimize this spread so that it can be neglected for low-resolution spectrometers.

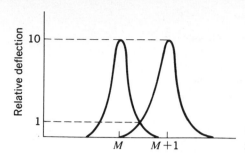

Fig. 21-3 Illustrating the definition of resolution in mass spectrometry.

Rearrangement of Eq. (21-1), with omission of E_0, gives

$$v = \sqrt{2V\frac{e}{m}} \qquad (21\text{-}2)$$

which shows that the ions move with velocities determined to a good degree of approximation by their charge-to-mass ratio e/m.

RESOLUTION

The term *resolution* is unfortunately not always used in the same sense. We shall follow the most widely used definition: The resolution is the ratio $M/\Delta M$, where ΔM is the difference in mass numbers which will give a valley of 10 percent between peaks of mass numbers M and $M + \Delta M$ (Fig. 21-3), where the two peaks are of equal height. Resolution is considered satisfactory if $\Delta M \gg 1$. Thus, if this criterion is met for masses up to 600 and 601, the *unit resolution* is said to be 600.

The resolution is not generally uniform over the whole range of masses which can be detected, becoming poorer at higher mass numbers. For some instruments, unit resolution will be specified as the same as the maximum observable mass, and can be expected to be better at lower masses. In others, denoted *high-resolution* instruments, the unit-resolution figure may be much larger than the maximum observable mass, which makes it possible to distinguish species in which the difference in mass is correspondingly less than one mass unit.*

MASS ANALYZERS

There are several possible ways to separate ions according to mass number. Only the more widely used methods will be described here.

Passage of charged particles through a magnetic field results in a circular trajectory of radius given by

$$r = \frac{mv}{eB} \qquad (21\text{-}3)$$

* In quadrupole spectrometers the resolution is proportional to the mass number.

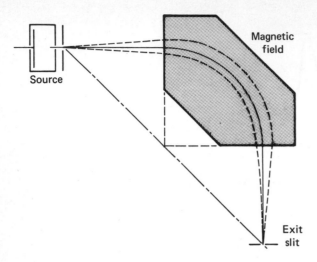

Fig. 21-4   Magnetic sector analyzer.

where B is the magnetic field strength.   It can easily be shown on geometric grounds that a homogeneous beam of ions diverging from a slit can be brought to focus by a magnetic field in the shape of a sector* (Fig. 21-4). This is the basis of the *magnetic sector* mass spectrometer.   As shown in Eq. (21-1), the ions emerging from the source have (nearly) uniform kinetic energy.   In order to focus, Eq. (21-3) shows that they must have equal momentum (mv).   The radius of the trajectory is given by a combination of these two equations:

$$r = \frac{1}{B} \sqrt{2V \frac{m}{e}} \tag{21-4}$$

This equation shows that every species of ion, characterized by a particular value of m/e, will follow its own curve.   The spectrometer is provided with an exit slit, as shown in Fig. 21-4, that will isolate those ions whose path has just the right radius of curvature.   According to Eq. (21-4), for the selected ions, the m/e ratio equals $k(B^2/V)$, where k is a constant of the apparatus.   Hence a range of masses can be scanned by varying either B or V, keeping the other constant.

The resolution of the sector mass analyzer as described above is limited because of the spread of kinetic energies, and also because the boundaries of the magnetic field cannot be sharply defined.   It can be

---

* The geometric construction requires that the two slits and the apex of the sector be collinear, as shown in the figure.   Sectors of 60° and 90° are used in commercial instruments.

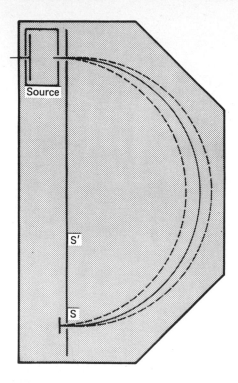

Fig. 21-5 180° magnetic sector analyzer. The source and the exit slit S are immersed in the magnetic field. S' indicates the region where other m/e ions may focus.

improved by correcting either of these features. The magnetic boundary effect can be removed through use of a sector of 180°, with both source and detector immersed in the magnetic field (Fig. 21-5).

DOUBLE-FOCUS INSTRUMENTS

The beam of ions can be rendered isoenergetic by means of an electrostatic sector (Fig. 21-6). The beam assumes a circular trajectory in passing through the annular space between two concentric cylindrical electrodes. The radius r and the applied field E determine the energy of the particles that can pass, according to the relation

$$\frac{1}{2} mv^2 = \frac{eEr}{2} \tag{21-5}$$

This electric sector, sometimes called an *energy filter*, is used in conjunction with a magnetic sector, forming a *double-focus* spectrometer, i.e., one that focuses the ions both in energy and in mass.

There are two double-focus designs in common use. One of these, the *Nier-Johnson* design (Fig. 21-7), makes use of 90° sectors with an

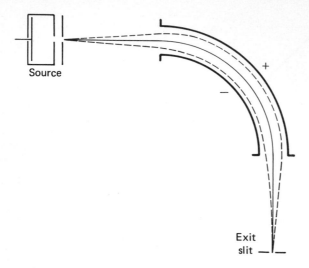

Fig. 21-6   Radial electrostatic sector.

intermediate slit.   The other, using *Mattauch-Herzog* geometry (Fig. 21-8),
utilizes an electric sector with an angle of 31°50′.   At this angle, the effect
of the sector is to collimate the ion beam, analogous to forming an optical
virtual image at infinity.   This means that all ions enter the magnetic
field at normal incidence, which minimizes the edge effect.   The magnetic
sector has an angle of 135°, which brings each ionic species to a focus
at the far boundary of the field.   An important feature of this design
is that all masses are focused simultaneously along a plane surface, the
focal plane, which makes possible direct photographic recording.   This is
not true of any other common design of mass spectrometer.

Double-focus spectrometers are capable of very high unit resolution—up to 50,000 or higher, compared to about 5000 as the maximum
resolution for a single-focus sector spectrometer.

QUADRUPOLE MASS ANALYZERS

The quadrupole analyzer is a device in which ions can be resolved according
to their m/e ratio without the need of a heavy magnet.   It consists of
four metallic rods, precisely straight and parallel, so positioned that the
ion beam shoots down the center of the array (Fig. 21-9).   Note that
a circular orifice rather than a slit is used as an inlet port.   Diagonally
opposite rods are connected electrically, the two pairs to opposite poles
of a dc source and also to a radiofrequency oscillator.

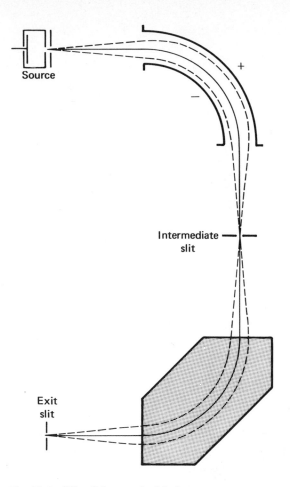

Source

Intermediate slit

Exit slit

Fig. 21-7   Nier-Johnson double-focus mass spectrometer.

Neither the dc nor the ac field has any effect on the forward motion
of the ions, but lateral motion will be produced by these interacting fields.
This can be analyzed (3) in terms of the coordinate system of Fig. 21-10.
If the rods were of symmetric hyperbolic cross section, then the potential
ϕ at any point (x,y) would be given as a function of time t by the equation

$$\phi = \frac{(V_{dc} + V_0 \cos \omega t)(x^2 - y^2)}{r^2} \tag{21-6}$$

where V_{dc} is the applied direct potential, V_0 is the amplitude of the alternating voltage of frequency ω radians per second, and r is the dimension

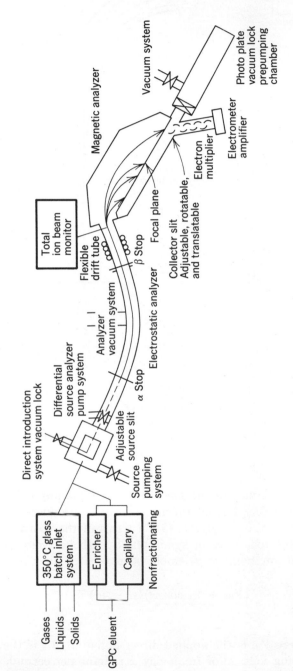

Fig. 21-8 The Mattauch–Herzog design of a high-resolution mass spectrometer.

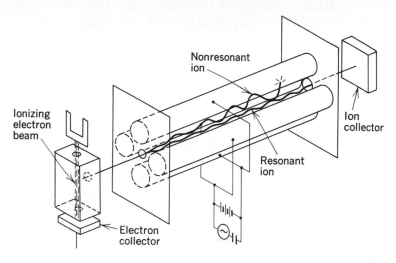

Fig. 21-9 Quadrupole mass spectrometer, schematic. Scanning in this device is accomplished by varying the radiofrequency or the magnitude of the rf and dc voltages. [*Research and Development* (17).]

defined in the figure. This relation holds to a good approximation even when the hyperbolic rods are replaced by less expensive cylindrical ones.

The force acting laterally on an ion of unit charge e is obtained by differentiating with respect to x and y:

$$F_x = -e\frac{\partial \phi}{\partial x} = -e\frac{(V_{dc} + V_0 \cos \omega t)2x}{r^2}$$

$$F_y = -e\frac{\partial \phi}{\partial y} = +e\frac{(V_{dc} + V_0 \cos \omega t)2y}{r^2}$$

$$(21\text{-}7)$$

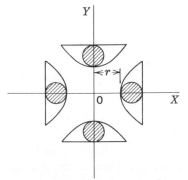

Fig. 21-10 Geometry of the rods in a quadrupole mass spectrometer. The z axis, along which the ions move, is perpendicular to the plane of the paper.

Then, since acceleration can be equated to the ratio of force to mass, we can write as the equations of motion for an ion of mass m and charge e,

$$\frac{d^2x}{dt^2} + \frac{2}{r^2}\frac{e}{m}(V_{dc} + V_0 \cos \omega t)x = 0$$

$$\frac{d^2y}{dt^2} - \frac{2}{r^2}\frac{e}{m}(V_{dc} + V_0 \cos \omega t)y = 0$$

(21-8)

These latter equations indicate that the motions of the ions will have a periodic component of frequency ω, but will also be dependent on the e/m ratio. A further mathematical treatment will show that for V_{dc}/V_0 < 0.168 there is only a narrow range of frequencies for which the ionic trajectories are stable (i.e., not divergent) with respect to both the x and y coordinates; outside this range the ions will collide with one or the other set of rods. Maximum resolution is obtained with the V_{dc}/V_0 ratio approaching as closely as possible the limiting value 0.168. If the ratio is allowed to become larger than this, no frequency can be found for which a stable trajectory will result, no matter what the e/m value. Mass selection can be achieved alternatively by keeping the frequency fixed and varying the dc and radiofrequency potentials while maintaining their ratio accurately constant. Quadrupole spectrometers are useful up to several hundred mass numbers. Unit resolution of the order of 400 to 500 can be achieved.

TIME-OF-FLIGHT MASS ANALYZERS

The instruments previously described produce a steady ion beam for any given setting of the controls. It is possible, however, to apply the accelerating potential intermittently and to cut up the beam accordingly into pulses. This permits sorting out the ions by their velocities, which is tantamount to mass sorting, without the need of a magnetic field. An apparatus for doing this, the time-of-flight spectrometer (3–5), is shown schematically in Fig. 21-11. An electron beam ionizes the incoming gas sample as in other spectrometers. An accelerating potential of the order of 100 V is applied to the *ion focus grid* in the form of a voltage pulse lasting 1 μs or less and repeating a few thousand times per second. This positive pulse accelerates the ions out through the grid where they are picked up by the field of the *ion energy grid*. All ions receive the same energy; therefore the velocity they acquire is proportional to $\sqrt{e/m}$ [cf. Eq. (21-2)]. The ions are then allowed to drift at constant velocity in a field-free region (40 cm or so) to a detector. The time of transit through the drift space is $T = k\sqrt{m/e}$, where k is a constant depending on the distance and the parameters of the ion gun. The value of k is not far from unity if T is

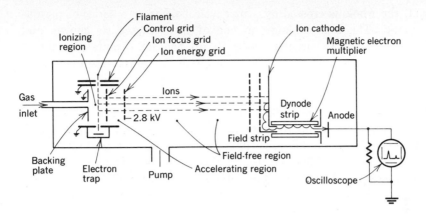

Fig. 21-11   Schematic diagram of the time-of-flight mass spectrometer.   (*CVC Products, Inc.*)

taken in microseconds. A singly charged N_2 molecule accordingly has a flight time of approximately $\sqrt{28} = 5.30$ μs, whereas for a singly charged O_2 molecule, $T = \sqrt{32} = 5.66$ μs, and for a Xe ion, $T = \sqrt{132} = 11.50$ μs. (The detector will be described subsequently.)

Time-of-flight spectrometers have been built with mass ranges up to 1200 amu, and unit resolutions of about 500.

DETECTORS

Ion detectors are of three types: (*1*) a simple collector known as a Faraday cup, (*2*) an electron multiplier, and (*3*) a photographic plate.

The *Faraday cup* is the least elaborate, and also the least sensitive of these detectors. It is used in many spectrometers in which extreme sensitivity is not needed, because of its convenience and lower cost. It consists of an insulated conductor directly connected to an electrometer amplifier. It is cup-shaped to lessen the likelihood of escape of secondary electrons liberated by ion impact.

The *electron multiplier* is essentially similar to the photomultiplier as used in ultraviolet and visible detection, but has the primary cathode optimized for detection of ions rather than photons. Being immersed in the spectrometer vacuum, it does not need the glass envelope of the photomultiplier. Because of its internal amplification, it is perhaps 1000 times more sensitive than the Faraday cup.

The Bendix Corporation has developed a novel type of electron multiplier, shown in Fig. 21-11, especially for use with time-of-flight mass spectrometers. It consists of a pair of glass plates coated with a high-resistance metallic film. An electric gradient is impressed along the lengths of these plates, and the whole is crossed by a magnetic field from a series of permanent magnets. The ions impinge on one of these plates, as shown, and emitted electrons follow circular paths to hit the same plate at another spot, a process that is repeated many times along the plate, until eventually the amplified stream of electrons strikes an anode connected to the readout device.

Photographic detection is only usable with Mattauch-Herzog instruments. Because the photoplate integrates the ion signal over a period of time, it is capable of greater sensitivity than even the electron multiplier. It can also make more effective use of the high-resolution capability of the double-focus spectrometer. The plates are processed by the usual photographic techniques and read with the aid of a densitometer.

DATA PROCESSING

The peaks appearing in a mass spectrum are often extremely sharp and numerous, necessitating a fast-response analog recorder or analog-to-digital conversion system. The most familiar readout is by a multiple-trace photographic recording oscillograph, giving a record such as that of Fig. 21-12. The several traces represent the same signal with differing exactly known amplification. The height of each peak can be measured on the trace that gives the largest deflection within the limits of the paper.

Time-of-flight spectrometers scan a complete spectrum in a few hundred microseconds with every pulse of ions received by the detector, so even an oscillograph is not fast enough for direct recording. A cathode-ray oscilloscope, being free of the inertia of a moving system, is suitable. For a permanent record, the oscilloscope screen can be photographed, or a sampling recorder can be used. This latter synthesizes an analog record by responding to successively delayed points on successive spectra, somewhat like the principle of a stroboscope for visualizing rapid rotary motion.

Further sophistication in handling mass spectral data requires digitizing. Small dedicated computers are available that can detect the presence of each peak during a scan, sense the corresponding field and accelerating voltage, compute the mass number, and print it out along with the peak height on a digital printer or feed it into a further computational facility if required. Similar computer methods can be used to analyze the information collected on photoplates, by processing the data from the densitometer.

SAMPLE HANDLING

There are a number of methods for introducing samples into the ionization source, the choice depending on physical properties such as melting and boiling points. Many spectrometers will accept samples introduced in more than one way. If the sample is a gas or a volatile liquid, it is best handled

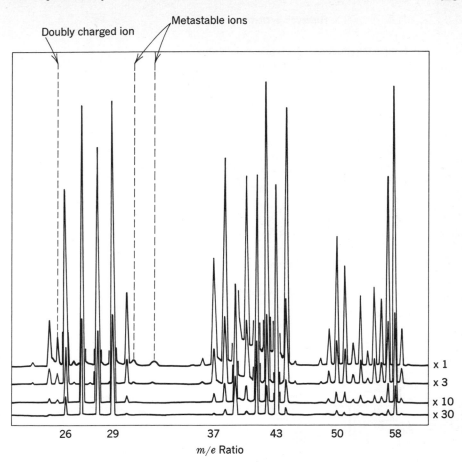

Fig. 21-12 The mass spectrum of *n*-butane, as recorded on a magnetic sector mass spectrometer. Simultaneous tracings by galvanometers of different sensitivities provide a dynamic range up to 30,000. (*E. I. Du Pont de Nemours & Company.*)

by allowing it to diffuse into a previously evacuated glass or metal bulb which communicates with the source via a tiny orifice called a *pinhole* or *molecular leak* (Fig. 21-13).

The molecular leak may consist of one or more tiny needle holes in a thin gold membrane. The holes must be small compared to the mean free path of the gas molecules at the expected pressures; a diameter of 0.01 mm is about right. This ensures conformity to Graham's law of effusion: Gases will pass through an orifice inversely as the square roots of their molecular weights. Since most of the molecules (those that escape ionization) follow the same law in leaving the ion source, the relative partial pressures of different gases will be the same within the source as they were in the external reservoir, an essential condition for quantitative

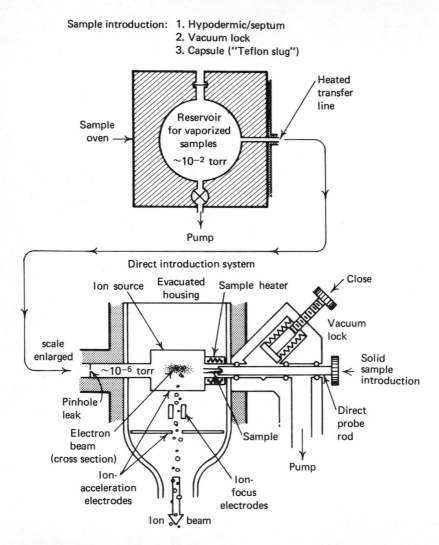

Reservoir system

Sample introduction: 1. Hypodermic/septum
                       2. Vacuum lock
                       3. Capsule ("Teflon slug")

Fig. 21-13 Two sample-introduction systems.   Above, by a heated reservoir and diffusion orifice; below, right, by a sample probe inserted through a vacuum lock. [*W. A. Benjamin* (8).]

analysis.   The expansion flask will contain enough sample to run as many duplicate spectra as may be needed.

Organic solids that cannot be handled as vapors are introduced directly into the source by means of a *probe*, typically a stainless-steel rod, 6 mm in diameter and 25 cm long, bearing a tiny cup at its tip to hold the sample (Fig. 21-13).   The probe is inserted through a vacuum lock.   A milligram or even less is sufficient sample to give a satisfactory spectrum,

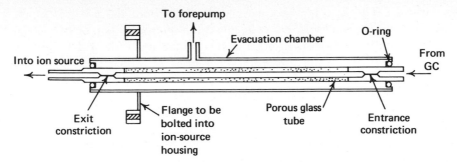

Fig. 21-14 A pressure-reduction system linking a mass spectrometer to a gas chromatograph. [*Analytical Chemistry* (18).]

but it does not last very long before being pumped down to vanishing amounts. More refractory substances must be handled with a spark source or laser vaporization, or other special arrangement.

An important adjunct to the sample-handling system is a permanently connected gas chromatograph (6). The mass spectrometer then provides a means of identifying GC peaks as they are being eluted. Chromatographs were first used with time-of-flight spectrometers, but now are commonly installed with quadrupole and sector instruments. A number of companies manufacture combined GC–mass spectrometers, while others provide a GC as an option.

Unless special high-capacity vacuum pumps are available, the GC carrier gas must be prevented from entering the mass spectrometer. With helium or hydrogen as carrier this can be accomplished by passing the gas stream through a porous tube (Fig. 21-14). The low-molecular-weight carrier diffuses readily through the tube to be pumped away, while the heavier sample molecules continue on.

Spectrometers designed for GC work are often fitted with an additional electrode just outside the ion source, positioned to intercept a constant small fraction of the total ion beam before mass dispersion. This is to be connected directly, after amplification, to an auxiliary strip-chart recorder, to serve as a GC ionization detector.

FRAGMENTATION

It is instructive to consider the sequence of events within an electron-impact ion source as the electron energy is increased through its useful range. Let us assume that an organic compound with several carbon atoms is present in the gas phase. The first ionization potential of such a compound is between 8 and 12 eV,* so only when this energy is reached will any ions

* The ionization potential of a compound or element is the energy in electron volts necessary to remove the least strongly held electron to an infinite distance. In electron-impact ionization, this may be taken as the accelerating potential in the electron gun.

be formed. The first ion to appear will result from the removal of one electron from the molecule, hence is called the *molecular ion,* or *parent ion.* As the potential is increased, other bonds in the molecule will be broken, each characterized by an *appearance potential* at which it first becomes evident. Hence mass spectrometry can provide information about ionization potentials and relative bond strengths.

The majority of mass spectra are run with a 70-V electron beam, energetic enough to break any bonds likely to be present. Therefore each compound yields a characteristic series of fragments, called its *fragmentation* or *cracking pattern.* Such patterns are commonly described both by listing and by bar-graph presentation, as in Fig. 21-15, the mass spectrum of *n*-butane. Note that the molecular ion ($m/e = 58$) gives a relatively small peak, while a peak at 43 is the most intense. (The most intense peak in a spectrum is called the *base peak;* it may or may not also be the molecular peak.) This indicates, of course, that a fragment of mass 43 is more resistant to further bond cleavage than the parent molecule itself. In this case it was formed by the loss of a fragment of mass $58 - 43 = 15$, clearly a methyl group, so the base peak is due to the $C_3H_7^+$ ion. The small peak at mass 59 is due to the presence of ^{13}C and ^2H in their natural abundances, and is an example of an *isotope satellite* peak. An occasional small peak at a half-integral mass number, such as that at 25.5 in the figure, is due to a doubly charged ion (mass 51); the large peak at 29 may include both $C_4H_{10}^{2+}$ and $C_2H_5^+$ ions.

Mass spectra, even with only moderate resolution, such as those discussed above, can provide a wealth of information about a compound. The rules for deriving elemental composition and structure are much too detailed to include here. There are many books devoted to this subject (7–9).

High-resolution spectrometry permits additional information to be deduced, by observation of *mass defects,* the difference between a true atomic or molecular weight and the nominal, whole number, value. For example, the atomic weights of ^1H, ^{12}C, ^{14}N, and ^{16}O are, respectively, 1.00782522, 12.00000000, 14.00307440, and 15.99491502 (10). To four decimal places, this means that the molecular weight of a CO fragment is less than that of N_2 by 0.0012 mass units, a difference of about 40 ppm. The molecular weight of C_2H_4, on the other hand, is greater than that of N_2 by about 850 ppm. Through measurements with this high degree of precision, the elemental compositions of all fragments can be determined unequivocally.

The technique required for high accuracy mass number determination, if done manually, is much too time-consuming. Computer assistance is a practical necessity.

A few broader-than-normal maxima are often observed superimposed upon the usual spectrum, particularly with double-focus instruments. These are caused by *metastable ions,* formed in the ion source, that sponta-

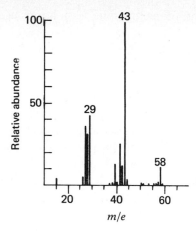

Fig. 21-15 The mass spectrum and relative abundances for *n*-butane.
[*W. A. Benjamin* (8).]

m/e	Relative abundance	*m/e*	Relative abundance
1	1.1	37	1.0
2	0.10	38	1.8
12	0.13	39	12.
13	0.26	39.2 m	0.4
14	0.96	40	1.6
15	5.3	41	27.
16	0.12	42	12.
25	0.46	43	100.
25.1 m	0.11	44	3.3
25.5 d	0.36	49	0.40
26	6.1	50	1.2
27	37.	51	1.0
28	32.	52	0.26
29	44.	53	0.74
30	0.98	54	0.19
30.4 m	0.14	55	0.93
31.9 m	0.20	56	0.72
		57	2.4
		58	12.
		59	0.54

m "Metastable" peak
d Doubly charged ions

neously decompose during their passage through the spectrometer. The new fragments formed when a metastable ion breaks down must include at least one positive ion and, as its mass is less, it will apear displaced in the spectrum. The resulting peak tends to be diffuse, because the decomposition events for identical ions will not all occur at the same position in the apparatus. Observations of metastables can be of use in organic mechanism studies.

QUALITATIVE ANALYSIS

The identification of unknown materials requires certain assignment of mass numbers to peaks on a chart. Some mass spectrometers have precise mass marker devices built in, or such can be purchased separately in the form of the digitizing computers mentioned previously. Without these, mass assignments can be difficult, becoming less certain as the mass increases. For a magnetic deflection instrument in which the magnetic field is scanned at a constant rate, Eq. (21-4) tells us that mass number increments are equal for equal increments in the field, so that identification of two peaks will calibrate a chart. However, it is often necessary to scan the field exponentially, or (e.g., with permanent-magnet instruments) to scan the accelerating potential. In such cases, no simple distance measurement on the chart will suffice, and a calibration with known compounds must be used. At the lower end of the scale, peaks due to the components of air (Fig. 21-16) will identify mass positions. For a convenient spot check, mercury vapor can be used, giving a characteristic pattern at $m/e = 198$ to 204. For a more inclusive calibration, *perfluorokerosine* (PFK) is often suitable. McLafferty (11) has published a "partial" list of 73 fragment peaks from this mixture of fluorocarbons. Peaks appear at 69 (CF_3), 93 (C_3F_3), 105 (C_4F_3), 131 (C_3F_5), and homologous peaks greater by increments of 50 (CF_2). PFK can be run immediately preceding or following an unknown, with no change in controls between, or it can be mixed with the unknown. The latter method, used primarily with high-dispersion instruments, is possible because the mass defects of fluorine-containing compounds are different from others, hence do not exactly overlap.

QUANTITATIVE ANALYSIS

The fragmentation patterns of the components of a mixture are additive; hence mixtures can be analyzed if spectra for the several components, run under the same conditions, are available. The calculation involves a set of n simultaneous equations in n unknowns, for a mixture of n components. A desk calculator or computer, depending on the size of n, will simplify the computation.

With components present to more than about 10 mole percent, precision and accuracy can well be such that errors (90 percent confidence level) are less than ± 0.5 mole percent (12). Mixtures of this order are

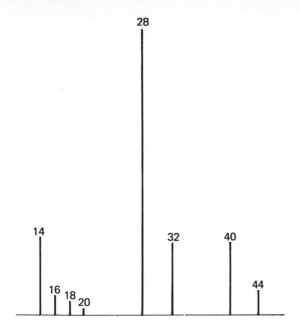

Fig. 21-16   The mass spectrum of air.

better handled with the aid of a GC inlet system.   GC provides quantitation of each of its peaks, while the mass spectrometer can be used to identify them, and many times to determine whether they are formed by more than a single constituent.

The mass spectrometer can be used in tracer studies, utilizing compounds enriched with stable isotopes that are in low natural abundance, such as ^2H, ^{13}C, ^{33}S, and ^{40}K, among many others.   Some isotopes should be avoided because of potential interference; thus ^{58}Fe would be a poor choice for an Fe tracer, as it is indistinguishable from ^{58}Ni except at highest resolution (~30 ppm).   The experimental technique of *isotope dilution* is often used in tracer studies; it is described in detail in Chap. 23, as it is more widely used in connection with radioactive isotopes (13).

Figure 21-17 shows an interesting example of the combined use of GC and a quadrupole mass spectrometer.   The sample is a mixture of the two optical isomers of the *N*-acetyl derivatives of the amino acid alanine, in which the L form is labeled with an atom of deuterium.   These isomers can be separated on a chromatographic column of a special optically active material.   The upper trace is the normal signal from the GC detector. From this trace alone, there is no way of knowing which peak is due to which compound.   The lower trace shows a mass scan repeated at ap-

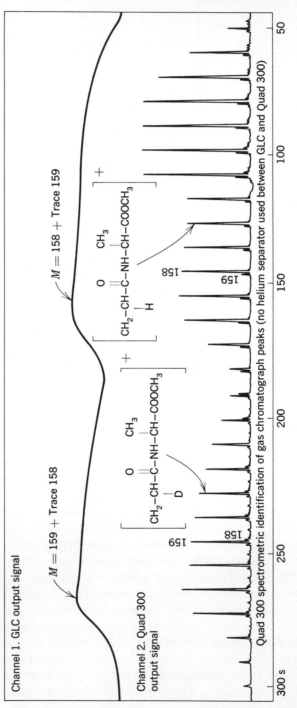

Fig. 21-17 Combined use of GC and mass spectrometry in the study of optical isomers. (*Courtesy of Dr. B. Halpern, and of Electronic Associates, Inc.*)

proximately 10-s intervals, which shows unequivocally that the isomer with the deuterium atom corresponds to the *second* GC peak.

A number of abridged mass spectrometers are manufactured for special purposes. A portable model designed to respond only to helium is widely used as a *leak detector* for vacuum systems. It is connected temporarily to the system under test, and a fine jet of helium played around any joints or other locations suspected of leaking. One advantage is that several leaks can be located without the necessity of repairing one before searching for the next.

Another application, for which a small quadrupole unit is suitable, is a *residual gas analyzer*. For this purpose, the mass spectrometer amounts to a very sensitive pressure gage that can determine not only the total residual pressure in an evacuated chamber, but also the partial pressure of each gas present.

CHEMICAL IONIZATION

The gas-phase chemical production of ions in the source chamber of a mass spectrometer was brought to general attention by a review article in 1968 (14). Suppose one wishes to obtain a mass spectrum of *n*-octadecane, $C_{18}H_{38}$. Conventional electron impact will give mostly fragments with 4, 5, and 6 carbon atoms; ions with 17 or 18 carbons do not appear. However, mixing the sample with a large excess (perhaps $10^4:1$) of methane will change the pattern profoundly. Now the principal ion is $C_{17}H_{37}^+$, and there are smaller amounts of ions with successively fewer carbon atoms. Also found will be CH_5^+ and $C_2H_5^+$ from the methane.

Clearly the probability of octadecane molecules being hit by electrons in the primary ionization process is far less than that of methane molecules being hit. Most of the observed ions of high mass must therefore result from ion-molecule reactions between, for example, CH_5^+ and n-$C_{18}H_{38}$. The methane serves as a source of reactant ions.

Mass spectrometric observations on ions produced in high-pressure (100 Pa ≈ 1 torr) methane alone show considerable amounts of CH_5^+ and $C_2H_5^+$ ions, which together represent almost 90 percent of the total ions present. The remainder is made up of $C_2H_3^+$, $C_2H_4^+$, $C_3H_5^+$, and $C_3H_7^+$. (In low-pressure methane none of these is significant, and we find only CH_4^+, CH_3^+, etc.) These ions result from such reactions as

$$CH_4^+ + CH_4 \rightarrow CH_5^+ + CH_3$$
$$CH_3^+ + CH_4 \rightarrow C_2H_5^+ + H_2$$

In the presence of octadecane, the predominant secondary reactions are H^- abstraction processes such as

$$CH_5^+ + C_{18}H_{38} \rightarrow C_{18}H_{37}^+ + CH_4 + H_2$$

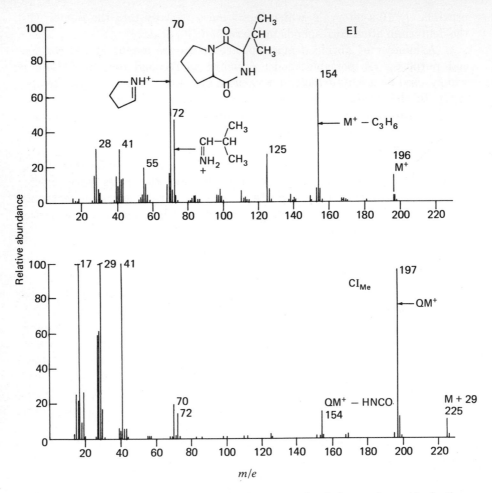

Fig. 21-18   Mass spectra of a diketopiperazine by conventional electron-impact ionization
(EI) and by chemical ionization (CI).   The reagent gas in the CI spectrum was methane.
Note the great intensity (hence sensitivity) of the $(M + 1)^+$ peak, designated QM^+ for
"quasi-molecular" ion.   The lesser peak at 225 represents addition of $C_2H_5^+$ and serves
to verify that the molecular ion must be 196.   This is an example of a computer-recon-
structed spectrum.   (*Finnigan Corporation.*)

Thus the mass spectra from chemical ionization sources are markedly
different from conventional spectra, and produce a different kind of information about ionization processes.   The structure of the source chamber
for chemical ionization and the accompanying vacuum system must be
specially designed to permit maintaining a pressure ratio of perhaps 10^5
from the inside to the outside of the exit slit.   Figure 21-18 shows the
mass spectra of the same compound as produced by electron-impact and
chemical ionization sources.

ION-CYCLOTRON RESONANCE (ICR) (15)

One other mass analyzer remains to be described briefly, a type particularly useful for study of ion-molecule reactions. Ions are formed by electron impact in a region of crossed magnetic and electric fields (Fig. 21-19), causing them to follow cycloidal trajectories, drifting away from the point of formation in a direction perpendicular to both fields with a velocity $v_d = E/B$, where E and B are the electric and magnetic field strengths. Ions are detected, somewhat as in NMR, by subjecting them to a weak high-frequency ac field, and noting resonant frequencies v_c at which energy is absorbed. This frequency is given by the relation

$$v_c = \frac{B}{2\pi} \frac{e}{m} \tag{21-9}$$

(This is the same equation describing the motion of protons in a cyclotron ion accelerator, whence the name ICR.) Typically B is large (up to approximately 1.5 T), to give frequencies in the range 10^5 to 10^6 Hz, and E is small (approximately 0.2 V $\cdot$ cm^{-1}), to keep v_d small.

Because of the cycloidal trajectory and moderate speed, the ions may cover an effective path length of tens or hundreds of metres in a space 10 or 15 cm long, within a period of several milliseconds. (Compare this

Fig. 21-19 ICR cell used for study of ion-molecule reactions. [*Analytical Chemistry* (15).]

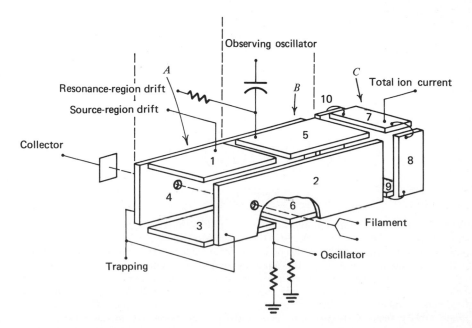

to residence times of microseconds and path lengths of 0.5 cm in a typical ion source for a conventional mass spectrometer.) This means that ion-molecule collisions have a probability of occurring perhaps 1000 times greater than in the chemical ionization source previously described. It follows that operation at lower pressure is practicable, with less stringent requirements on the vacuum pumps.

A less elaborate form of the ICR spectrometer, lacking the drift potential, is known as the *omegatron*. In it, resonant ions are collected at a detector electrode. This is used today only as a residual gas analyzer.

PROBLEMS

21-1 An organic compound is analyzed for its N content by isotope dilution (19). A measured amount of the compound containing ^{15}N in place of ^{14}N is added. After conversion of all the nitrogen to N_2, a mass spectrometer shows the following peak heights:

m/e	28	29	30
Height	978.5	360.6	52.5

Calculate the percent of the nitrogen which is ^{15}N.

21-2 An isotope dilution method for the determination of C in submilligram quantities with ^{13}C tracer has been reported by Boos et al. (20). The sample is mixed with a portion of succinic acid, $C_4H_6O_4$, which contains about 30 atom percent ^{13}C. The mixture is oxidized to CO_2 and H_2O, and the resulting CO_2 examined in a mass spectrometer. The ratio of mass 45 to 44 (corrected for the natural isotopic composition of O) is taken as the $^{13}C/^{12}C$ ratio, designated r. In natural C, the abundance of ^{13}C is 1.11 percent and ^{12}C 98.9 percent, which must be taken into account. The equations are as follows:

$$^{13}C_S = W_T \left(\frac{4}{119.3} \right) \left(\frac{r_T}{r_T + 1} \right) + W_S \left(\frac{X_C}{12.01} \right) (0.0111)$$

$$^{12}C_S = W_T \left(\frac{4}{119.3} \right) \left(\frac{1}{r_T + 1} \right) + W_S \left(\frac{X_C}{12.01} \right) (0.989)$$

$$r_S = \frac{^{13}C_S}{^{12}C_S}$$

where $^{12}C_S$ and $^{13}C_S$ represent the numbers of milligram atoms of the respective isotopes present in the mixed sample, r_S is the observed ratio, W_T and W_S are the weights in milligrams of tracer and sample, r_T is the ratio for pure tracer compound oxidized in the same manner, and X_C is the quantity sought, the weight fraction of C in the unknown. (a) Explain the above equations, and from them derive an expression for X_C in terms of W_S, W_T, and r_S. (b) In a particular analysis, 0.156 mg of sample and 0.181 mg of tracer were taken. The ratio r_S was found to be 0.206. The tracer contained 31.41 percent of its carbon as ^{13}C. Calculate the percent C in the sample.

REFERENCES

1. Roboz, J.: "Introduction to Mass Spectrometry," chap. 6, Wiley-Interscience, New York, 1968.
2. Chait, E. M.: *Anal. Chem.,* **44**(3):77A (1972).
3. Farmer, J. B.: In C. A. McDowell (ed.), "Mass Spectrometry," chap. 2, McGraw-Hill, New York, 1963.
4. Wiley, W. C.: *Science,* **124**:817 (1956).
5. Wiley, W. C., and I. H. McLaren: *Rev. Sci. Instrum.,* **26**:1150 (1955).
6. Karasek, F. W.: *Anal. Chem.,* **44**(4):32A (1972).
7. Ref. 1, pt. II.
8. McLafferty, F. W.: "Interpretation of Mass Spectra," 2d ed., W. A. Benjamin, Reading, Mass., 1973.
9. Shrader, S. R.: "Introductory Mass Spectrometry," Allyn and Bacon, Boston, 1971.
10. Wapstra, A. H., and N. B. Gove: *J. Nuclear Data,* **9**:267 (1972).
11. McLafferty, F. W.: *Anal. Chem.,* **28**:306 (1956).
12. Ref. 1, p. 319.
13. Ahearn, A. J. (ed.): "Trace Analysis by Mass Spectrometry," Academic, New York, 1972.
14. Field, F. H.: *Acct. Chem. Res.,* **1**:42 (1968).
15. Henis, J. M. S.: *J. Am. Chem. Soc.,* **90**:844 (1968); *Anal. Chem.,* **41**(10):22A (1969).
16. Elliott, R. M.: In C. A. McDowell (ed.), "Mass Spectrometry," chap. 4, McGraw-Hill, New York, 1963.
17. Lichtman, D.: *Res. Dev.,* **15**(2):52 (1964).
18. Watson, J. T., and K. Biemann: *Anal. Chem.,* **37**:844 (1965).
19. Grosse, A. V., S. G. Hindin, and A. D. Kirshenbaum: *Anal. Chem.,* **21**:386 (1949).
20. Boos, R. N., S. L. Jones, and N. R. Trenner: *Anal. Chem.,* **28**:390 (1956).

Chapter 22

Thermometric Methods

Many of the analytical methods discussed in other chapters have significant temperature coefficients, but in general their measurement does not itself provide analytical information. In the present chapter we shall consider a number of methods in which some property of the system is measured as a function of the temperature. It will help to clarify the relations between these to list them here for reference (Table 22-1) (1). There are other possible thermometric methods in addition to those listed, which are less used at present and will not be discussed in detail.

THERMOGRAVIMETRIC ANALYSIS (TGA)

This is a technique whereby the weight of a sample can be followed over a period of time while its temperature is being changed (usually increased at a constant rate). Several examples of thermograms obtained by this method are shown in Fig. 22-1 (2). Curve 1 shows the weight of a precipitate of silver chromate, collected in a filtering crucible. The initial drop in weight represents the loss of excess wash water. Just above 92°C the weight becomes constant and remains so to about 812°C. From there to 945°C oxygen is lost. The loss in weight shows that the decomposition

Table 22-1 Thermoanalytical methods

Designation	Property measured	Apparatus
Thermogravimetric analysis (TGA)	Change in weight	Thermobalance
Derivative thermogravimetric analysis (DTG)	Rate of change of weight	Thermobalance
Differential thermal analysis (DTA)	Heat evolved or absorbed	DTA apparatus
Calorimetric DTA	Heat evolved or absorbed	Differential calorimeter
Thermometric titration	Change of temperature	Titration calorimeter

proceeds according to the reaction $2Ag_2CrO_4 \rightarrow 2O_2 + 2Ag + Ag_2Cr_2O_4$. The residue is thus a mixture of silver and silver chromite. It follows that the silver chromate precipitate, if used for a gravimetric chromium analysis, may be dried anywhere in the plateau region between about 100 and 800°C, say at 110°C. Laboratory directions in older textbooks specified exactly 135°C.

The balance is calibrated, preferably each time it is used, by placing a known weight on the pan to give a reference mark, such as that in the upper right corner of Fig. 22-1.

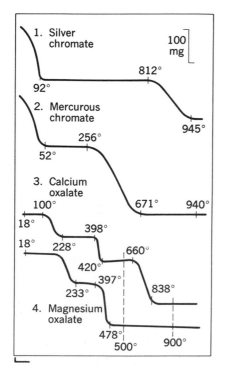

Fig. 22-1 Examples of curves taken with a thermobalance. [*American Elsevier* (2).]

Curve 2 of the same figure shows a heating curve for mercurous chromate. This compound is stable between about 52 and 256°C and then decomposes according to the equation $Hg_2CrO_4 \rightarrow Hg_2O + CrO_3$. The mercurous oxide is lost by sublimation, and the chromium trioxide remains at constant weight above 671°C. Because of the high atomic weight of mercury, the precipitate of mercurous chromate provides a particularly favorable gravimetric factor for the determination of chromium. It had previously been the practice to ignite the precipitate under a hood and weigh the chromium trioxide. Thermogravimetric study shows this procedure not only to be unnecessary, but to result in lowered precision.

Much of the reported work in thermogravimetry has been directed toward the establishment of optimum temperature ranges for the conditioning of precipitates for conventional gravimetric analysis, as the preceding examples suggest. The technique, however, has a greater potential than this.

Consider, for example, curves 3 and 4 of Fig. 22-1. A significant difference in behavior is observed between the oxalates of calcium and magnesium, which permits their simultaneous determination. Calcium oxalate loses its carbon and excess oxygen in two steps, $CaC_2O_4 \rightarrow CaCO_3 + CO$ and $CaCO_3 \rightarrow CaO + CO_2$, whereas the magnesium compound does not pass through the carbonate stage, $MgC_2O_4 \rightarrow MgO + CO + CO_2$. The ranges of stability are:

Compound	°C	Compound	°C
$CaC_2O_4 \cdot H_2O$	Up to 100	$MgC_2O_4 \cdot 2H_2O$	Up to 176
CaC_2O_4	226–398	MgC_2O_4	233–397
$CaCO_3$	420–660	MgO	480 and up
CaO	840 and up		

Thus, at 500°C, calcium carbonate and magnesium oxide are stable, whereas at 900°C both metals exist as the simple oxides. Comparison of the weights of a mixed precipitate at these two temperatures will permit calculation of both the calcium and the magnesium content of the original sample.

Another example is the analysis of copper-silver alloys based on the relative stabilities of the nitrates (Fig. 22-2) (2). $AgNO_3$ is stable up to 473°C, where it starts to lose NO_2 and O_2, leaving a residue of metallic silver above 608°C. $Cu(NO_3)_2$, on the other hand, decomposes in two steps to the oxide CuO, which is the stable form up to at least 950°C. A binary alloy can be analyzed by successive weighings at 400 and 700°C in a short time (perhaps 30 min) with an accuracy of about ±0.3 percent.

The limiting temperatures of the various segments of the thermograms such as those of Figs. 22-1 and 22-2 cannot be considered reproducible

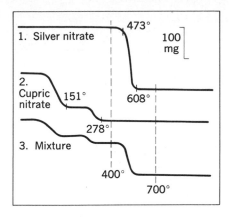

Fig. 22-2 Examples of curves taken with a thermobalance. [*American Elsevier* (2).]

without qualification. The thermogravimetric method, as ordinarily carried out, is a dynamic one, and the system is never at equilibrium. Hence the temperatures of distinctive features on the curves are somewhat different as observed on different instruments, or on the same instrument at different rates of temperature scanning, or with different size samples, etc.

THERMOBALANCES

There are more than a dozen manufacturers of thermobalances, and some of them produce several models. The weighing mechanism may be a modification of a single- or double-pan balance, an electronically self-balancing device, a torsion balance, or a simple spring balance. Several models have an electric furnace for heating the sample, located beneath the balance; the crucible is suspended within it by a long platinum wire. This design requires stringent precautions against convection effects interfering with the balance. Some designers have preferred to place the furnace above the balance, with the crucible supported at the top of a rod extending upward from the balance beam.

All thermobalances intended for precise work have provision for automatic recording of weight either against time or, with an XY recorder, directly against temperature. If plotted against time, either the temperature must be programmed to increase at a steady rate, or a second pen must be arranged to plot the temperature-time relationship. An example of the latter is seen in Fig. 22-5.

There are also a number of nonrecording thermobalances, though not usually dignified by that name, intended principally for the determination of superficial moisture in bulk materials. The sample is heated, as by an infrared lamp, while on the pan of a balance specially designed to minimize errors produced by air currents. Readings are taken manually

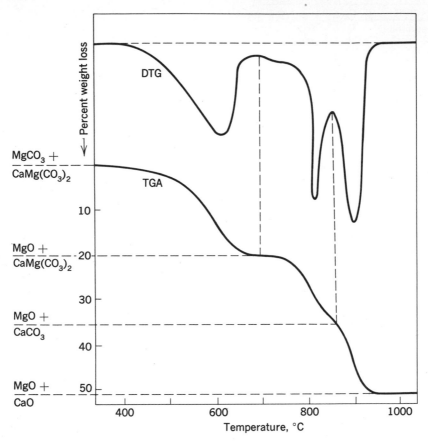

Fig. 22-3 Relation between TGA and DTG curves for the pyrolysis of mixed calcium and magnesium carbonates. [*Hungarian Scientific Instruments* (3).]

until no further change is seen. A precision of a few tenths of a percent water in a 1-g sample can be obtained in a few minutes.

DERIVATIVE THERMOGRAVIMETRIC ANALYSIS (DTG)

There is sometimes an advantage in being able to compare a thermogram with its first derivative, as in Fig. 22-3 (3). The plateau in the thermogram at 700°C is clear enough, but the shoulder at about 870°C could not have been pinpointed without the derivative curve.

Several commercial thermobalances are provided with electronic circuits to take the derivative automatically. A two-pen recorder permits a convenient direct comparison of the two curves.

DIFFERENTIAL THERMAL ANALYSIS (DTA)

This is a technique by which phase transitions or chemical reactions can be followed by observation of the heat absorbed or liberated. It is especially suited to studies of structural changes within a solid at elevated temperatures, where few other methods are available.

In a typical apparatus, one set of thermocouple junctions (Fig. 22-4) is inserted into an inert material, such as aluminum oxide, which does not change in any manner through the temperature range to be studied. The other set is placed in the sample under test. With constant heating, any transition or thermally induced reaction in the sample will be recorded as a peak or dip in an otherwise straight line. An endothermic process will cause the thermocouple junction in the sample to lag behind the junction in the inert material, and hence develop a voltage; an exothermic event will cause a voltage of opposite sign. It is customary to plot exotherms upward and endotherms downward, but this convention is not universally followed.

Since the usual mode of operation is to supply heat to the samples, endothermic events are more likely to occur than exothermic, and the latter when they are observed are often caused by secondary processes.

As an example of this approach, consider the curves of Fig. 22-5 (4). Curve 1 is essentially the same as curve 3 of Fig. 22-1; the slight differences result from differing instrumental conditions. Curve 2 represents the same type of experiment, but the sample is in an atmosphere of CO_2 rather than in air; as should be expected, no difference is evident until the decomposition of $CaCO_3$, which now requires a higher temperature.

Curve 3 is a differential thermogram (DTA curve) also showing the decomposition of calcium oxalate in an atmosphere of CO_2. It is seen that the three points of weight loss correspond to three endothermic pro-

Thermocouple wells

Fig. 22-4 Simple arrangement for DTA.

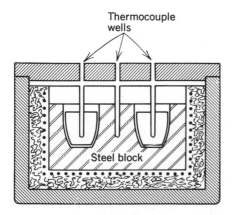

Steel block

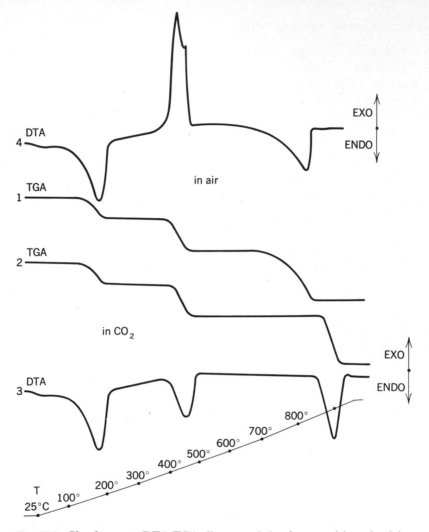

Fig. 22-5 Simultaneous DTA-TGA diagram of the decomposition of calcium oxalate monohydrate in air and in carbon dioxide. [*Plenum* (4).]

cesses: it requires energy to break the bonds in the successive elimination of H_2O, CO, and CO_2. By contrast, the second peak in curve 4, where the atmosphere is air, is sharply exothermic, but corresponds to the same weight loss.   The explanation for the difference lies in the exothermic burning of CO in air at the temperature of the furnace.

Comparison of Figs. 22-3 and 22-5 reveals that there is a degree of similarity between the curves obtained in DTG (derivative thermograms) and DTA (differential thermograms). DTG can give information

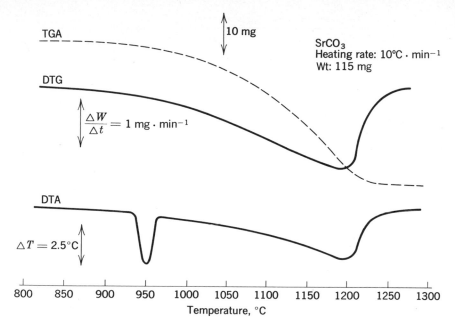

Fig. 22-6 Decomposition of $SrCO_3$ in air, showing the DTA carbonate decomposition endotherm overlapping the endothermic rhombic-hexagonal crystalline transition. (*Mettler Instrument Corporation.*)

only about weight changes, whereas DTA will reveal changes in energy, regardless of constancy or change in weight. In Fig. 22-5, if DTG curves had been included, they would have been identical up to about 700°C, even though the DTA curves vary greatly. Figure 22-6, on the other hand, shows a situation in which a pronounced endotherm in DTA (at 950°C) fails to show up at all in either TGA or DTG; this is evidence of a crystalline transition, from rhombic to hexagonal modifications of $SrCO_3$, which, of course, does not involve a change in weight.

DTA APPARATUS

There are many companies in the DTA field. Their products vary with respect to such parameters as sample size, temperature range, selectable scanning rates, precision, convenience, and cost. Figure 22-7 shows schematically the essential parts of a typical DTA apparatus, which includes provision for bathing the samples with a controlled atmosphere; the gas can be made to flow *through* the bed of particulate sample, thereby flushing away any gaseous product of decomposition.

In this apparatus, as well as in some thermobalances, provision is made for detecting and analyzing effluent gases, by GC or other means.

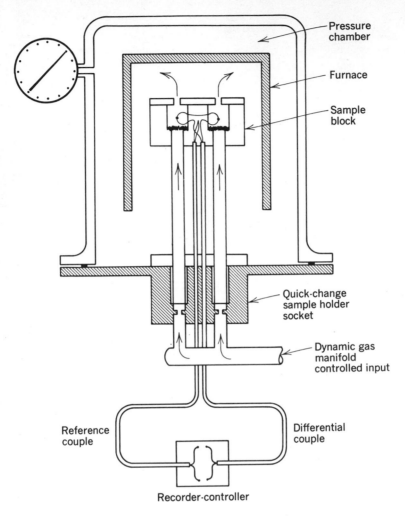

Fig. 22-7 DTA apparatus, schematic. (*Tracor, Inc.*)

Several companies have combined DTA and TGA capabilities into a single instrument. The record is obtained on a two-pen recorder.

SCANNING CALORIMETRIC DTA

Conventional DTA as just described is capable of giving good qualitative data about the temperatures and signs of transitions, but it is difficult or impossible to obtain quantitative data—the heat of transition if the purity is known, or the amount of a constituent in a sample if the heat of transition is known. This difficulty arises from uncontrollable and frequently unknown factors such as the specific heat and thermal conductivity of the sample both before and after the transition. The rate of heating, placement of thermocouples, and other instrumental parameters will also affect the areas beneath endotherms or exotherms.

Quantitative results can be achieved by converting the sample compartment in a DTA apparatus to a differential calorimeter (5). This has been done in three different ways by three different companies. Perkin-Elmer manufactures an instrument called a Differential Scanning Calorimeter (DSC), in which the calorimeter is of the isothermal type (6). Each sample holder (unknown and reference) is provided with its own resistive heater. When the differential thermocouple starts to register a voltage, an automatic control loop sends just enough power into whichever of the two samples is the cooler, to counteract the trend and keep the two temperatures within a very small fraction of a degree of equality. A second electronic control loop forces the temperature of the reference (hence effectively of both) samples to increase linearly with time. The recorder traces out the electric power which has to be delivered to one or the other sample to maintain isothermal conditions. The resulting thermogram resembles conventional DTA, but the area beneath a peak is an *exact* measure of the energy supplied to the unknown sample to compensate for an endothermic event, or to the reference material to equal the energy emitted in the unknown when an exothermic event takes place. Differences in thermal conductivity, heat capacity, etc., are now irrelevant.

Technical Equipment Corporation* has chosen an adiabatic approach in designing their comparable unit, the Deltatherm Dynamic Adiabatic Calorimeter (DAC) (7). In an adiabatic calorimeter there must be no passage of heat between the sample and its surroundings so, instead of a reference sample, the DAC has a massive copper block, with the sample in a central cavity but thermally isolated from its walls. The block is heated at a constant rate, and the amount of power required to maintain the sample at the same temperature as the surrounding walls is recorded. This device gives comparable precision in the measurement of heats of transition, but the adiabatic system facilitates the determination of specific heats as well.

The third instrument in this category is the Du Pont DTA apparatus with accessory calorimeter. This is also adiabatic, but makes use of a reference sample. The reference temperature actuates the x axis of an xy recorder, while the temperature differential between the two samples controls the y input. The area beneath a DTA peak under these conditions is an accurate measure of the heat of transition, independent of specific heat and other variables; specific heats can also be determined.

A scanning calorimeter provides a convenient means for the precise determination of impurity limits in high-purity organic compounds, through observation of the melting-point depression (8, 9). Experimental and

* Denver, Colo.

calculational details are given in the references. The standard deviation of the results is reported to be ± 4 percent of the impurity measured.

THERMOMETRIC TITRATIONS (10)

Since practically all chemical reactions are accompanied by a heat effect, it is possible to follow the course of a reaction by observing the heat liberated or absorbed. Such a measurement can be made by titration in a small Dewar flask or a beaker nested in Styrofoam. The temperature can be read with a thermistor thermometer.

Thermometric titration can readily be automated, and several firms manufacture suitable equipment. In most models titrant is added at constant rate by a syringe pump or its equivalent, and the resistance of the detector monitored on a strip-chart recorder.

A large variety of titrations have been followed with success by the thermometric method. These include neutralizations of both strong and weak acids, precipitations, redox reactions, and complex formations. Any solvent can be employed; besides water, work has been reported in acetic acid, carbon tetrachloride, benzene, and nitrobenzene, and in the fused eutectic of lithium and potassium nitrates. Reported precision is usually better than ± 1 percent standard deviation, sometimes much better.

It is essential that the two reacting solutions do not differ appreciably with respect to extraneous materials that would contribute noticeable heat effects either by reaction with each other or with the solvent (heat of dilution).

It has been pointed out by Jordan (10) that thermometric titrimetry constitutes one of a very few methods of titration not based solely on consideration of the free-energy change ΔG, hence on the equilibrium constant of the reaction. The quantity measured is ΔH, not ΔG, in the familiar thermodynamic equation $\Delta H = \Delta G + T \Delta S$. Hence thermometric titrations may give useful results even if ΔG is zero or positive. Two examples will indicate the value of this approach.

The neutralization of boric acid follows the equation $H_3BO_3 + OH^- \rightleftharpoons H_2BO_3^- + H_2O$. Given the first ionization constant $k_a = 5.8 \times 10^{-10}$ (at 25°C), it follows that the standard free-energy change of this reaction is $\Delta G° = -6.5$ kcal·mol^{-1}, corresponding to a neutralization constant $K_n = K_a/K_w = 5.8 \times 10^{-4}$. This may be compared to hydrochloric acid, where the only reaction to be considered is $H^+ + OH^- \rightleftharpoons H_2O$, $K_n = 1/K_w = 1 \times 10^{14}$, and $\Delta G° = -19.2$ kcal·mol^{-1}. Thus boric acid cannot be titrated successfully by any method that depends on the pH (such as potentiometric or photometric techniques), because the hydrogen-ion activity is determined by the equilibrium constant.

However, it so happens that the entropy term $T \Delta S°$ is -3.7 kcal · mol^{-1} for boric acid, as against $+5.7$ for hydrochloric acid. These

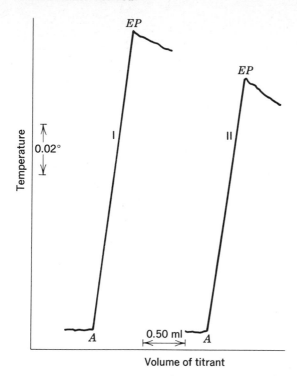

Fig. 22-8 Typical thermometric titration curves, in 0.01 *F* aqueous solutions. (I) Hydrochloric acid and (II) boric acid, titrated by NaOH. *A* indicates the start of the titration, and *EP* the end point. [*Record of Chemical Progress* (11).]

combine to give enthalpies of neutralization of -10.2 kcal · mol^{-1} for boric and -13.5 kcal · mol^{-1} for hydrochloric, showing that they can be titrated thermometrically with equal ease. This is illustrated in Fig. 22-8 (11).

Another interesting situation arises in the titration of calcium and magnesium with EDTA. The stability constants of the chelates differ by less than two orders of magnitude, so that titration based on an indicator can give only the sum of the two ions. However, the entropy of the reaction of Mg^{2+} with EDTA is twice that for the Ca^{2+} reaction. This not only gives a distinct difference in ΔH values, but it actually changes the sign: ΔH° is $+5.5$ kcal · mol^{-1} for the magnesium reaction, and -5.7 kcal · mol^{-1} for calcium. A titration curve for a mixture of the two is given in Fig. 22-9 (10).

It is possible to improve the precision (or to work at greater dilution) by electronic compensation of errors due to heat transfer to the surroundings. This can be accomplished with the operational amplifier circuit of Fig. 22-10 (12). The unbalance voltage from the thermistor bridge is

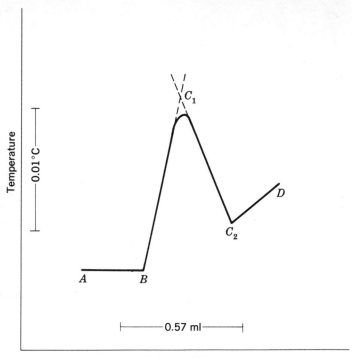

Fig. 22-9 Titration of a mixture of 0.25 mmol each of Ca^{2+} and Mg^{2+} ions with EDTA. The titration was started at B and showed a calcium end point extrapolated to C_1 and a magnesium end point at C_2. C_2 to D represents excess reagent. [*Wiley-Interscience* (10).]

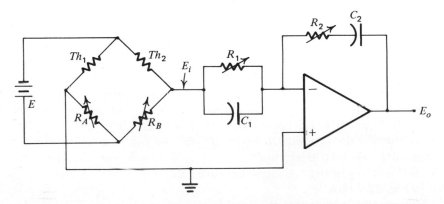

Fig. 22-10 An operational amplifier circuit for use in thermometric titration. Th_1 and Th_2 are thermistors; R_A and R_B are zero-adjusting resistors. For a detailed explanation of the function of the RC combinations, see the reference. [*Analytical Chemistry* (12), *adapted.*]

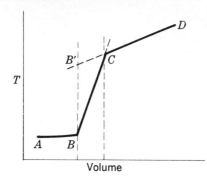

Fig. 22-11 Idealized thermometric and enthalpometric titration curve.

impressed on point E_i. Proper adjustment of the two resistors R_1 and R_2 will produce the required corrections, so that the voltage appearing at E_0 will closely approximate an ideal titration curve like that of Fig. 22-11.

Thermometric titrimetry can give information in addition to the usual stoichiometry. Figure 22-11 represents a generalized titration curve. Segment AB is a pretitration trace on the recorder, to establish a base line; the titration is started at point B. In the vicinity of C the trace often shows curvature rather than a sharp angle, and this can be related to the equilibrium constant (hence to ΔG°) for the reaction. From C to D the curve commonly rises as shown, corresponding to the heat of dilution of the reagent; it may, however, slope downward, indicating either that the dilution is endothermic (which is unusual) or that the reagent is cooler than the contents of the titration vessel. The temperature change that can rightly be ascribed to the reaction is ΔT, from B to the extrapolated intersection of CD with the zero-time ordinate. ΔT is proportional to the total amount of heat liberated, $Q \equiv -N \Delta H$, where N is the number of moles reacting, and ΔH is the heat absorbed per mole. The constant of proportionality is the heat capacity of the vessel and its contents, which we will designate as k in kilocalories per degree, so that

$$\Delta T = \frac{Q}{k} = -\frac{N \Delta H}{k} \qquad (22\text{-}1)$$

The quantity k can be determined by a simple calibration step, for example, by a known amount of electrical heating. Then N can be determined if ΔH° is known, or vice versa, if we make the assumption that ΔH and ΔH° do not differ materially. Note that the calorimetric determination of N does not require a standardized titrant, and is not strictly a titration. This procedure is called *enthalpimetry*.

It follows from Eq. (22-1) that the temperature change ΔT in a given reaction is determined only by the number of moles of substance reacting. This makes possible an extremely simple analytical procedure known as *direct-injection enthalpimetry* (DIE) (10, 13). A small volume of concentrated reagent is added in a single dose to a specified volume of a dilute solution of the sample to be titrated, and the resulting temperature change noted. It is important that the reagent be added in considerable excess, and that it be injected, with stirring, as nearly instantaneously as possible. Typically the reagent might be 300 µl of a 1 *M* solution, for a sample of the order of 100 µmol in 25 ml. Figure 22-12 shows a few examples. The reagent solution need not be standardized, but the whole procedure must be calibrated with known samples. The electronic

Fig. 22-12 Typical injection enthalpograms. (I) HCl treated with excess NaOH; (II) H_3BO_3 with excess NaOH; (III) Mg^{2+} with excess EDTA; (IV) Pb^{2+} with excess EDTA. Sample quantities were of the order of 0.1 mmol in 25 ml. At time *IT* 0.300 ml of 1 *M* reagent was injected. ΔE is the extrapolated unbalance potential of the thermistor bridge. [*Wiley-Interscience* (10).]

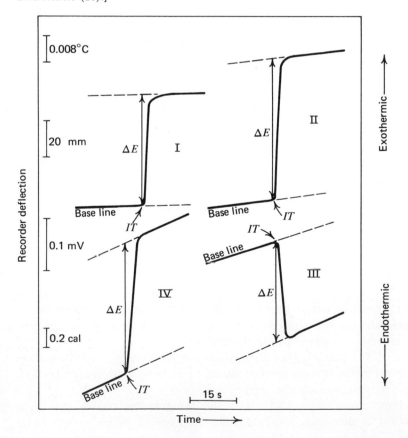

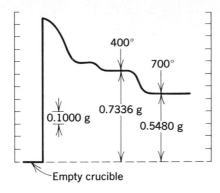

Fig. 22-13   Thermobalance trace of a mixture of copper and silver nitrates.

compensation technique mentioned previously can be applied to DIE as well (12).

PROBLEMS

22-1   A sample of a copper-silver alloy is dissolved in nitric acid in a small crucible, the crucible placed in the thermobalance, and its temperature gradually raised to 750°C. The trace obtained (weight against time) is shown in Fig. 22-13, together with the weights of residue as read from the graph. Calculate the composition of the alloy in terms of percent by weight.

22-2   Microcosmic salt, $Na(NH_4)HPO_4 \cdot 4H_2O$, upon heating, first liberates four water molecules, and then another water and a molecule of NH_3, to end up as sodium metaphosphate, $NaPO_3$. Sketch curves which you might expect from a study of this substance (*a*) with a thermobalance, and (*b*) by DTA.

22-3   What is the end product of pyrolysis of $SrCO_3$, as shown by Fig. 22-6?

REFERENCES

1.   Wendlandt, W. W.: "Thermal Methods of Analysis," Wiley-Interscience, New York, 1964.
2.   Duval, C.: "Inorganic Thermogravimetric Analysis," 2d ed., American Elsevier, New York, 1963.
3.   Paulik, F., J. Paulik, and L. Erdey: *Hung. Sci. Instrum.,* **1**:3 (1964).
4.   Vaughan, H. P., and W. G. Wiedemann: In P. M. Waters (ed.), "Vacuum Micro-balance Techniques," Plenum, New York, 1965.
5.   Wilhoit, R. C.: *J. Chem. Educ.,* **44**:A571, A629, A685, A853 (1967).
6.   Watson, E. S., M. J. O'Neill, J. Justin, and N. Brenner: *Anal. Chem.,* **36**:1233 (1964).
7.   Dosch, E. L.: *Instrum. Soc. Am. Conf. Proc. Preprint* No. 2.6-5-64 (1964).
8.   Plato, C., and A. R. Glasgow, Jr.: *Anal. Chem.,* **41**:330 (1969).
9.   Plato, C.: *Anal. Chem.,* **44**:1531 (1972).
10. Jordan, J.: In I. M. Kolthoff and P. J. Elving (eds.), "Treatise on Analytical Chemistry," pt. I, vol. 8, chap. 91, Wiley-Interscience, New York, 1968.
11. Jordan, J.: *Record Chem. Progr.,* **19**:193 (1958).
12. Nakanishi, M., and S. Fujieda: *Anal Chem.,* **44**:574 (1972); **46**:119 (1974).
13. Wasilewski, J. C., P. T. -S. Pei, and J. Jordan: *Anal. Chem.,* **36**:2131 (1964).

Chapter 23

Radioactivity as
an Analytical Tool

Those properties of atomic nuclei that are of analytical significance include nuclear masses and spins, both considered in previous chapters. Also important is the ability of some nuclei to undergo specific reactions. Such reactions include spontaneous radioactive decay accompanied by release of particulate and electromagnetic radiation, on the one hand and, on the other, addition reactions in which the nucleus combines with a neutron or another particle. A related topic is Mössbauer spectroscopy, based on resonance absorption of gamma photons by certain nuclei.

RADIOACTIVITY

Although it may be assumed that the reader has a knowledge of the rudiments of radioactive phenomena, it will be appropriate to summarize those principles briefly for reference purposes.

Nearly all known elements exist in several isotopic forms. Many of these isotopes do not occur in nature, but can be formed artificially by various procedures from suitable isotopes of the same or other elements. Most artificially prepared isotopes and many naturally occurring ones are unstable, in that their nuclei tend to disintegrate spontaneously with the

Table 23-1   Particles produced in radioactive decay

Particle	Symbol	Mass*	Charge†	Penetrating power	Ionizing power
Electron	β^-	5.439×10^{-4}	-1	Medium	Medium
Positron	β^+	5.439×10^{-4}	$+1$	Medium	Medium
Alpha particle	α	3.9948	$+2$	Low	High
Neutron	n	1.0000	0	Very high	Nil
Photon (gamma ray)	γ	0	0	Very high	Very low

* In units of 1.673×10^{-27} kg.
† In units of 1.60240×10^{-19} C.

ejection of energetic particles or, in some cases, with the emission of radiant energy. The other product of the disintegration is a residual nucleus slightly lighter in mass than before. This is the phenomenon we call *radioactivity.*

Several different types of particles are ejected by radioactive substances. Those of importance for our purposes (see Table 23-1) are the electron (negative), the positron (or positive electron), the alpha particle, and the neutron. The emission of these particles is frequently, but not always, accompanied by the radiation of energy as gamma rays. Another mode of radioactive decomposition sometimes encountered is the spontaneous capture by the nucleus of an electron from the K level (or, less frequently, from the L or higher levels). This process, known as *electron capture,* is most commonly evidenced by the emission of the characteristic x-rays produced by electrons from higher energy levels falling in to fill the vacancy created by the capture.

The particles and radiations from different radioactive nuclei vary widely in their energy content and in the frequency with which they are produced. Both of these properties are characteristic of the particular isotope which is disintegrating, and hence their measurement under suitably standardized conditions will serve to prove the presence of that isotope.

The frequency of occurrence of atomic disintegrations is related to a constant characteristic of each active isotope, namely, its *half-life,* the time required for any given sample of the isotope to be reduced to one-half its initial quantity. This varies among the known active materials from millionths of a second to millions of years. The half-lives at either extreme, of course, cannot be measured directly, but are inferred from other evidence. Isotopes of very short life cannot be useful for analytical purposes, simply because any experiment takes a considerable amount of time, and these isotopes disappear too quickly. On the other hand, isotopes of very long life are difficult to apply, because the disintegrations are too infrequent. Isotopes useful for analytical applications are those with half-lives roughly between a few hours and a few thousand years. If the experiments can be carried out quickly in the same laboratory where the radioactive material is prepared, the lower limit can be reduced to perhaps 10 min. The length of any experiment generally cannot exceed about 10 times the half-life of the isotope employed.

A selection of radioisotopes which have been found useful in analytical applications is given in Table 23-2. Radioisotopes may be useful as sources of radiations, or as tracers to follow some reaction or process and assist in determining its extent. Before turning to these applications, we shall consider the methods of detection and measurement.

DETECTORS OF RADIATIONS

The detectors described in Chap. 10 for use with x-rays are also applicable to the detection of radioactivity. Gamma rays are of course physically indistinguishable from x-rays. Particulate radiations (except neutrons) have less ability to penetrate solids, hence will suffer loss in passing through the walls or surface layers of some types of detectors, even though once in the sensitive region they might well be easily detectable.

Photographic detection is used primarily for mapping the distribution of radioactive materials on a solid surface, a process known as *autoradiography* or *radioautography*.

SCINTILLATION COUNTERS

When a ray or particle strikes a suitable fluorescent material, a tiny flash of visible light is emitted. Counting such flashes therefore provides a measure of the number of incident particles or photons. The circuitry used for this purpose is similar to that described in Chap. 3 for photon counting in the ultraviolet.

Scintillators can be either solids or liquids. The most used solid is thallium-activated sodium iodide. It is often shaped as in Fig. 23-1, to accept samples inserted in a cylindrical well. This provides a high efficiency of collection of radiations. Good optical contact between the scintillator and photomultiplier is essential, and the surrounding walls must be highly reflective to minimize loss of light. The whole assembly is shielded with a dense metal, usually lead, to reduce background interference.

Also available are high-efficiency units in which a pair of scintillators encloses the relatively small sample as in a sandwich, so that essentially all the emitted radiation enters the detector. These are called 4π detectors.

For quantitative measurements of isotopes which emit low-energy beta particles, such as ^{14}C, ^{35}S, and especially ^3H (tritium), *a liquid scintillator* is to be preferred, into which the active compound can be directly

Table 23-2   Radioisotopes used in analysis*

Isotope	Type of decay†	Half-life	Energy of radiation, MeV	
			Particles	Gamma transitions
^{3}H	β^-	12.26 years	0.0186	None
^{14}C	β^-	5720 years	0.155	None
^{22}Na	β^+ (90%)	2.58 years	0.545	1.27
	EC (10%)			
^{32}P	β^-	14.3 days	1.71	None
^{35}S	β^-	87 days	0.167	None
^{36}Cl	β^-	3.0×10^5 years	0.714	None
^{40}K	β^- (89%)	1.27×10^9 years	1.32	1.46
	EC (11%)			
^{42}K	β^-	12.36 h	3.55 (75%)	1.52 (25%)
			1.98 (25%)	
^{45}Ca	β^-	165 days	0.255	0.32
^{51}Cr	EC	27.8 days	. . . .	0.32 (8%)
^{55}Fe	EC	2.60 years	. . . .	None
^{59}Fe	β^-	45 days	0.460 (50%)	1.29, 1.10
			0.27 (50%)	
^{57}Co	EC	270 days	. . . .	0.122, 0.0144, 0.136
^{60}Co	β^-	5.26 years	0.32	1.333, 1.173
^{65}Zn	EC (97.5%)	245 days	0.33	1.11
	β^+ (2.5%)			
^{85}Kr	β^-	10.6 years	0.67	(γ)
^{90}Sr	β^-	29 years	0.54	None
^{90}Y	β^-	64 h	2.27	(γ)
^{95}Zr	β^-	65 days	0.36, 0.40	0.72, 0.76
^{95}Nb	β^-	35.1 days	0.16	0.77
^{110}Ag	β^-	253 days	0.085 (58%)	0.44, 2.46
^{119m}Sn	IT	250 days	. . . .	0.065, 0.024
^{131}I	β^-	8.06 days	0.60 (87.2%)	0.364 (80.9%) (others)
			(others)	
^{137}Cs	β^-	30 years	0.51 (92%)	0.662 (from ^{137m}Ba
			1.17 (8%)	daughter)
^{133}Ba	EC	7.2 years	. . . .	0.360, 0.292, 0.081,
				0.070
^{140}La	β^-	40.2 h	1.34 (70%) (others)	0.49, 0.82, 1.60 (others)
^{147}Pm	β^-	2.65 years	0.225	(γ)
^{170}Tm	β^-	127 days	0.97 (76%)	0.084
			0.88 (24%)	
^{203}Hg	β^-	47 days	0.21	0.28
^{198}Au	β^-	2.70 days	0.96 (others)	0.412 (others)
^{204}Tl	β^- (98%)	3.80 years	0.76	None
	EC (2%)			
^{210}Pb	β^-	22 years	0.015, 0.061	0.046

* Data selected from extensive tabulation by Friedlander, Kennedy, and Miller (1).
† EC = electron capture; IT = internal transition.

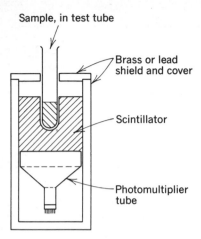

Sample, in test tube

Brass or lead
shield and cover

Scintillator

Photomultiplier
tube

Fig. 23-1 Well scintillation counter, schematic.

incorporated. This ensures maximum efficiency in the production of scintillations from betas. A number of organic compounds will act as scintillators when dissolved in suitable solvents. These include anthracene, p-terphenyl, 2,5-diphenyloxazole (PPO), α-naphthylphenyloxazole (NPO), and phenylbiphenyloxadiazole (PBD). Of these, PPO is the most effective, but its emitted radiation is ultraviolet. It is customary to mix with PPO a *secondary scintillator*, which translates through a fluorescent mechanism the ultraviolet scintillations into the visible range. The most used secondary scintillator is 1,4-bis-2-(5-phenyloxazolyl)-benzene (POPOP) or its dimethyl derivative (dimethyl-POPOP). A recommended solution contains 5 g·l^{-1} PPO and 0.3 g·l^{-1} dimethyl-POPOP in toluene.

When working with such low-level activity, special precautions must be taken to distinguish those pulses which are produced by scintillations from spurious pulses arising in the photomultiplier, caused either by stray or cosmic radiation, or by shot-effect noise. This can be accomplished by employing two identical photomultipliers looking at the same scintillator and connected in a *coincidence circuit* (Fig. 23-2). This consists of an electronic unit called an *AND-gate,* which will transmit a signal to the counting equipment only when it receives *simultaneous* pulses from both photomultipliers. The shot noise, being random, will occasionally produce simultaneous signals and, if necessary, this source of undesired counts can be reduced still further either by a triple coincidence circuit with three photomultipliers, or by reducing the thermal motion of electrons by refrigeration.

Scintillation counting is inherently proportional, in that the energy in each flash of light is determined by the energy of the particle which originated it. This proportionality can be maintained through the detector and amplifier to the ultimate record. NaI(Tl) gives a considerably larger

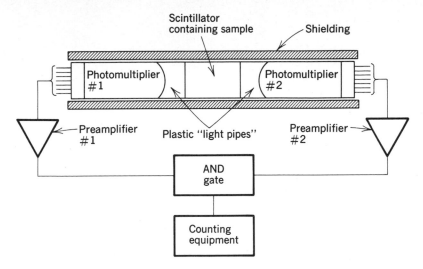

Fig. 23-2 Liquid scintillation counting assembly, schematic. The sample, photomultipliers, and preamplifiers may be refrigerated to decrease electrical noise.

pulse than a liquid scintillator for the same excitation, but is not able to count as fast by a factor of approximately 1000 (2).

GAS-IONIZATION DETECTORS

As indicated in Table 23-1, all types of nuclear radiation (except neutrons) produce significant ionization in materials into or through which they penetrate. (It is this ionization which is the direct cause of radiation damage to living material.) Ionization due to radiation is most easily measured in gases.

Let us consider the phenomena which occur in a gas-filled glass vessel provided with two electrodes, one of which is a metal tube perhaps 2 cm in diameter and 10 cm in length, and the other a wire passing along the axis of the cylinder (Fig. 23-3). Let the central wire be connected through a high resistance R_1 to the positive terminal of a variable-voltage dc supply, while the outer electrode is held at ground potential, as is the negative of the power supply. The positive electrode is connected also, normally through a capacitor C, to the input of an amplifier (indicated in Fig. 23-3 as an operational amplifier). The output of the amplifier is made to operate a deflecting meter or an electromechanical counting register. Let us subject the tube to a constant small source of energetic beta particles, which we will assume to be few enough per second to cause individual pulses of ionization. We will now increase the applied potential

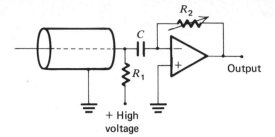

Fig. 23-3 Elementary circuit for an ionization chamber.

gradually from zero to several thousand volts. At small voltages (region *A* of Fig. 23-4) (3) the ions produced are accelerated only slowly by the electric field, and many of them recombine to form neutral molecules before they can reach the electrodes. As the potential is increased, the number of ions per pulse reaching the electrodes also increases, until a condition of saturation is attained, where essentially *all* the ions formed are discharged at the electrodes, and the observed size of the pulses is constant over a region of 100 V or more (region *B*). This is called the *ionization chamber* region.

Upon increasing the voltage beyond this region, at *C*, the pulse size again increases, because the ions are accelerated sufficiently to cause secondary ionization by collision with other molecules of the gas. In this portion of the curve, the height of a pulse is still dependent on the energy of the original ionizing event. The pulse height is now a *multiple* of that caused by the primary ionization; hence this part of the curve is called

Fig. 23-4 Number of ions collected as a function of applied voltage. Curve *a*, alpha particles; curve *b*, beta particles. [*Van Nostrand* (3).]

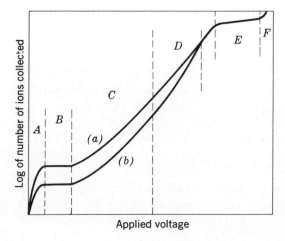

the proportional region, and a device operating under these conditions is called a *proportional counter.*

Increasing the voltage into region D produces greatly increased secondary ionization, so that proportionality is lost. At E, a plateau is attained where over a band of 100 or 200 V all pulses have equal magnitude, regardless of the energy of the ionizing particle. This is the Geiger region, and the counter used in this manner is a *Geiger counter* (sometimes called a Geiger-Müller, or G-M counter). Beyond this plateau (at F) the tube breaks into a continuous glow discharge.

Thus detection and counting of pulses are possible in three modes, corresponding to B, C, and E. The relative advantages and disadvantages of each can be summarized as follows.

The ionization chamber has the advantage of requiring only a low voltage (100 to 200 V), but the current it passes is very small (perhaps 10^{-8} A) and therefore requires either a high-gain amplifier or a highly sensitive electrometer. In practice, it is most useful for pulse work with the highly ionizing alpha particle. With beta and gamma particles it is ordinarily restricted to the measurement of relatively intense beams of radiation, where the current is continuous rather than pulsed. For this application, the capacitor C of Fig. 23-3 is removed, and the amplifier must be of the direct-coupled type which can respond to slow changes in a minute direct current.

The proportional counter operates at a much higher potential, perhaps 1000 to 2000 V. As with the ionization chamber, the size of pulse is proportional to the number of ions produced by the primary particle, but the ion-multiplication process results in internal amplification which may be as great as 10^3. The external amplifier therefore need not have unusually high gain, but if observed pulse heights are to reflect the energy of the original particle, it must be linear (i.e., distortion-free). A counter operated in this region has an extremely short recovery time, and can count more than 10^5 pulses per minute. A well-regulated voltage supply must be provided.

The Geiger counter takes a somewhat higher voltage (though still in the 1000- to 2000-V range), but the pulses are so large that little or no amplification is required. This favorable situation is often more than offset by the fact that the pulses are all uniform in size and therefore give no information about the energy of the ionizing particles. Another disadvantage is the slow speed which limits counting rates to about 10^4 per minute.

Geiger counters are used primarily in portable rate meters and survey instruments, where simplicity of the associated circuits is more important than high accuracy. Pulse-height discrimination is not possible, but it is easy to distinguish between beta and gamma radiation by an aluminum shield, as this will exclude betas without affecting gamma rays.

The operating regions for Geiger and proportional counters must be determined by plotting the counting rate, in counts per minute, for a fixed source of radiation, against applied volts. Precautions must be taken to prevent the positive ions which are accelerated toward the outer (negative) electrode from causing the emission of secondary electrons. Such electrons would move toward the central wire, produce secondary ionization of the gas, and a continuous discharge would result. This can be avoided by adding to the filling gas a few percent of an organic compound (alcohol or methane) or a halogen. The positive ions transfer their excess energy to molecules of the additive rather than to the electrode.

A modification of the proportional counter that is particularly convenient, especially for weakly penetrating radiations, is the *flow counter*, in which the gas (e.g., argon–10 percent methane) is allowed to flow slowly through the tube during use. This obviates the slow deterioration of the organic additive and also permits a sample to be placed directly inside the counter, often an important advantage.

A convenient mounting arrangement for an end-window counter is shown in Fig. 23-5. The sample is mounted on a stainless-steel disk called a *planchet*, and placed on a plastic shelf beneath the face of the counter tube. The shelf can be located at several distances from the counter to ac-

Fig. 23-5 Iron shield and sample holder for planchet counting. Any type of end-window detector can be installed. The overall height is about 40 cm, and the weight 90 kg. (*Radiation Counter Laboratories, Inc.*)

commodate samples of varying strength.   Aluminum filters can be inserted as desired.

SEMICONDUCTOR DETECTORS

Excellent proportional detection can be achieved with detectors of silicon or germanium (2, 4).   Crystals of these elements are prepared so as to have a sensitive volume within which absorbed radiation will displace electrons from their normal locations.

One such technique requires the preparation of a p-n (or n-p) junction* on or just barely beneath the open surface of the crystal.   The upper and lower surfaces are then made conducting with thin, deposited metal films.   The resulting diode is connected electrically in the reverse-biased mode, so that electrons are pulled away from the junction on the n side, while positive holes are pulled away on the p side.   This procedure causes the formation of a *depletion region,* the thickness of which is variable from 0 to about 1 mm by variation of the applied voltage.

According to a second technique, ions of lithium are diffused or *drifted* into the crystal under the influence of an electric field; the lithium has the effect of "cleaning up" or compensating the natural charge carriers, so that when a field is applied across the treated crystal the depletion layer will be many times deeper, perhaps as thick as 1 cm.

It is necessary that lithium-drifted germanium detectors be operated at liquid-nitrogen temperature (77 K), because the lithium ions are sufficiently mobile at room temperature to destroy the geometry of the depletion layer.

These detectors are used in a manner quite analogous to gas-filled ionization chambers.   Radiation absorbed in the depletion region produces pairs of electrons and holes which are accelerated toward the respective electrodes, forming a current proportional to the energy of the ionizing particle or photon.   A sensitive and noise-free electrometer amplifier is required, because the currents are of only a few microamperes.

The thin depletion layers of less expensive surface-junction detectors are sufficient for the detection of beta and heavier particles, which cannot penetrate deeply, but the thicker layer of a lithium-drifted detector, together with the larger absorption coefficient of germanium as compared to silicon, makes the germanium-lithium detector by far the best for gamma spectroscopy.

BACKGROUND

There is a significant amount of radiation always present in the atmosphere, so that even in the absence of a sample any detector of radioactivity

* The nature and properties of semiconductor junctions will be discussed in more detail in Chap. 26.

will show a finite response. This radiation is due in part to the natural radioactivity of the surroundings, and in part to cosmic rays. By shielding the counter with 5 to 10 cm of lead, it can be reduced markedly to perhaps 15 to 20 counts per minute. A much higher background count, or a sudden increase, may indicate accidental contamination of the immediate surroundings with radioactive matter, or it may indicate incipient failure of the counter itself. All measurements of activities must be corrected for background count before any use or interpretation can be undertaken.

ELECTRONIC SCALERS

Since scintillation, proportional, and Geiger counters all deliver their information as pulses, a facility must be provided for counting these pulses. An electromechanical counter is satisfactory for slow counting, less than 10 to 100 counts per second, but has too much mechanical inertia to permit faster operation. For faster counting a decimal scaling circuit is useful. This is a special amplifier that will transmit to its output terminals only every *tenth* pulse that appears at its input. As many of these units can be operated in cascade as are needed to express the data with sufficient precision. The scaling circuits are directly connected to a numerical read-out display.

Scalers are usually provided with a timing circuit, so that the count can be continued for a predetermined length of time. Alternatively, the time required to reach a predetermined number can be measured. Radioactive decay is statistical in nature, that is, the exact number of atoms which will distintegrate and eject particles in any particular second is governed by the laws of probability. Observed counts are significant only when a large enough number has been accumulated to permit valid statistical analysis. The most convenient criterion is the *standard deviation σ*, sometimes called the *root-mean-square deviation*. It can be shown that, for radioactivity measurements when the half-life of the decaying isotope is long compared to the duration of the experiment, the standard deviation is simply the square root of n, the total number of counts. One is thus justified in expressing the results of the experiments as $n \pm \sqrt{n}$. This explains the advantage in determining the time for a predetermined count: It results in uniform precision.

The background count is also statistical in nature. If we let σ_s represent the standard deviation of the sample count, σ_b that of the background, and σ_t that of the sample with background (total), then it can be shown that

$$\sigma_s = (\sigma_t{}^2 + \sigma_b{}^2)^{1/2} \tag{23-1}$$

There are two ways of handling this: The background count can be run long enough that $\sigma_b \ll \sigma_t$, and hence can be neglected, or the ratio of counting times t_t for the sample and t_b for the background which will give the best precision in the least time can be estimated by the relation

$$\frac{t_t}{t_b} = \left(\frac{R_t}{R_b}\right)^{\frac{1}{2}} \tag{23-2}$$

where the R's refer to the respective count rates, which need only be known roughly from a preliminary measurement (5).

As an example, suppose in an experiment the approximate activities for sample and background are R_t = 1000 counts per minute and R_b = 40 counts per minute. The time ratio should be $t_t/t_b = ({}^{1000}\!\!\!/_{40})^{\frac{1}{2}}$ = 5. If a precision of 1 percent standard deviation is desired, the total count must be 10,000 counts, R_t = 10,000 $\pm$ $\sqrt{10{,}000}$ = 10,000 $\pm$ 100, which means counting for 10 min. Therefore the background count must be taken for at least one-fifth of 10 min, or 2 min. The value of σ_s is given by $(10^4 + 80)^{\frac{1}{2}}$ = 100.4 counts per 10 min, so the final result is R_s = (1000 − 40) $\pm$ 10 = 960 $\pm$ 10 counts per minute.

Another source of error is due to the occurrence of *coincidences*, here defined as two pulses coming so close together that the counter does not have sufficient time to recover from one pulse in time to respond to the next. The recovery time varies considerably from one type of detector to another. If the recovery time is designated by τ, the observed counting rate by R, and the true pulse rate by R', the governing relation is

$$R' = R + R^2\tau \tag{23-3}$$

It must be realized that in general only a fraction of the radiation from any sample can enter the counter (unless, of course, the sample is *inside* the counter). With a single end-window counting tube, the geometric efficiency must be less than 50 percent, and may be much less. With thin-walled and dipping counters, similarly low efficiencies must be expected. The figure can be raised to much better than 50 percent by making use of the 4π geometry previously mentioned. However, an efficiency of less than 50 percent is often tolerated, because the layout of the equipment may be much simpler. It is always important to maintain constant geometry for determinations which are to be compared with one another. This is facilitated by suitable design of sample-holding equipment such as that shown in Fig. 23-5.

Precautions must be taken against the effects of partial absorption of radiations by the sample itself or by other materials which may be present, such as solvent or filter paper.

PULSE-HEIGHT ANALYSIS

Ionizing particles and photons vary widely in energy content, as illustrated in Table 23-2. Individual active species can often be identified by observation of these energy values. One method is by the use of *standard absorbers*—known thicknesses of aluminum or copper for less energetic particles, or lead for gamma rays. The *half thickness*, i.e., the thickness of metal which reduces the activity to half its value, can be translated into energy by means of predetermined calibration curves.

A more elegant method is by the observation of pulse heights from a semiconductor, proportional gas, or scintillation counter (4). The instrument for this application is the multichannel analyzer described in Chap. 10. Figure 23-6 shows the spectrum of the radiations from ^{131}I, and Fig. 23-7 a similar spectrum of a mixture of gamma emitters, both taken with multichannel analyzers.

In high-speed counting, error can arise from *dead time*. This is the summation of the time duration of pulses, during which the detector is not ready to receive another pulse. Trouble from this source is particularly likely to arise in gamma-ray spectroscopy with a multichannel analyzer. It may be compensated without the need of cutting the sample size by means of a *pulser* (6–8). A series of pulses from a 10-kHz oscillator is fed to the analyzer through an electronic switching system, such that pulses are counted only during the time that the detector is receptive. The pulses

Fig. 23-6   Radiation spectrum of [131]I.   (*Nuclear-Chicago Corporation.*)

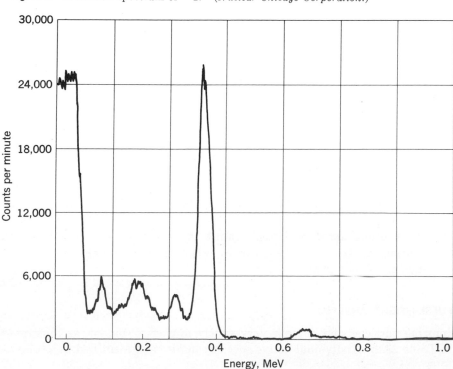

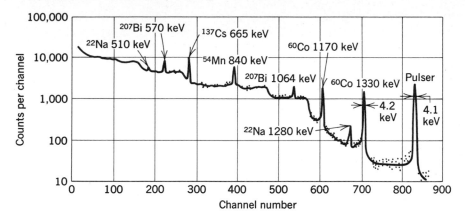

Fig. 23-7   Gamma-ray spectrum taken on a multichannel analyzer with a lithium-drifted germanium detector.   (*Isotopes, Inc.*)

are recorded as a peak in a channel at the high-energy end of the spectrum. The size of this peak measures the effective on-time of the detector, hence can provide a correction factor applicable to all the gamma-ray peaks.   A pulser peak is visible at the right-hand end of the spectrum of Fig. 23-7.

For quantitative results, the area beneath each peak must be measured accurately, a procedure often complicated by lack of a well-defined base line.   Covell has shown (8) on a firm statistical basis that a valid area measure can be obtained by a procedure illustrated in Fig. 23-8.   In this figure each vertical bar represents one analyzer channel.   A line AB is drawn intersecting the curve an equal number of channels below and above the peak (at channels $P - 5$ and $P + 5$ in the figure), the number of channels selected depending on their width.   The area enclosed between the curve and this line, measured with a planimeter, is proportional to the intensity of the corresponding radiation.

NEUTRON COUNTING

Since neutrons are uncharged, they do not produce ionization in a gas by any direct process.   They may however be detected in a counter filled with gaseous BF_3, because neutrons react very readily with the ^{10}B nucleus to produce ^7Li and alpha particles.   The resultant alphas trigger the counter in the usual way.   Neutrons can be detected in an analogous manner by a scintillation counter: A boron compound is added to an alpha-sensitive scintillator.

Several different types of neutron spectrometers have been designed. Low-energy neutrons can be diffracted in a crystal spectrometer very much as can x-rays, since their deBroglie wavelength ($\lambda = h/mv$) is in the x-ray range.   Other methods involve selection according to velocities, whereby

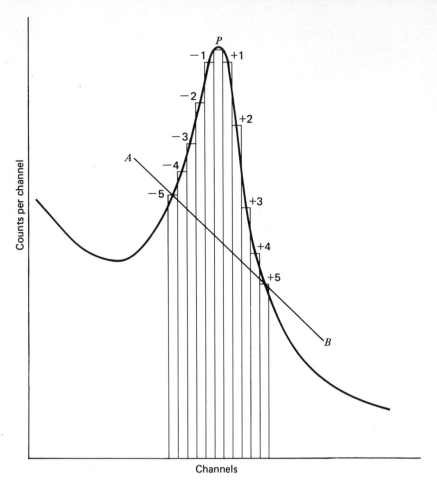

Fig. 23-8 Curve of a single species, from a gamma-ray spectrum, showing method of quantitative evaluation.

all neutrons which pass over a measured course in an evacuated tube in a given length of time are observed. Then scanning of the time for a fixed distance will yield a spectrum. The details of these instruments cannot be described in the space available here.

The selective absorption of neutrons of varying energy content is an extremely powerful analytical tool but, unfortunately for general application, it requires a source of neutrons of such high flux that only a nuclear reactor can provide them. A less powerful source can be applied to the absorptiometric determination of a few elements which have unusually high absorptive power (*cross section*), particularly B, Cd, Li, Hg, Ir, In, Au, Ag, and several lanthanides. An extensive introduction to the subject has been published by Taylor, Anderson, and Havens (9). More recent data on nuclear cross sections are available (1).

ANALYTICAL APPLICATIONS OF RADIOACTIVE SOURCES

As pointed out previously, gamma sources can often be substituted to advantage for conventional x-ray sources.

The absorption of alpha and beta rays is potentially useful for analytical purposes. Because of their low penetrating power, alpha rays are best suited for the analysis of gases. This process has been studied by Deisler, McHenry, and Wilhelm (10), who mounted an alpha-ray source (polonium, ^{210}Po, in an aged preparation of radium-D, ^{210}Pb) inside an ionization chamber which also served as the sample container. Under conditions of constant applied potential and constant gas pressure, the current through the chamber is a function only of gas composition. In favorable cases, binary gas mixtures can be analyzed with a precision of ±0.2 to 0.3 mole percent, by reference to a calibration curve prepared from measurements on known mixtures. An instrument based on this principle is manufactured by Mine Safety Appliances Company, Pittsburgh, Pa., for the detection of such toxic gases as $Ni(CO)_4$ and tetraethyl lead (TEL) in air in the parts-per-billion range.

The absorption of beta rays has also been applied to analysis (11, 12), this time primarily with liquids, though the principle is also applicable to solids and gases. It can be shown on theoretical grounds that the absorption of beta rays by matter is almost entirely due to electron-electron collisions. Elementary considerations show that hydrogen has a greater number of electrons per unit of weight than any other element, by a factor of at least 2. This renders the method particularly sensitive to the presence of hydrogen. Tritium and other beta emitters are used in electron-capture GC detectors, as described in Chap. 19.

RADIOACTIVE TRACERS

The ease with which the presence of active isotopes can be detected and the precision with which they can be measured, even in very small quantities, lead to a variety of analytical procedures of great versatility. The most important general techniques are activation analysis and isotope dilution.

ACTIVATION ANALYSIS

A great many elements become radioactive when bombarded with energetic particles such as protons, deuterons, alpha particles, or neutrons. The resulting activity can provide data for quantitative analyses (3). As an example, we shall consider activation by exposure to neutrons of *thermal* velocities, i.e., with kinetic energy less than about 0.2 eV.

The neutron is captured by the atomic nucleus to give a larger nucleus with the same positive charge, an isotope of the same element. This new nucleus is in many cases unstable and spontaneously decomposes by the emission of a particle or a gamma ray; in other words, it is radioactive. The active isotopes formed in this way from the various elements vary widely in half-life, and in many instances can be identified by this constant, along with other pertinent information, such as the gamma-ray energy spectrum.

This phenomenon can be used for analysis by subjecting a sample to neutron bombardment either in a nuclear reactor in which uranium is undergoing fission, or by other means. Radioactivity will be induced in each of the elements present that are capable of being so activated. The intensity of radioactivity in the sample can then be plotted against the time to give a *decay curve*. This curve is generally complex in nature, being the summation of the activities of all the active elements present. The half-life of the longest-lived component can be determined from the later portions of the curve after more transient substances have virtually vanished. The activity due to this element can then be subtracted point by point ("stripped away") from the readings from shorter times. Then the next longest-lived substance can similarly be identified and stripped away, then the next, and so on.

For example, an alloy of aluminum was analyzed by this method (13). A small, carefully cleaned and weighed sample of the metal (about 25 mg) was placed in an atomic reactor for a period of 5 min. The subsequent activity of the sample was followed with a standard counting apparatus for about 60 h. The results are plotted in Fig. 23-9. The activity due to the aluminum was very intense but of short duration (2.3-min half-life). It was therefore convenient to wait for ½ h after irradiation before commencing measurements with the counter, so that practically all the active aluminum had time to decay away and thus not interfere with subsequent measurements.

By reference to the figure, it is seen that the activity (on a logarithmic scale) is a linear function of time after about 40 h. This straight-line portion extrapolated backward gives the activity at any time due to an isotope with a half-life that can be read from this graph as about 15 h. This is identified from tables of isotopes as ^{24}Na with a known half-life of 15.0 h. Further, the plot shows a linear region extending from 1 to 18 h which must be due to an isotope of half-life close to 2.5 h. This was found to be ^{56}Mn (2.58 h).

Rather than plotting complete decay curves, it is possible to obtain the desired results by running gamma-ray spectra at selected times following irradiation. Anders (6) has described a typical assembly for the simul-

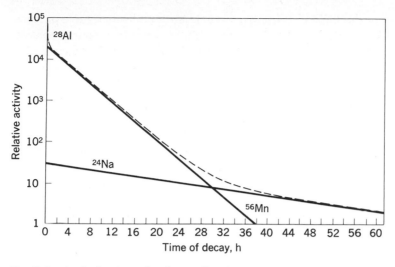

Fig. 23-9 Analysis of an aluminum alloy by neutron activation. [*Analytical Chemistry* (13).]

taneous determination of several elements. In his system the sample is irradiated for 5 min in a reactor at a flux of about 10^{12} neutrons per square centimetre per second, and then counted at five specified times ranging from 3 min to several hours after removal from the reactor. The data are processed and plotted by computer.

The sensitivity of analysis by means of neutron activation depends on the intensity of the activating source, on the ability of the element sought to capture neutrons (called the *neutron-capture cross section*), and on the half-life of the induced activity. The governing relation (3) is:

$$A = N\sigma\phi \left[1 - \exp\left(- \frac{0.693t}{T_{\frac{1}{2}}} \right) \right] \tag{23-4}$$

where A = induced activity at the end of the period of irradiation, in disintegrations per second, N = the number of atoms present of the isotope being activated, σ = the neutron-capture cross section in square centimetres, ϕ = the flux in neutrons per square centimetre per second, t = the irradiation time, and $T_{\frac{1}{2}}$ = the half-life of the product. ($T_{\frac{1}{2}}$ and t must be in the same units.) Quantitative analyses are rarely based on calculations from this equation, as sufficiently reliable data are seldom available for σ, ϕ, and $T_{\frac{1}{2}}$; as a further complication ϕ may not be homogeneous and may vary with time. For practical purposes, standard samples are irradiated simultaneously with unknowns, and the analysis carried out by simple comparison.

If a nuclear reactor is available for activation, in favorable cases as little as 10^{-10} g of an element can be detected. For less powerful neutron sources, the method is limited to those elements which have particularly favorable nuclear properties. For example, a neutron source consisting of 25 mg of Ra mixed with 250 mg of Be, producing a usable flux of about 100 neutrons per square centimetre per second, will activate only Rh, Ag, In, Ir, and Dy, but can offer a convenient and accurate method for the determination of these elements, even in trace amounts. A source composed of ^{124}Sb and Be with a flux of 10^3 to 10^4 will activate about 19 elements. A nuclear reactor may have a flux as great as 10^{14}, and will activate nearly all the elements heavier than O, but with greatly differing sensitivities. For some elements (In, Re, Ir, Sm, Eu, Dy, Ho, Lu, V, As, Sb) activation analysis appears to be capable of greater sensitivity than strictly chemical analysis, whereas for others (Fe, Ca, Pb, Bi, Zn, Cd, Na, K) activation is no better than, or distinctly inferior to, chemical methods (14).

Activation can also be produced by the bombardment of samples by *protons*. The high-energy protons required are obtained from a particle accelerator, such as a cyclotron (15, 16). This procedure is especially useful for rapid analyses using short-lived isotopes. The proton-capture cross section of many nuclei differs from σ, the neutron cross section, and this permits analysis of some matrix materials that would mask activation by neutrons.

Activation by electromagnetic radiation is also possible. Gamma rays from ^{124}Sb have been used for activation of ^{9}Be, and from ^{24}Na for measurement of deuterium in water (17). These sources produce the most energetic gammas of any isotopes with a long enough lifetime to be practical, and ^{9}Be and ^{2}H are the only nuclei that can be activated by them. For wider applicability, *Bremsstrahlung* radiation produced by the absorption of high-energy electrons in a high-atomic weight target must be used. (This is the same type of radiation seen as continuous background in x-ray tube emissions, Fig. 10-1, but at much higher energy.) Photons of 20 to 40 MeV are needed. For some elements (e.g., F, Fe, and Pb) the sensitivity is better than for neutron activation. In other cases, matrix interferences are eliminated, making the method attractive.

Neutron activation is far more widely used than other methods of activation. The reports of a conference (18) on the subject held in 1968 include descriptions of about 150 diverse applications in such fields as biomedical, forensic, archeological, geochemical, environmental, and industrial chemistry.

ISOTOPE DILUTION

This is a technique which is suitable where a compound can be isolated in a pure state but with only a poor yield. A known amount of the same substance containing an active isotope is added to the unknown and thoroughly mixed with it. A sample of the pure substance is then isolated from the mixture, and its activity is determined. A simple calculation then gives the quantity of the substance in the original material.

Consider a solution which contains W grams of a compound which is to be determined. To the solution is added a portion of the same compound, which is *tagged* with a radioactive atom; the added portion weighs w grams and has an activity of A counts per minute and a specific activity $S_0 = A/w$. After thorough mixing, g grams of this compound is isolated in a pure state and found to have an activity of B counts per minute and specific activity $S = B/g$. Now the total amount of activity (assuming loss by decay to be negligible) must be the same after mixing as before, or

$$S_0 w = (W + w)S \qquad (23\text{-}5)$$

from which it follows that

$$W = g\,\frac{A}{B} - w \qquad (23\text{-}6)$$

If the added material is highly active, then the amount added w can be very small relative to W, and Eq. (23-6) reduces to

$$W = g\,\frac{A}{B} \qquad (23\text{-}7)$$

Suppose it is required to determine the amount of glycine in a mixture with other amino acids. Glycine can be isolated chemically, but only with a low yield, which makes isotope dilution an appropriate technique. We start by synthesizing or obtaining commercially a sample of glycine which contains an atom of ^{14}C in perhaps one in every million of its molecules. A 0.500-g portion of this active preparation (specific activity, corrected for background, is 25,000 counts per minute per gram) is mixed with the unknown. From the mix is obtained 0.200 g of pure glycine with an activity of 1250 counts per 10 min. The background is 100 counts per 5 min. The data may be summarized as follows:

$w = 0.500$ g

$S_0 = 25{,}000$ counts per minute per gram

$A = wS_0 = 12{,}500$ counts per minute

$g = 0.200$ g

$B = \dfrac{1250}{10} - \dfrac{100}{5} = 105$ counts per minute from which

$$W = g \frac{A}{B} - w = 0.200 \frac{12,500}{105} - 0.500$$

$$= 23.3 \text{ g}$$

which is the required weight of glycine in the sample. If w were neglected (as in Eq. 23-7), the resulting error in this particular example would be about 2 percent, whereas the counting error, calculated as detailed in a previous paragraph, is about 3.5 percent.

As a further illustration, let us examine the work of Salyer and Sweet (19) on the electrogravimetric determination of Co in steel or other alloys. The reason for using isotope dilution was that Co, deposited anodically as Co_2O_3, is apt to form a poorly adherent layer; this precludes a conventional gravimetric determination, but loss of some particles of oxide during washing and drying is not objectionable in isotope dilution. Other shortcuts are permissible, such as the substitution of centrifuging for quantitative filtration and washing. A standard curve was prepared by the addition of equal aliquots of ^{60}Co to samples containing various amounts of pure Co and then electrodepositing Co_2O_3 in a standardized manner. The unknown was fortified by an aliquot of ^{60}Co immediately on dissolution of the sample. Chemical treatment was required in order to remove elements which might interfere with the electrolysis. Co was then deposited, the deposit weighed and counted, and the Co in the original sample ascertained by reference to the standard curve. Standard deviations varied from 0.005 to 0.025 percent.

The use of a standard curve in this way tends to eliminate some sources of error, in much the same way that a blank determination does in many types of analysis. However, in many applications the curve is dispensed with, and the answer obtained directly from Eq. (23-6) or (23-7).

RADIOMETRIC ANALYSIS

This term has been applied to analytical procedures in which a radioactive substance is used indirectly to determine the quantity of an inactive substance. An excellent example (20) is the determination of chloride ion by precipitation with radioactive silver, ^{110}Ag. From the diminution of the activity of a silver nitrate solution following removal of silver chloride precipitated by the unknown sample, the chloride content is easily calculated. A sample computation is given in the original paper. Extremely small quantities of chloride could be determined with greater precision by measuring the activity of the precipitate formed.

MÖSSBAUER SPECTROSCOPY (1, 21, 22)

This term designates a study of the phenomenon of resonance fluorescence of gamma rays. It is comparable to resonance fluorescence in optical regions, but involves *intranuclear* rather than electronic energy levels. An important characteristic of this radiation, under optimum conditions of measurement, is the extreme sharpness of the lines. The resonance gamma ray of ^{67}Zn, for example, has a width at half height of only 4.8×10^{-11} eV, but with a photon energy of approximately 93 keV, less than 1 part in 10^{15}. This may be compared with the Zn $K\alpha$ x-ray, which has a half-height width of 4.7×10^{-8} eV for a photon of 8.6 keV, or about 1 part in 10^{11}.

The most extensively studied element is Fe, because the nuclear energy levels of one isotope, ^{57}Fe, are most easily accessible. This nuclide has a metastable energy level 14.4 keV above the stable ground state, and it is the gamma rays emitted in the transition between these levels that are readily absorbed by normal ^{57}Fe nuclei. The excited nuclei are obtained by spontaneous decay of ^{57}Co (267-day half-life). So the source material in a Mössbauer experiment with Fe is made of ^{57}Co plated on a suitable support. The absorber can be ordinary Fe in any chemical form, as the natural abundance of ^{57}Fe, about 2.2 percent, gives adequate sensitivity.

Since the frequency bands are so narrow, extremely slight changes in the energy states of the absorbing nuclei can shift the frequency at which absorption can occur by more than the width of the line of the primary radiation, so that no absorption will take place. The effect of the state of chemical combination on the nuclear levels can be of just this order of magnitude. Such a *chemical shift* can be observed and measured by imposing a translational motion on either the emitter or absorber in such a way that the resulting Doppler shift will exactly compensate for the chemical shift. The required motion turns out to be of the order of a few millimetres per second, hence is easily realizable in practice.

Figure 23-10*a* shows schematically a typical apparatus for Mössbauer spectrometry. Since relative motion of source and absorber is required, either may be made movable, but as the absorbing sample must often be refrigerated, it is more convenient to move the source. A motor designed to produce linear reciprocating motion is driven by a signal generator programmed to give constant *acceleration*, first in one direction and then the other (Fig. 23-10*b*). With this type of motion, a whole range of *velocities* is covered in each cycle of operation; the *displacement* follows a parabolic curve.

The signal from the detector is fed through a single-channel analyzer, the function of which is to restrict the response to the single resonant gamma ray. From there it passes to a multichannel analyzer, synchronized

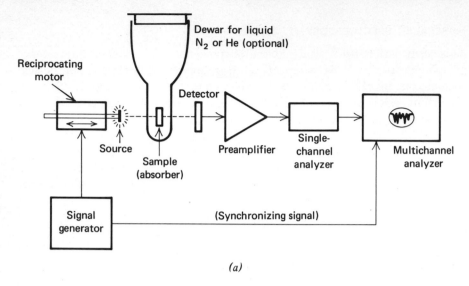

(a)

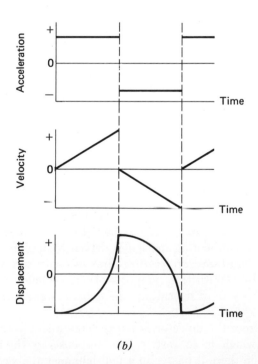

(b)

Fig. 23-10   *(a)* Schematic of a Mössbauer spectrometer; *(b)* time sequences.

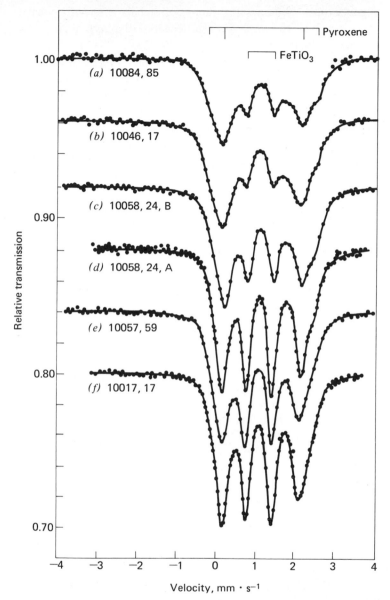

Fig. 23-11   A typical series of Mössbauer spectra: six samples of dust and rocks from the moon showing varying proportions of pyroxene (an iron-bearing silicate mineral) and FeTiO₃.   [*American Laboratory* (23).]

with the signal generator. This assigns each channel to a specific narrow increment of velocities, so the built-in oscilloscope displays a response as a function of source velocity.

Figure 23-11 shows the general form of a Mössbauer spectrum (23). The peaks often show *hyperfine* splitting into two, four, or six parts. In the case of ^{57}Fe, six peaks are seen at low temperatures, as a result of the existence of two major energy states corresponding to $I = \frac{1}{2}$ and $I = \frac{3}{2}$, each of which is split by the electric and magnetic fields existing within the nucleus.

Mössbauer spectra can give information about valence states and crystal structures of any compounds or alloys containing the elements to which it is sensitive. A major drawback is the practical restriction to a very few elements. By far the greatest amount of work has been done on ^{57}Fe, with ^{61}Ni and ^{119}Sn following. There are about 32 elements in which the effect has been observed, and 17 more in which it is to be expected. Many of these elements owe their response to low-percentage isotopes, are themselves in low abundance, or are difficult to work with for one reason or another. The importance of Fe, Ni, and Sn chemistries is enough to ensure this technique a place among instrumental methods.

SAFETY PRECAUTIONS

The small amounts of radioactive substances needed for analytical experiments with tracers do not generally present radiation hazards which are difficult to guard against. Beta-active isotopes are safe if kept in ordinary glass or metal containers; when out of such containers, they should be handled with tongs, and the operator should wear rubber or plastic gloves. Gamma emitters may require more extensive shielding, perhaps a few centimetres of lead, depending on the photon energy of the specific isotope. Pipetting of these solutions by mouth is never permissible.

A survey meter should always be available so that cleanup of accidental spillage can be checked. Often the hazard to the experimenter is less real than the chance of contaminating the laboratory so that the background count is increased unduly. Most active tracer materials are as safe to use in the laboratory as such more familiar materials as benzene and silver nitrate. Safety precautions are no more exacting, merely of a different type.

Active materials in larger than tracer quantities do, of course, require more elaborate safety measures, descriptions of which are readily available elsewhere.

PROBLEMS

23-1 A mixture is to be assayed for penicillin by the isotope dilution method. A 10.0-mg portion of pure radioactive penicillin, which has an activity of 4500 counts per minute per milligram, as measured on a particular counting apparatus, is added to the given specimen. From the mixture it is possible to isolate only 0.35 mg of pure crystalline penicillin. Its activity is determined on the same apparatus to be 390 counts per minute per milligram. (Background corrections have been applied.) What was the penicillin content of the original sample in grams?

23-2 According to the *reverse isotope dilution* method, a weight w of inactive compound is added to a preparation containing an unknown amount W of an active form of the same compound, which has a known specific activity S_0. A sample is then isolated in pure form, weighed and counted, just as in normal isotope dilution. Show mathematically that Eq. (23-6) applies equally to this case.

23-3 A method for the simultaneous determination of U and Th in minerals (24) requires (1) measurement of combined U and Th by radioactivity, and (2) determination of the Th/U ratio by x-ray emission spectroscopy. The combined radioactivity is expressed as percent equivalent uranium, namely, that amount of uranium in pitchblende necessary to give an equal activity. The Th/U ratio is taken as the ratio of peak heights for the x-ray lines: Th $L\alpha$ and U $L\alpha$. The U and Th contents are given by the relation $x + 0.2xy$ = percent equivalent U, where x is the weight percent of U and y is the Th/U weight ratio. For a particular counting apparatus 1 percent equivalent U corresponds to 2100 counts per minute above background. A 1.000-g sample of a monazite sand, when prepared and counted according to the standard procedure, gave 2780 counts per minute (corrected for background). X-ray examination gave peak heights of 72.3 scale divisions for Th $L\alpha$, and 1.58 divisions for U $L\alpha$. Compute the U and Th contents of the sample in terms of weight percent.

23-4 In a study of solubilities of slightly soluble salts, it is necessary to determine the concentrations of oxalate solutions in the parts-per-million range. This analysis is to be carried out radiometrically by precipitation of $^{45}CaC_2O_4$. Calculate the oxalate concentration (in parts per million) in a sample from the following: A standard solution is prepared which is 0.680 F in $CaCl_2$ and has an activity of 20,000 counts per minute per milliliter (corrected). To a 100.0-ml sample of trace oxalate solution is added 5.00 ml of the standard solution. No precipitate is visible, beyond a slight turbidity. A few drops of $FeCl_3$ solution is added, and the solution made alkaline with ammonia to precipitate $Fe(OH)_3$. The precipitate is collected on a small filter paper by suction, washed once, dried, and counted. The counting apparatus is known by prior experiment to have a 30.0 percent efficiency for ^{45}Ca beta rays. The time required for a preset count of 6000 is 18.60 min. The background is 150 counts per 5 min. (The efficiency correction need not be applied to the standard solution.)

23-5 A method has been described (25) for the determination of oxidation products of propane by the reverse isotope dilution method. The sample to be oxidized is propane-2-^{14}C. Among the products was found a considerable quantity of 2-propanol-^{14}C. To the mixture of products was added a measured amount of inactive 2-propanol, and a portion isolated by conventional fractionation. The following data were obtained. (The symbol μCi stands for *microcurie*, an alternative unit for activity.)

Quantity of propane-2-^{14}C	10 mmol
Specific activity of propane	72.8 μCi $\cdot$ mmol^{-1}
Inactive propanol added	16 mmol
Specific activity of propanol	5.8 μCi $\cdot$ mmol^{-1}

Compute the percent of propane converted to propanol.

23-6 The beta radiation from an active source is to be measured with a Geiger counter. The maximum uncertainty permitted is ±1 percent. Counts recorded at the end of successive 5-min periods for sample and background are as follows:

Time, min	0	5	10	15	20	25	30
Background, counts per minute	0	127	249	377	502	672	793
Sample, counts per minute	0	2155	4297	6451	8602	10,749	12,907

(*a*) What is the minimum time over which the count must be taken to give the required precision? (*b*) How long would the background have to be counted? (*c*) What is the actual corrected count in counts per minute, with precision limits?

REFERENCES

1. Friedlander, G., J. W. Kennedy, and J. M. Miller: "Nuclear and Radiochemistry," 2d ed., Wiley, New York, 1964.
2. Steyn, J. J., and S. S. Nargolwalla: In J. Krugers (ed.), "Instrumentation in Applied Nuclear Chemistry," p. 117, Plenum, New York, 1973.
3. Lyon, W. S., Jr. (ed.): "Guide to Activation Analysis," Van Nostrand, Princeton, N.J., 1964.
4. Prussin, S. G., J. A. Harris, and J. M. Hollander: *Anal. Chem.*, **37**:1127 (1965).
5. Choppin, G. R.: "Experimental Nuclear Chemistry," Prentice-Hall, Englewood Cliffs, N.J., 1961.
6. Anders, O. U.: *Anal. Chem.*, **41**:428 (1969).
7. De Wispelaere, C., J. Op de Beeck, and J. Hoste: *Anal. Chem.*, **45**:547 (1973).
8. Covell, D. F.: *Anal. Chem.*, **31**:1785 (1959).
9. Taylor, T. I., R. H. Anderson, and W. W. Havens, Jr.: *Science,* **114**:341 (1951).
10. Deisler, P. F., Jr., K. W. McHenry, Jr., and R. H. Wilhelm: *Anal. Chem.,* **27**:1366 (1955).
11. Jacobs, R. B., L. G. Lewis, and F. J. Piehl: *Anal. Chem.,* **28**:324 (1956).
12. Smith, V. N., and J. W. Otvos: *Anal. Chem.,* **26**:359 (1954).
13. Boyd, G. E.: *Anal. Chem.,* **21**:335 (1949).
14. Meinke, W. W.: *Science,* **121**:177 (1955); *Anal. Chem.,* **30**:686 (1958).
15. Debrun, J. -L., D. C. Riddle, and E. A. Schweikert: *Anal. Chem.,* **44**:1386 (1972).
16. Bankert, S. F., S. D. Bloom, and G. D. Sauter: *Anal. Chem.,* **45**:692 (1973).
17. Lutz, G. J.: *Anal. Chem.,* **43**:93 (1971).
18. DeVoe, J. R., and P. D. LaFleur (eds.): "Modern Trends in Activation Analysis," 2 vols., *Natl. Bur. Stand. Spec. Publ.* 312 (1969).
19. Salyer, D., and T. R. Sweet: *Anal. Chem.,* **28**:61 (1956); **29**:2 (1957).
20. Hein, R. E., and R. H. McFarland: *J. Chem. Educ.,* **33**:33 (1956).
21. Greenwood, N. N., and T. C. Gibb: "Mössbauer Spectroscopy," Chapman and Hall, London, 1972.
22. Cohen, R. L.: *Science,* **178**:828 (1972).
23. Muir, A. H., Jr., R. M. Housley, R. W. Grant, M. Abdel-Gawad, and M. Blander: *Am. Lab.,* **2**(11):8 (1970).
24. Campbell, W. J., and H. F. Carl: *Anal. Chem.,* **27**:1884 (1955).
25. Clingman, W. H., Jr., and H. H. Hammen: *Anal. Chem.,* **32**:323(1960).

Chapter 24

Automatic Analyzers

AUTOMATION VERSUS MECHANIZATION

Many of the instruments described in previous chapters of this book are highly mechanized, meaning that built-in electrical and mechanical devices relieve the operator of explicit care of many details. Such mechanization permits much more effective utilization of the analytical capabilities of the instrument than could be attained by purely manual operation. This is especially true where the method involves scanning one variable while continuously measuring another.

As an example consider an infrared spectrophotometer. Even if the electromechanical aids usually taken for granted were omitted and all work done manually, equally valid spectra could be obtained. An experienced operator would need at least 5 to 10 min to set the prism or grating at the desired wavelength, adjust the slit width, measure the transmittance of both blank and sample, and record the data in a notebook. This would have to be repeated many hundreds of times to cover the 2 to 20 μm range. The data would have to be calculated point by point, and then plotted on graph paper. It might take several days to complete one high-resolution spectrum. Clearly, the mechanization of such an instrument permits not only great savings of time, but also greater freedom from error.

Of course an automatic recording spectrophotometer will have its own potential sources of error. Imperfections and malfunction of mechanical parts can produce both systematic and random errors that may be difficult to track down and eliminate. But instrumental errors of this sort can generally be reduced by careful design and skillful construction, and the residual inaccuracies further diminished by overall calibration procedures. Human error that may enter into a nonmechanized procedure is aggravated by the tedium involved in many repetitions of the same type of measurement, hence is minimized in the operation of an automatic instrument.

The term *automation*, however, is usually taken to mean more than mere automatic operation of an instrument. It refers to a *system*, often consisting of one major instrument, such as a spectrophotometer, together with various other components to enable it to examine numerous samples in close succession and record the results. This type of automatic analysis is valuable whenever a large number of similar determinations must be carried out on a routine basis. Much impetus in this direction has stemmed from the needs of clinical laboratories, but automated instruments are also extensively used in quality control laboratories in chemical, pharmaceutical, metallurgical, and other industries.

In principle, a machine can be designed to carry out any procedures that a human operator might do. Mechanical operations, such as dissolving, pipetting, and diluting, can be performed by a machine with better efficiency than by a human. That this is also true of those crucial steps involving decision making is not so obvious. An experienced chemist or technician, by observing the progress of a reaction, can determine when it has reached completion and decide accordingly what the next step should be. A machine can be programmed to do this too, by making quantitative measurements followed by logical decisions based on those measurements. The machine has the advantage over the human that the correctness of its judgment will not falter because of fatigue and tedium.

On the other hand, the machine does not excel at recognizing unusual occurrences that do not follow the expected patterns. A successful automatic analyzer must be able to recognize these unusual responses, at least to the extent that it will call the operator's attention to anything out of the ordinary.

Another requirement of an automatic analysis system is that of bookkeeping. There must be no discrepancy between the sequence of samples fed into the analyzer and the sequence of answers coming out. This is particularly imperative in multistep analyses, where a number of samples may be in progress simultaneously.

Most automated analytical systems are built around one major instrumental approach. Sample handling, reagent addition, and other features are available if required, often supplied as discrete modules. In most cases the chemistry of specific analytical procedures is adapted from standard nonautomated methods. Minor changes are often made for reasons of convenience, so that all procedures to be used on a given instrument will fit into the same format in such features as temperature and time required for color development.

In the remainder of this chapter we will examine a few representative examples of automated instrumental systems.

METTLER INSTRUMENT CORPORATION

Mettler has designed an automated analytical system based on titration. Any type of reaction capable of giving a potentiometric end point with ion-selective, pH, or redox electrodes can be implemented. Motor-driven syringe burets of several different capacities are easily interchanged. The controls are flexible, so that the titration can be stopped at the end point, or a continuing curve can be plotted on a recorder well past the end point. Alternatively, data can be displayed in either analog or digital form, or printed out. Direct connection to an electronic data processor is available.

In a fully extended system, samples are weighed (on an electronic balance) into a series of up to 44 beakers. These are transported sequentially by a pneumatic drive device to the titration head, where each is titrated. The results are calculated automatically and printed out on paper tape.

AMERICAN MONITOR CORPORATION*

The Programachem automates spectrophotometric analyses (primarily clinical) within the wavelength range 325 to 800 nm, with a specially designed grating spectrophotometer. As many as 89 samples can be loaded into the instrument at one time, and analyzed at an average rate of 350 per hour. As each liquid sample is rotated to the operating position, a probe dips into it and withdraws an accurately reproducible aliquot which it then transfers to a reaction tube. Here it is diluted, and reagent is added to develop color. Absorbance measurement at a preselected wavelength completes the analysis. The results are computed internally and printed out on a paper strip.

Reagent solutions for as many as 25 different tests are stored in the Programachem cabinet, and selected according to instructions punched into a program card. In common with most automatic analyzers, this one is intended for applications in which a large number of similar determinations are required, but where it may be desirable to change over quickly to a different analysis. The operator can run a series of blood glucose determinations, for example, and then change to cholesterol merely by inserting

* Marketed by Fisher Scientific Company.

the proper program card and initiating a purge cycle to rid the system of the previous reagents.

TECHNICON INSTRUMENTS CORPORATION

The designers of the Technicon AutoAnalyzer have abandoned conventional procedures in sample handling, with the result that their equipment is quite different in appearance from most others. The heart of the system is a multiple-channel peristaltic pump arranged to force sample, diluent, reagents, and even air through plastic tubing to accomplish the necessary chemical and physical procedures on the flowing material. The final analytical measurement is usually spectrophotometric, or less frequently fluorometric or flame photometric (1, 2). Most analyses suggested by the manufacturer are of clinical nature, but the equipment can be used in industrial laboratories as well.

To illustrate the operation of the AutoAnalyzer, the procedure for the determination of calcium in blood serum will be described in some detail. Figure 24-1 shows the pertinent apparatus, and Fig. 24-2 the sche-

Fig. 24-1 General view of a typical assembly of the Technicon AutoAnalyzer. From right to left the units are: sample turntable, peristaltic pump, thermostated bath containing mixing coils, colorimeter, digital printer, and recorder. (*Technicon Instruments Corporation.*)

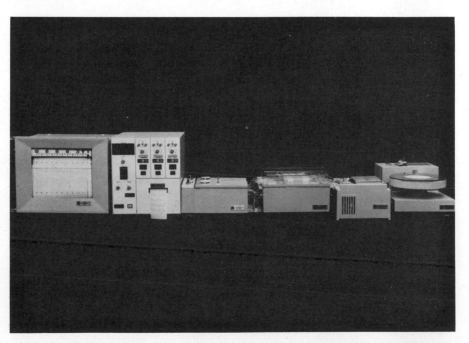

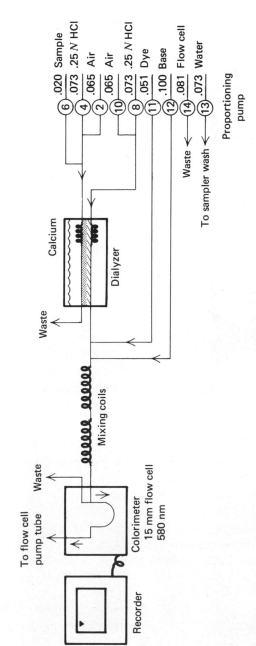

Fig. 24-2 Technicon AutoAnalyzer flow sheet for calcium determination. (*Technicon Instruments Corporation.*)

matic flow diagram. In the latter figure, the nine numbered horizontal lines at the right represent utilized channels in the standard 14-channel peristaltic pump; the related decimal fraction is the inside diameter (in inches) of the correct plastic tubing. In operation, dilute HCl is pumped through channel 4 and mixed with air (channel 2) at a T connection. The flow rates are such that the acid stream is segmented by air bubbles spaced at about 1-cm intervals. The effect of the bubbles is to promote mixing within the liquid segments and to limit drastically interaction between successive samples. The sample (serum or plasma) is pumped through channel 6 to join with the HCl stream. The next step is dialysis. The acidified sample is passed through a long spiral passage provided with a septum made of a sheet of cellophane. In an identical passage on the other side of the septum flows a similarly segmented stream of HCl from pump channels 8 and 10. Ionic components of the sample (including Ca^{2+} and Mg^{2+}) diffuse into the second stream, while protein and other large organic molecules cannot pass the barrier and are pumped to waste. At this point, a reagent solution containing the dye cresolphthalein and 8-quinolinol is added, followed immediately by a solution of 0.5 N diethylamine.* Sample and reagent are mixed thoroughly in a mixing coil, and then flow into the colorimeter. Here the liquid is separated from air bubbles, and a portion of it pumped through a cuvet and channel 14 of the pump to waste. Channel 13 pumps water to rinse the sampling device. The entire procedure is capable of 60 determinations per hour. Each sample is preceded by a wash cycle to ensure removal of traces of the preceding sample. Standard solutions are interspersed at intervals for calibration. The results are read out on a strip-chart recorder or digital printer.

E. I. DU PONT DE NEMOURS & COMPANY, INSTRUMENT DIVISION

The Du Pont Automatic Clinical Analyzer follows an entirely different, unique plan. A set of reagents for a single determination are premeasured by the manufacturer and encapsulated in a special plastic kit or pack, which also serves as the reaction chamber and cuvet for the ultimate photometric analysis. Packs for certain tests contain individual disposable chromatographic columns to isolate specific constituents or molecular-weight fractions.

A separate pack is used for each test performed on a sample. Each pack contains both the name of the test for identification by the operator and a binary code to instruct the analyzer. The instrument is programmed by insertion of the appropriate pack or packs behind each sample cup in the input tray.

The analyzer automatically injects the exact amount of sample and diluent into each pack in succession, mixes the reagents, waits a preset

* The 8-quinolinol virtually eliminates interference due to Mg^{2+}.

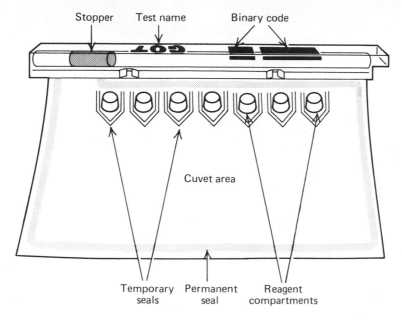

Stopper   Test name   Binary code

Cuvet area

Temporary   Permanent   Reagent
  seals   seal   compartments

Fig. 24-3 Plastic analytical pack, showing seven reagent pouches, three of them loaded. (*E. I. Du Pont de Nemours & Company.*)

length of time, forms a precise optical cell within the transparent pack walls, and measures the reaction product photometrically. These operations are controlled and monitored by a built-in, solid-state, special-purpose computer, and are performed under precisely regulated conditions within the instrument. The computer calculates the concentration value for each test and prints out a separate report sheet for each sample. This report contains all the test results on that sample, along with the patient identification.

Figure 24-3 shows one of the analysis packs. As many as seven reagents are located in inner sealed pouches, to be broken open at the appropriate points in the cycle. A short chromatographic column can be inserted along the top of the pack if required. After automatic injection of the sample followed by diluent, the pack is clipped to a moving chain to be transported from station to station through the apparatus shown in Fig. 24-4. After being heated to 37°C, the pack is pressed between jaws so shaped as to break the first four of the seven reagent pouches, mixing their contents with the sample. The pack then passes through five delay stations, to give adequate time (about 3 min) for full color development to occur, before the release of the remaining three reagents. The pack next moves to the photometer station, where it is pressed between quartz plates to form a cuvet exactly 1 cm thick. The photometer uses interfer-

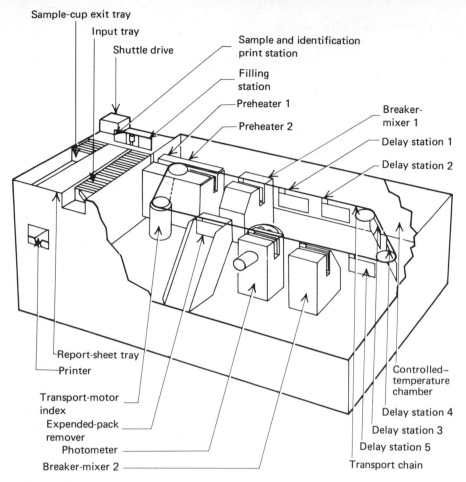

Fig. 24-4 Cutaway plan of the Du Pont ACA instrument. (*E. I. Du Pont de Nemours & Company.*)

ence filters to measure absorbance within the available range of 340 to 600 nm. The expended pack is automatically discarded.

GeMSAEC* CENTRIFUGAL ANALYZER

The analytical system designated by this acronym is built around a specially designed centrifuge (3). All necessary dilutions and additions of reagents, as well as spectrophotometric measurement, are carried out while the centrifuge is in motion. Figure 24-5 (4, 5) shows (*a*) a reagent and

* Acronym for the original sponsors, the Institute of General Medical Sciences and the U.S. Atomic Energy Commission; the work was carried out at the Oak Ridge National Laboratory.

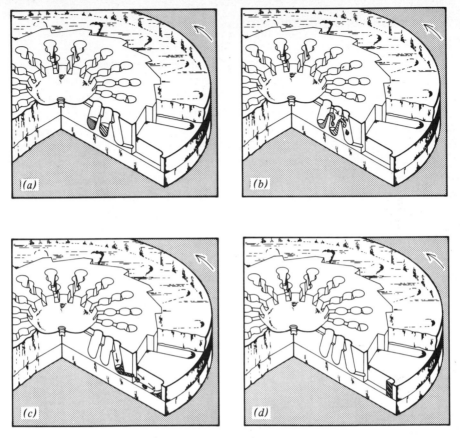

Fig. 24-5 Principle of operation of the GeMSAEC centrifugal analyzer. (*a*) Reagent and sample loaded in place; (*b*) same, in process of transfer; (*c*) same, being further transferred into optical position, shown in (*d*). [*Analytical Chemistry* (4).]

sample preloaded into cavities in a plastic rotor. Part (*b*) depicts the first result of spinning the rotor; the liquids are mixed and transferred to a third cavity. In (*c*), the mixed solution is transferred further to a position where a light beam can pass vertically through the solution. In another form of rotor, a reagent can be added equally to all stations while the centrifuge is rotating.

The photomultiplier detector sees a continuing series of pulses, each corresponding to one sample space in the rotor, separated by dark intervals. Standards and blanks are included in each loaded rotor (17 or more positions). A minicomputer corrects for the dark current, calculates concentrations, and averages the results over many rotations of the centrifuge.

The GeMSAEC equipment is now available commercially, and is finding many applications (6).

COMPUTER-CONTROLLED INSTRUMENTS

The preceding systems are intended for fast repetitive analyses wherein the same procedure is applied to many samples in sequence. Automation is also applicable to more complex analytical situations in which each sample presents its own requirements (7). The objective is to utilize the capabilities of a computer to optimize the variable parameters of the analytical instrument, thereby achieving more efficient error-free operation. The computer is also used to process the data resulting from the experiment.

Generally the electronic portions of the analytical instrument must be designed with computer operation in mind, so that all-electronic controls, with no moving parts, as far as possible take the place of knob-operated variable resistors and panel switches. The computer is wired directly to these control points, as well as to the detector, and is programmed to make the necessary adjustments to optimize the instrument.

The computer for this application is often a small, dedicated type, but in some installations control is exercised through remote terminals to a large, general-purpose computer. Further discussion of computers, especially regarding their interfacing with laboratory instruments, will be found in Chap. 27.

REFERENCES

1. Skeggs, L. T., Jr. (ed.): "Automation in Analytical Chemistry," Technicon Symposia, Mediad, New York, 1966.
2. Skeggs, L. T., Jr.: *Anal. Chem.*, **38**(6):31A (1966).
3. Anderson, N. G.: *Science*, **166**:317 (1969).
4. Scott, C. D., and C. A. Burtis: *Anal. Chem.*, **45**:327A (1973).
5. Tiffany, T. O., C. A. Burtis, J. C. Mailen, and L. H. Thacker: *Anal. Chem.*, **45**:1716 (1973).
6. Pesce, M. A., *J. Chem. Educ.*, **51**:A521 (1974).
7. Spinrad, R. J.: *Science*, **158**:55 (1967).

Chapter 25

General Considerations in Analysis

We have now completed a survey of some of the most useful analytical methods available to the chemist. This rather imposing array of techniques may well appear confusing. We must give some consideration to the problem of choosing the most appropriate method for any analytical problem that may arise.

Suppose that you, as an analytical chemist, are asked to devise a procedure for the quantitative determination of substance X. Here is a checklist of some of the questions you might ask before undertaking the task:

1. What is the matrix or host material in which the desired substance is found?
2. What impurities are likely to be present, and in approximately what concentrations?
3. What range of quantities can be expected for X?
4. What degree of precision and accuracy is required?
5. What reference standards are available?
6. Is the analysis to be performed in the laboratory, in a plant location, or in the field?

7. How many samples are expected per day?
8. Is it essential that the answers be obtained quickly? If so, how quickly?
9. To what extent is long-term reliability required (as for continuous unattended operation), and to what extent can it be traded off to lower the cost of equipment?
10. In what physical form is the answer desired (automatic recording, printed or punched tape, written report, etc.)?
11. What special or unusual facilities are available which might affect the selection of a method (e.g., an atomic reactor)?

It may happen that some compromise is necessary. High precision may not be compatible with speed, for example. In many instances, personal preference may well prove the deciding factor. Thus colorimetric and polarographic methods may be made to yield about the same accuracy with similarly dilute samples, and the time consumed and the cost of apparatus are comparable. The analyst is then free to choose the method with which he is more familiar. Many of the more generally applicable methods of analysis are listed in Table 25-1, with comments designed to help in the selection of a procedure for various kinds of samples.

SENSITIVITY AND DETECTION LIMITS

The sensitivity S of an analytical method or instrument may be defined as the ratio of the change in the response R to the change in the quantity or concentration C that is measured:

$$S = \frac{dR}{dC} \quad \text{or} \quad \frac{\Delta R}{\Delta C} \tag{25-1}$$

This quantity is highly dependent on experimental conditions. The maximum sensitivity of which a method is capable is generally quoted in terms of *limits of detection*, defined as that quantity (or concentration) of a substance for which an analytical signal just disappears as the amount present approaches zero. In most cases the signal must be considerably larger to permit unequivocal identification, and larger yet for a quantitative determination.

It is difficult to generalize about the relative ultimate sensitivities of various methods, since such data differ widely from one element or type of compound to another. Karasek (1) has given tentative figures for several methods (Table 25-2), and Morrison (2) has compiled a valuable comparison of the sensitivities of several analytical methods as applied to all the elements for which data were available (Table 25-3). One finds from the compilation, for example, that the element europium can be detected

Table 25-1   The comparative applicability of various analytical procedures

Type of sample	Procedure	Application
1. Alloys, ores	a. Spectrography	General; rapid
	b. Electrodeposition	General; slower; less-expensive apparatus
	c. Colorimetry	More specific; especially for lesser constituents
	d. Activation	Specific; less convenient except in special cases
	e. X-ray absorption	Where sought element and impurities vary widely in atomic weight
	f. X-ray fluorescence	General; rapid
2. Traces of metal ions	a. Colorimetry	These are of comparable sensitivity and accuracy; highly specific
	b. Nephelometry	
	c. Fluorimetry	
	d. Polarography	Less specific
	e. Stripping analysis	Specific and highly sensitive
	f. Conductivity	Solutions; nonspecific
3. Gaseous mixtures	a. Gas chromatography	General; some specificity
	b. Gravimetric	Especially for carbon dioxide or water
	c. Volumetric (Orsat, etc.)	Mixtures; to determine several constituents
	d. Manometric (Warburg)	Evolution or uptake; small samples
	e. Infrared absorption	Routine assay for single component
	f. Mass spectra	General; expensive apparatus
4. Mixtures (complete separation not required)	a. Infrared spectra	Especially for organic compounds
	b. Raman spectra	
	c. X-ray diffraction	Crystalline solids
	d. Isotope dilution	Analysis for single component
	e. Mass spectra	For simple volatile compounds
	f. NMR	For liquids
5. Mixtures (separation procedures)	a. Ion exchange	For ionic materials
	b. Countercurrent distribution	Must be partially soluble in each of two immiscible liquids
	c. Partition chromatography	
	d. Adsorption chromatography	Chiefly for organic compounds
	e. Electrodeposition	For metallic cations
6. Surface composition	a. Electron spectroscopy	Nonvolatile solids
	b. Electron-beam microprobe	
	c. Infrared, with ATR	Any solids with covalent bonding

Table 25-2 Detection and identification limits for analytical methods* (g)

Method	Detection limits	Identification limits
Gas chromatography	10^{-6}–10^{-12}	
Infrared spectrophotometry	10^{-7}	10^{-6}
Ultraviolet spectrophotometry	10^{-7}	10^{-6}
NMR (time-averaged)	10^{-7}	10^{-5}
Mass spectrometry (batch inlet)	10^{-6}	10^{-5}
Mass spectrometry (direct probe)	10^{-12}	10^{-11}
GC-MS combination	10^{-11}	10^{-10}

* From F. W. Karasek (1).

in a quantity as small as 0.5 pg by neutron activation, but only to 1 ng by flame emission, and 100 ng by AA. On the other hand, iron is determinable by activation only to 5 μg, and by flame emission to 3 ng.

The data in this table can be taken as a general guide in the selection of an appropriate method, but exact detection limits will vary greatly with changes of experimental detail. No allowance has been taken for preconcentration procedures, and these will often permit marked extension of the limits. Time-averaging procedures likewise can increase sensitivity.

The limiting minimum concentration C_m can be defined in terms of the signal-to-noise ratio S/N, where S is the magnitude of the desired signal, and N is the spurious noise signal resulting from the random error inherent in the system (3). The required value of C_m is that concentration for which the S/N ratio is given by:

$$\frac{S}{N} = \frac{t\sqrt{2}}{\sqrt{n}} \tag{25-2}$$

where t is the Student-t statistic which can be obtained from handbook tables, and n is the number of *pairs* of readings taken (i.e., one reading for the blank or background, and one for the sample). See the reference for the derivation and further significance of this relation.

PRECISION

This quantity is measured inversely by the relative standard deviation s. The smaller the value of s, the better the precision. It is intimately related to the *accuracy*, which is the closeness of agreement between the observed result and the known or "true" value.

The precision of measurement can be improved by repetition with suitable statistical treatment of the data. A procedure which has a similar effect is titration. In an instrumental titration one has the opportunity to (and in many situations must) take a whole series of measurements, both before and after the end point; drawing a smooth curve through these points has about the same effect on the overall precision as would be obtained by taking the same number of individual readings on a solution without titrating. (It must be remembered, of course, that the information obtained with and without titration may not be, even ideally, the same, but may pertain to different equilibrium states.)

The precision obtainable from one method as compared with another is often affected by the form of the response curve, apart from the inherent ability of the instrument to detect signals. Figure 25-1 shows two titration curves corresponding to the same reaction, the titration of carbonate by strong acid in the presence of a large concentration of bicarbonate (4). The potentiometric curve is essentially useless, but the photometric titration (at wavelength 235 nm, where the carbonate ion absorbs but bicarbonate does not) shows an excellent end point, obtained by extrapolating two straight-line segments.

COMPARISON WITH STANDARDS

The majority of the analytical methods that have been discussed involve the comparison of a physical property of the unknown with the corresponding property of a standard or a series of standards containing the same material in known amount. This may be achieved by means of a calibration curve, which is a plot of the magnitude of the physical property against the concentration of the desired constituent (or some simple function of the concentration, such as its logarithm or reciprocal). In some instances the shape of the curve is predicted by theory (Beer's law, the Ilkovič equation, etc.), and it may be more convenient to perform a calculation based on the equation than to employ a calibration curve. This is the case, for example, in the determination of a cation by the measurement of a half-cell potential: The Nernst equation will give the desired information directly, but it actually represents a curve which can be drawn for the graphical comparison of unknowns with the standard solution from which $E°$ was originally evaluated (Fig. 13-1).

Another general procedure for comparison of unknowns and standards is to bracket the unknown between two suitably close standards, one slightly below and one slightly above it with respect to the quantity measured. This finds application in optical comparators in which the intensity

Table 25-3   Absolute detection limits for the elements by various techniques* (ng)

Element	Absorption spectrophotometry	Ultraviolet–visible fluorescence	Atomic absorption	Flame emission photometry	Neutron activation	Spark source mass spectroscopy	Emission spectroscopy Dc arc	Emission spectroscopy Copper spark	Emission spectroscopy Graphite spark
Ag	5	...	0.001†	3	0.01	0.2	0.75§	200	0.5
Al	0.5	0.5	0.01¶	4	1	0.02	30§	20	3
Ar	...	...	...	...	...	0.03			
As	10	...	1**	100	0.1	0.06	600	500	100
Au	5	...	5	50	0.05	0.2	100	40	0
B	50	40	...	5	...	0.01	2§	10	0.3
Ba	100	...	30	3	5	0.2	4	10	
Be	8	1	10	20	...	0.008	0.13§		5.9§
Bi	600	...	0.2†**	100	50	0.2	4	0.2	5
Br	10	...	...	...	0.5	0.1			
C	...	...	...	...	...	0.01			
Ca	100	3	8×10^{-5}¶	0.2	100	0.03	0.18§	20	
Cd	3	2000	1×10^{-3}¶	20	5	0.3	78§	100	20
Ce	40	...	...	200	10	0.1	100	40	
Cl	100	...	...	...	1	0.04			
Co	3	20	2	8	0.5	0.05	4.9§	40	5
Cr	7	...	8×10^{-3}¶	0.5	100	0.05	8.3§	20	1
Cs	...	30	3	0.4	50	0.1	300	300	
Cu	2	...	5×10^{-3}¶	1	0.1	0.08	0.2§	100	0.5
Dy	...	...	300	4	0.0001	0.5	50	40	
Er	...	...	300	8	0.1	0.5	50	30	
Eu	4000	...	100	1	0.0005	0.2	30	2	
F	...	1	...	...	100	0.02	...	10	
Fe	50	20	0.01¶	3	5000	0.05	30	80	3
Ga	200	1	50	2	0.5	0.09	37§	60	
Gd	40	...	10,000	10	1	0.5	50	20	
Ge	4000	2000	200	500	0.5	0.2	16.7§	...	675§
H	...	...	...	...	...	0.0008			
He	...	...	...	...	...	0.003			

Hf	10	50	300	0.4	100	4000		100	
Hg	10	500	100	0.6	1	100	0.2†		5
Ho		20	50	0.1	0.5	4	300		
I		80	67§	0.1	0.005	500	0.2†	40	10
In		500	300	0.3	0.01	1	0.3		50
Ir	56.5§	20	100	0.03	5	5000			6
K	0.3	5	10	0.1		0.01			
Kr		0.2	7.6§	0.1	0.1	20	0.2		200
La		100	100	0.006		0.01			
Li		10	0.13§	0.1	0.005	6	0.1	200	4000
Lu		3	10	0.03		0.8			3
Mg		5	100	0.05	50	0.5		0.2	
Mn			10	0.3	0.005	20	3 × 10⁻³¶		2
Mo		30	14.2§	0.01	10	0.01	0.02¶		5
N									10
Na		30	80	0.02	0.5		0.4		10
Ne									8000
Nb	253§	20	11.4§	0.02	0.5	200	5000	100	
Nd	1	30		0.08	10	10	5000		5
Ni		1		0.4	5	5	0.01¶		4000
Np		200						2	4
O			1000	0.07					
Os	10	2000	10	0.01	5	500			
P		200	200	0.4	50	40			50
Pa			80						100
Pb		30	300	0.03	1000	0.5	3 × 10⁻³¶		
Pd		50	30	0.3	0.05	5	30		6
Pr		30	300	0.3	0.05	50	15,000		10
Pt		2	200	0.1	5	60	20		4000
Pu				0.5					10
Rb		200		0.1	5	0.1	2		
Re		20	300	0.2	0.05	100	1000		200
Rh		200		0.09	0.05	10	20		50
Ru				0.03	1	10	20	1000	100
S	10	500	93.5§	0.03	500	2000		200	
Sb				0.2	0.5	30	0.1†		4

Table 25-3 Absolute detection limits for the elements by various techniques (ng) (Continued)

Element	Absorption spectrophotometry	Ultraviolet–visible fluorescence	Atomic absorption	Flame emission photometry	Neutron activation	Spark source mass spectroscopy	Emission spectroscopy		
							Dc arc	Copper spark	Graphite spark
Sc	8	100	200	3	1	0.04	30	0.8	
Se	200	2	1**	600	500	0.1	10,000		
Si	100	80	6000	700	5	0.03	6.6§	10	
Sm	1000	100	1000	10	0.05	0.5	50	40	
Sn	60	100	0.2¶	30	50	0.3	10	.....	914§
Sr	10	.....	2	0.2	0.5	0.09	5.7§	30	
Ta	40	.....	.....	1000	5	0.2	2730§	100	10
Tb	1000	3000	6000	10	5	0.1	300	100	
Te	400	20	2**	100	5	0.3	4000	300	
Th	50	.....	.....	8000	5	0.2	115§	40	
Ti	10	.....	0.06¶	20	.....	0.05	1.3§	10	3
Tl	20	30	5	3	1	0.2	30	200	50
Tm	.....	.....	300	8	0.5	0.1	50	10	
U	300	0.1	.....	200	0.5	0.2	1130§	200	
V	10	2500	0.1¶	10	0.1	0.04	6.4§	8	1
W	30	1000	250	200	0.1	0.5	244§	80	10
Xe	.....	.....	.....	.....	.....	0.4			
Y	1000	20	1000	10	.....	0.07	3	0.8	
Yb	.....	.....	100	2	0.1	0.5	50	5	
Zn	100	1000	1×10^{-3}¶	80	10	0.1	58§	200	10
Zr	50	20	.....	500	100	0.1	4.6§	20	3

* From Morrison and Skogerboe (2) except where noted.
† Data from AA with graphite rod source, from H. Massmann, quoted by H. Kaiser, in W. W. Meinke and B. F. Scribner (eds.), "Trace Characterization, Chemical and Physical," p. 164, NBS Monograph 100, Washington, 1967.
¶ Data from AA with graphite rod source, from D. R. Thomerson and K. C. Thompson, Am. Lab., **6**(3):53 (1974).
§ H. Kaiser, loc. cit., p. 163.
** Data from AA with a special hydride generator, from D. R. Thomerson and K. C. Thompson, loc. cit.

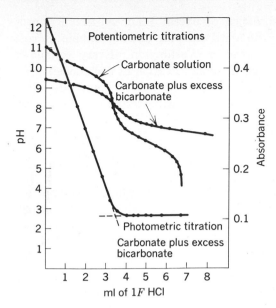

Fig. 25-1 Potentiometric and spec trophotometric titrations of 3.4×10^{-2} F sodium carbonate solution containing 1 F sodium bicarbonate. [*Analytical Chemistry* (4).]

of color is matched directly by eye, and also in the method of (differential) "ultimate precision" in spectrophotometry.

In all comparisons it is highly desirable that the standards duplicate the unknown as closely as possible. This principle results in the substantial reduction of systematic errors which have the same effect on all solutions. In some cases the precision can be greatly increased, since the full-scale span of the instrument can be applied to measuring the difference between two rather close magnitudes, rather than the distance of each magnitude from zero. This has been discussed in its application to photometric analysis in Chap. 3, but the principle is of wider applicability.

Closely related is the type of apparatus in which the comparison between standard and unknown is made directly in a single operation. Examples are the potentiometric concentration cell, the thermal conductivity detector in GC, and photometers and spectrophotometers that employ a balanced system of two light beams passing through two samples.

It must be remembered that comparison with standards cannot improve the *precision* of an analysis, but it may have an effect on the *accuracy*, which can never be better than the standards. The preparation and preservation of standards for extremely dilute solutions (micromolar to nanomolar) can be quite difficult. The walls of a glass vessel have a tendency to adsorb solute, and may reduce the concentration significantly below the intended value; this can be overcome in favorable cases by taking the precaution of rinsing out the vessel with some of the solution to be stored.

An important aid in the direction of overall standardization is provided by the extensive series of standard samples made available at cost by the National Bureau of Standards in Washington. Every sample is accompanied by a certificate bearing the concentration of each constituent, from the major elements down to those present in only a few thousandths of a percent. A great many determinations can be tested as to accuracy and precision by means of these samples.

STANDARD ADDITION

This is a very generally applicable method of implementing the comparison with a standard. It has been mentioned in the discussion of a few instrumental methods (polarography, for example), but it can readily be adapted to others. A reading is taken on the sample to be analyzed; then a measured quantity of standard is added to the sample with mixing, and the measurement is repeated. If the analysis is destructive, (as a titration usually is), then the standard must be added to a second aliquot.

In many circumstances this procedure will serve to identify the feature of the record that pertains to the desired material, and at the same time give the information needed for a quantitative analysis. The dilution of the sample by addition of the standard must be allowed for or shown to be negligible.

This technique has the great advantage that the standard and unknown are measured under essentially identical conditions. Even if the kind and quantity of other substances present are not known accurately, they can be taken as identical in the two measurements.

DATA PLOTTING

The precision with which data can be read from a graph, and hence the effective sensitivity of the method, can be greatly affected by the manner in which the data are displayed.

It will be instructive to consider a number of functions of the form $R = f(C)$, relating instrument response to a concentration term, which might describe the behavior of some analytical systems. In Fig. 25-2, curves are plotted for four such functions: (1) linear, $R = k_1 C$; (2) square-law or power series, $R = k_2' C^2 + k_2 C$; (3) reciprocal, $R = k_3/C$; (4) logarithmic, $R = k_4 \log C$. The derivative $\partial R/\partial C$ of each of these is also plotted. It is apparent from the figure that the reciprocal and logarithmic functions give very steep curves, hence greatest sensitivity, at low concentrations, while the square function gives greater sensitivity at higher concentrations. For the linear function, as expected, the sensitivity as defined above is constant over the whole range.

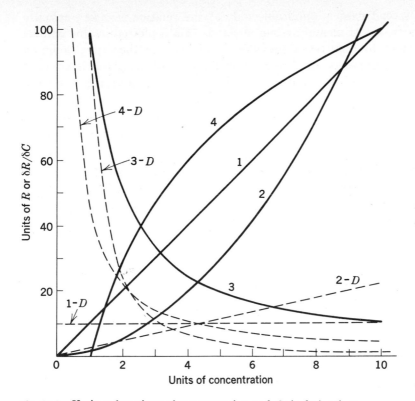

Fig. 25-2 Various functions of concentration and their derivatives.

(1)	$R = k_1 C$	(1-D) $\partial R/\partial C = k_1$	$k_1 = 10$
(2)	$R = k_2' C^2 + k_2 C$	(2-D) $\partial R/\partial C = 2k_2' C + k_2$	$k_2 = k_2' = 1$
(3)	$R = k_3(1/C)$	(3-D) $\partial R/\partial C = -k_3(1/C^2)*$	$k_3 = 100$
(4)	$R = k_4 \log C$	(4-D) $\partial R/\partial C = 0.435 k_4(1/C)$	$k_4 = 100$

* The negative sign is ignored for convenience in plotting.

The slopes in Fig. 25-2 are written as partial derivatives to emphasize the fact that there are likely to be other variables affecting the sensitivity. An example is spectrophotometric analysis, in which the sensitivity can be varied by a change in wavelength. The response curve will be of the same form, but the horizontal scale will be compressed or expanded.

It is often useful to plot data on log-log paper. There are two reasons for doing this: (1) to compress onto one set of coordinates data covering several orders of magnitude, and (2) to determine the constants k and n in a relation of the form $y = kx^n$. The first of these objectives is attained, for example, when one plots an absorption spectrum as $\log A$ against wavelength or wave number; there is no theoretical reason for doing this, as there is in the plotting of A $[= \log (1/T)]$ in place of T.

The second case is applicable only if the relation $y = kx^n$ actually prevails in the experimental system at hand. Data plotted in this manner in an attempt to discover a previously unknown mathematical relation may give the impression of a correlation more significant than is actually justified. Possible pitfalls and their avoidance have been vividly described by P. N. Rowe (5).

PROBLEMS

25-1 Identify several instrumental methods as corresponding to each of the functions plotted in Fig. 25-2, and show equations to verify your choices.

25-2 How would you best determine water quantitatively in each of the following circumstances (outline a procedure where possible):

(a) Water vapor in tanks of compressed H_2 or O_2
(b) Water dissolved in "pure" chloroform or ether
(c) Water content of the atmosphere in a sealed space vehicle
(d) Water collected in the bottom of a large gasoline storage tank

25-3 Devise an instrumental method for the determination of TEL (tetraethyl lead) and TML (tetramethyl lead) when either or both may be present in a gasoline.

25-4 One of the most difficult separations is between Zr and Hf. A result of the "lanthanide contraction" is that Zr and Hf atoms are almost precisely the same size which, together with their similar electronic structure, makes them chemically nearly identical. Suggest at least two tentative methods by which you might attack their separation. In each method, what data would you look for to evaluate the success of the method? What nonseparative analytical methods might be used to analyze mixtures of these elements?

25-5 In the colorimetric analysis of manganese and chromium in steel, the color of the ferric iron will cause interference unless suitable precautions are taken. It is possible to remove the bulk of the iron by ether extraction in the presence of hydrochloric acid, or by ion exchange following oxidation, or the iron may be complexed with citrate, tartrate, or other reagent to destroy its color. Alternatively, the iron may be allowed to remain in the solution and its absorbance corrected for by taking photometric readings before and after oxidation of the manganese and chromium; or combinations of these methods may be employed. Compare these procedures critically. Why is mercury-cathode electrolysis inapplicable as a separatory procedure?

25-6 In Chap. 10, mention was made of the use of sodium iodide crystals "activated" by the addition of a trace of thallous iodide, as luminescent crystals for the detection of x-radiation. Devise a nondestructive method for determining the amount of thallous iodide in sodium iodide crystals. Estimate the precision.

25-7 In the following table is given the average chemical composition of seawater (in millimoles per kilogram). Note that in certain instances the element or radical may be present in more than one form: Thus carbon dioxide may be present partly as dissolved gas and partly as carbonate or bicarbonate; several of the metals may be present in more than one oxidation state. Suggest appropriate analytical methods by which seawater may be analyzed for each of these elements or ions. Any procedure by which more than one ion can be determined simultaneously will, of course, be an advantage.

Cl	535.0	Br	0.81	NO_3^-	0.014	Ag	0.0002
Na	454.0	Sr	0.15	Fe	0.0036	NO_2^-	0.0001
SO_4^{2-}	27.55	Al	0.07	Mn	0.003	As	0.00004
Mg	52.29	F	0.043	P	0.002	Zn	0.00003
Ca	10.19	Si	0.04	Cu	0.002	Au	2.5×10^{-7}
K	9.6	B	0.037	Ba	0.0015	H	10^{-6}-10^{-7}
CO_2	2.25	Li	0.015	I	0.00035		

REFERENCES

1. Karasek, F. W.: *Anal. Chem.,* **44**(4):32A (1972).
2. Morrison, G. H., and R. K. Skogerboe: In G. H. Morrison (ed.), "Trace Analysis: Physical Methods," Wiley-Interscience, New York, 1965.
3. St. John, P. A., W. J. McCarthy, and J. D. Winefordner: *Anal. Chem.,* **39**:1495 (1967); W. J. McCarthy: in J. D. Winefordner (ed.), "Spectrochemical Methods of Analysis," p. 493 ff., Wiley-Interscience, New York, 1971; T. Coor: *J. Chem. Educ.,* **45**:A533 (1968).
4. Underwood, A. L., and L. H. Howe, III: *Anal. Chem.,* **34**:692 (1962).
5. Rowe, P. N.: *Chem. Tech.,* **4**:9 (1974).

Chapter 26

Electronic Circuitry for Analytical Instruments

The great majority of instrumental methods of analysis treated in this book require electrical circuits, and this, in all but the simplest cases, implies the need for electronics. So for a full understanding of analytical instruments, their limitations, and what can go wrong with them, some knowledge of electronics is essential. The brief account in this chapter and the next can only be considered a survey. No attempt is made to derive the fundamental mathematical relations; a more complete treatment can be found in numerous texts.

The heart of any electronic device is one or more components which act directly on electrons. This includes vacuum and gas-filled tubes in which electrons are liberated by thermionic emission from a hot cathode or by the photoelectric effect. It also includes a variety of solid-state components, particularly semiconductor diodes, transistors, and photocells. None of these active elements is self-sufficient; they all require more-or-less complex supplementary circuits composed of resistors, capacitors, inductors, meters, etc. In most cases a power supply is also required, which may be a selection of batteries or may be rectified alternating current. In addition there will frequently be incorporated other components for convenience or safety, such as switches, fuses, and pilot lights.

504

Electronics first became important to chemists in the 1930s with the introduction of the glass electrode for pH measurement, which made necessary the development of an electronic pH meter. Other electronic instruments soon followed, notably spectrophotometers which displaced the inconvenient and often inexact photographic observation of absorption spectra. During and since the Second World War, advancing electronic technology has made prior instruments obsolete in nearly all fields.

In the years since about 1960, another revolution has taken place, in which vacuum tubes have been entirely supplanted in new designs of laboratory instruments by the introduction of transistors. At first, "transistorization" was mostly a process of replacing tubes by transistors with a minimum of circuit changes—something like a literal translation of one language into another. More recently, as the idiomatic capabilities of transistors have become more fully understood, new circuits have been designed to fit existing needs. Present-day instrument engineers rarely choose vacuum tubes except for a few specialized applications.

Now another electronics revolution is in progress, wherein discrete transistors and associated components are being replaced by *integrated circuits* (ICs). Each IC may contain many transistors, along with capacitors and resistors, all fabricated simultaneously on a single tiny slab of highly purified silicon.

In this chapter vacuum tubes will scarcely be mentioned. If information about tubes and their applications is desired, the reader is referred to any of the many texts on general electronics.

SEMICONDUCTORS

A semiconductor is a solid substance which is intermediate between metallic conductors on the one hand and nonconducting insulators on the other. It is characterized by a relatively large *negative* temperature coefficient of resistance, whereas the coefficient is *positive* for metals; this provides a convenient criterion for distinguishing the two types of conductors. The most used semiconductors are silicon and germanium.

In a crystal of high-purity silicon, each atom is bound covalently to each of four other atoms (the diamond structure) and, since each atom has just four valence electrons, it is fully satisfied by this structure. To make the crystal useful for electronic purposes, it must be *doped*, that is, a trace of impurity added. The foreign atoms must be of such nature that they can replace some of the silicon atoms in the crystal lattice. If the impurity is a pentavalent element, such as arsenic or antimony, then each of its atoms possesses an extra valence electron beyond those needed for the covalent lattice bonds. The extra electrons are easily torn loose from their parent atoms by thermal energy, and then are free to wander at random throughout the lattice. The impurity atoms become unipositive ions imbedded in the crystal, and hence immobile. On the other hand, if the impurity is trivalent gallium, indium, or gold, then there will be a deficiency of one electron per atom. The spot where the electron is lacking is called a *hole*. Occasionally an electron in a normal covalent bond located near one of the impurity atoms will have sufficient thermal energy that its vibrations will bring it so close to the hole that it will escape completely from its previous berth and move into the hole. The result of such a process is that the hole has moved from one spot to another within the lattice. In a piece of silicon doped in this way, the holes appear to wander freely through the crystal in a fashion exactly analogous to the surplus electrons in a piece doped with arsenic or antimony. The mobility of the holes, however, is somewhat less than that of the electrons.

If an electric field is applied to doped silicon, a current will flow which is carried almost exclusively by the excess electrons or holes provided by the impurity.* Silicon in which the current carriers are electrons (negative) is called *n*-type silicon, whereas that in which the predominant carriers are holes (positive) is *p*-type silicon.

DIODES

A crystal diode is a two-terminal semiconductor device which has the ability to pass current with ease in one direction but to block it from flowing the other way. Hence it is useful as a rectifier.

The crystal diode is made from a small bar of silicon or other semiconductor (called a *wafer* or *chip*), part *p* type and part *n* type, as indicated schematically in Fig. 26-1. When it is connected to a source of potential so that the *n* region is negative and the *p* region positive, the dominant carriers in both sections tend to move toward the *p-n* junction. At the junction, electrons from the *n* side neutralize holes from the *p* side, and so current flows easily. In the reverse connection, with the *n* region positive and the *p* region negative, both holes and electrons are pulled away from the junction, leaving an intermediate region with very few carriers of either type, hence a high resistance.

In Fig. 26-2 are shown the characteristic current-voltage curves for silicon and germanium crystal diodes. For positive (i.e., forward) values of potential, the current follows approximately an exponential function of the voltage. For negative (reverse) potentials, the current is practically zero until (for silicon) a critical value E_z is reached, at which the current increases negatively until limited by the series resistance in the circuit. This critical point, called the *zener* or *breakdown* voltage, corresponds to the potential necessary to tear electrons out of covalent bonds in the crystal,

---

* It should be emphasized that germanium or silicon which is doped, even relatively heavily, is still from the *chemical* standpoint extremely pure. A controlled impurity of the order of 1 ppm will usually be adequate to impart the desired electrical properties.

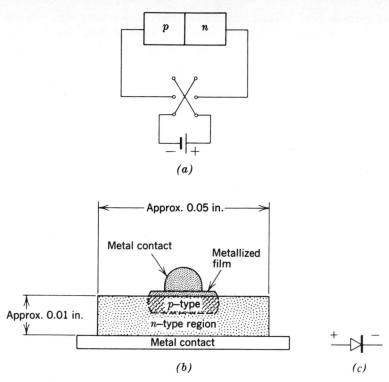

(a)

(b)

(c)

Fig. 26-1 A *p-n* junction diode. (*a*) Schematic; (*b*) physical structure of one form (the diode is manufactured by diffusing a *p*-type impurity into the *n*-type starting material); (*c*) symbol for a diode, showing polarity for forward bias, i.e., easy current flow. [(*b*), *Courtesy Education Development Center* (1).]

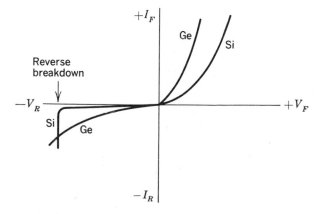

Fig. 26-2 Dc characteristics of germanium and silicon diodes.

thus creating pairs of positive and negative carriers which are swept toward the respective circuit connections. By varying the manufacturing techniques, silicon diodes can be prepared with zener voltages anywhere between about 2 and 200 V. Germanium diodes have a lower reverse resistance than silicon, and no well-defined zener breakdown potential.

Diodes are useful in two main areas. In the larger sizes they are employed as power rectifiers to convert alternating to direct current as a power source for other devices. Smaller diodes are applied to the rectification of ac signals; in this service they are often called *detectors* or *demodulators*.

BIPOLAR TRANSISTORS

The basic amplifying semiconductor device is the transistor, which exists in many modifications, the most prevalent being the bipolar junction transistor. This may be visualized with the aid of Fig. 26-3a, which shows schematically a wafer made up of two p-type segments separated by a

Fig. 26-3 A *p-n-p* junction transistor. (a) Schematic; (b) physical structure of one form. The *p*-type regions are obtained by alloying a metal containing large amounts of *p*-type impurity into the *n*-type semiconductor wafer. [(b), *Courtesy Education Development Center* (1).]

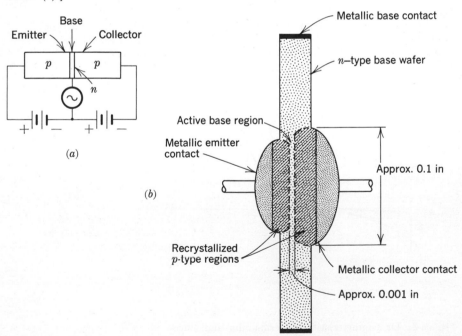

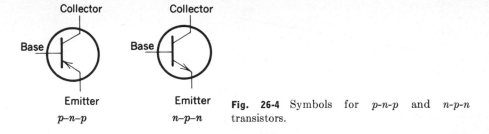

Fig. 26-4 Symbols for *p-n-p* and *n-p-n* transistors.

thin layer of *n* material. One way of making such a unit is to start with a wafer of relatively high resistance (i.e., lightly doped) *n*-type silicon, deposit bits of indium on each side, and then heat it to a high enough temperature to melt the indium (mp 155°C) and to cause it to diffuse into the silicon, converting *n* to *p* type for a region on each surface. The result resembles Fig. 26-3*b*, which shows a very thin portion of the original *n* silicon separating two *p* regions to which electric connections are made through the indium. For transistor action, the junction between the smaller *p* region, called the *emitter*, and the *n* material, called the *base*, is given a small forward bias, while the junction with the larger *p* region, the *collector*, is reverse-biased. Current can flow easily across the forward-biased emitter-base junction, but since the emitter region has more holes as carriers than the base has electrons, the major part of the emitter current is carried by holes. The base region, being both thin and lightly doped, is poor in electrons, so that most of the holes from the emitter diffuse across it, as they have a greater probability of being "collected" by the large collector junction than of combining with electrons in the base. Any current which may be injected into the transistor through the base connection will provide additional electrons to carry a portion of the emitter-base current, and hence will have a considerable effect on the number of holes available to flow into the collector. This is the basis of the amplifying ability of the transistor.

The transistor described above is of the *p-n-p* type, but it is equally possible to fabricate a unit with opposite characteristics, an *n-p-n* transistor. The two can be used in equivalent circuits, except that the polarities of all voltages and currents must be reversed. Figure 26-4 shows the standard symbols for the two types. The arrowhead on the emitter indicates the direction of easy positive current flow for the emitter-base junction.

Figure 26-5 gives a family of current-voltage characteristic curves for the collector circuit of a typical silicon *n-p-n* transistor, the 2N5182, showing the effect of the base current as a parameter. The ability of a transistor to amplify currents may be expressed as the ratio of a small change in collector current to the corresponding change in base current (at constant collector voltage). This quantity, denoted by β, lies between

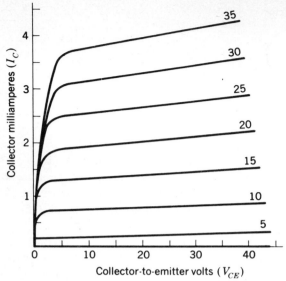

Fig. 26-5 Collector characteristics of a silicon transistor, type 2N5128. (*RCA Corporation, Harrison, N.J.*)

50 and 200 for common transistors. Figure 26-6, in which collector current is plotted against base current, is known as a *transfer plot*. The slope of this curve is the β of the transistor, in this case approximately 100 at $I_B = 20\ \mu A$.

A simple circuit to demonstrate the amplification possible with a transistor is given in Fig. 26-7. The circuit is powered from a single bat-

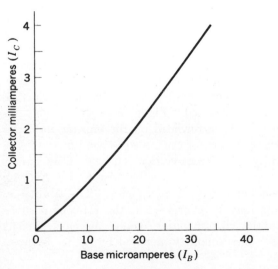

Fig. 26-6 Transfer plot for the 2N5128 transistor. (*RCA Corporation.*)

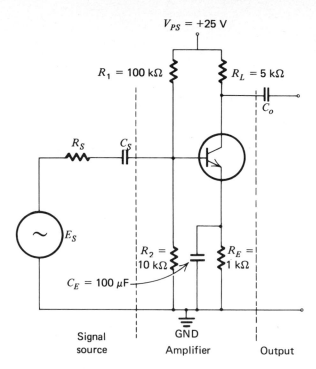

Fig. 26-7   A single-stage transistor amplifier.   The component values shown are illustrative only.

tery or rectifier at voltage V_{PS} relative to ground.   Resistors R_1, R_2, and R_E, known as *bias resistors*, are designed to provide a suitable dc potential between the base and emitter, so that when no signal is present the transistor will draw a small current, about the middle of its range.

The amplifying action of the transistor can be demonstrated as follows.   Suppose that an ac signal of 1-mV amplitude, originating in a source with a resistance of 1 kΩ, is led into the base of the transistor.   This produces an ac component of the base current equal to the E/R quotient for the source, namely, 10^{-3} V/10^{+3} Ω = 10^{-6} A or 1 μA.   Multiplying by β (100) shows that the resulting ac component of the collector current will be 100 μA.   This current, flowing through resistor R_L, the *load* resistor, produces an ac potential drop of $(100 \times 10^{-6}$ A) $(15 \times 10^{3}\Omega)$ = 1.50 V. Normally, the power supply is shunted by a large capacitance, so that the point marked V_{PS} is at ground potential so far as alternating current is concerned.   Hence the 1.50-V output signal will be seen at the output relative to ground.   The gain of this circuit (the ratio of output to input) can be calculated in terms of voltage, current, or power.   These gains

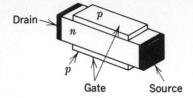

Fig. 26-8 Junction FET, n-channel type, schematic.

are, respectively,

$$G_V = \frac{1.50}{10^{-3}} = 1.50 \times 10^3$$

$$G_I = \beta = 100$$

$$G_P = G_V G_I = 1.5 \times 10^5$$

FIELD-EFFECT TRANSISTORS (FET)

This semiconductor device works on a principle somewhat different from that of the transistors we have previously discussed. Consider the sketch in Fig. 26-8. This device consists of a bar of n-type silicon, called the *channel,* with connections at both ends, called, respectively, the *source* and *drain.* The channel is sandwiched between layers of p material (connected together) called the *gate.* In some constructions, the gate is all in one piece, completely surrounding the central bar. Both n- and p-channel FETS are available.

The FET is classified as a *unipolar* transistor, as there is effectively only a single junction, whereas those transistors having two junctions (emitter-base and base-collector) are *bipolar.*

The operation of a FET can be followed by the aid of Fig. 26-9. If no voltage is applied to the gate, then current will flow unhindered through the channel, electrons passing from source to drain (the channel current is carried entirely by majority carriers). In its normal mode, the gate-to-channel junction is reverse-biased. This has the effect of depleting the area shown in dotted lines of electrons, hence increasing the resistance of the channel. This arrangement has the added feature that the reverse

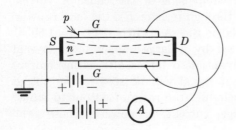

Fig. 26-9 Connections for the n-channel FET.

bias prevents the flow of appreciable current in the gate circuit.   The voltage applied to the gate may be several volts, say up to 10 or so.   Since
no current flows in this circuit, the gate characteristic is given in volts,
rather than in current units as in bipolar transistors which draw appreciable base currents.   The voltage impressed between drain and source may
be quite high (requiring some current-limiting component in the circuit).

The characteristic curves for a FET resemble those of Fig. 26-6, if
drain current (in milliamperes) is plotted against the drain-source voltage,
with the gate-source voltage as parameter.   The principal advantage of
the FET is its high input impedance, resulting from the reverse-bias condition.   This impedance may be in the range of tens or even hundreds of
megohms.

The *insulated-gate FET* (sometimes called MOSFET for metal oxide
semiconductor) is a modification of the FET, wherein a thin film of insulat-
ing material, usually silicon dioxide, separates the gate from the channel.
This eliminates the rectifying junction, so that the gate can be of either
polarity without drawing any current.   The electrostatic field between gate
and channel is still able to modify the distribution of holes or electrons in
the channel and so determine its resistance.   The MOSFET has the highest
input resistance of any transistor.   The conventional symbols for the several types of FETs are shown in Fig. 26-10.

There are many other semiconductor devices that find occasional use
in laboratory instruments.   One of the most widely used is the *thermistor*,

Fig. 26-10   Symbols for FETs.   (*a*) *n*-Channel FET;
(*b*) *p*-channel FET; (*c*) *n*-channel MOSFET; (*d*)
p-channel MOSFET.   *B* denotes a connection to the
bulk semiconductor material.

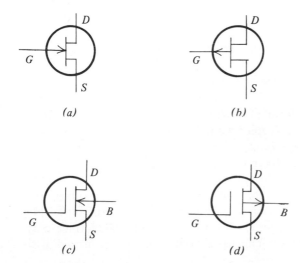

(a)   (b)

(c)   (d)

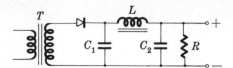

Fig. 26-11   Half-wave rectifier with filter.

a resistor with a large negative temperature coefficient, made of the sintered or fused oxides of transition metals.   It provides a sensitive temperature detector, though with rather less reproducibility than a resistance thermometer made of a metal such as platinum.   It can also serve in temperature-compensation circuits to counteract the positive coefficients of other components or to lower the bias at the base of a transistor if the temperature rises, thereby stabilizing the gain of the transistor.

ELECTRONIC CIRCUITS

The majority of modern laboratory instruments are powered from 115-V, 50- or 60-Hz ac lines.   Direct current is obtained either from batteries or by rectification with silicon rectifiers.

There are a number of possible connections for the rectifier diodes, each of which has merit for particular types of applications.   The simplest is the *half-wave* rectifier (Fig. 26-11).   The transformer T has two functions: It isolates the circuit from the ac supply line, and it permits a selection of the voltage level, as may be required.   The C_1-L-C_2 network constitutes a *filter* to reduce the residual ripple to a low value.   With the type of filter shown, the voltage of the dc output will be somewhat greater than the ac voltage at the transformer secondary.   If C_1 is omitted, the output will be lower.   For low-current requirements, the choke L may be replaced by a resistor.   The shunt resistor R, called a *bleeder*, ensures the discharge of the capacitors when the supply is turned off; it is particularly important in high-voltage units, as a safety measure.

The filter requirements are less stringent with a *full-wave* rectifier (Fig. 26-12).   In the circuit in (*a*), the output voltage approximates *half* that of the transformer secondary, while in the *bridge rectifier* (*b*) the full voltage is obtained.

Figure 26-13 shows one form of *voltage doubler* rectifier which gives an output about twice the transformer voltage.   The two capacitors each charge to the secondary voltage on opposite half-cycles of the alternating current, but in such a sense that their voltages add together for the output.

High voltages at low currents (a few kilovolts at less than 1 mA) are required for ionization chambers, counters, photomultiplier tubes, and cathode-ray tube applications.   Such requirements can be met by means of a half-wave rectifier with a simple RC filter, but the transformer may be unduly bulky and expensive.   The same result may be achieved with

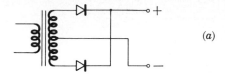

(a)

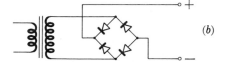

(b)

Fig. 26-12 Two forms of full-wave rectifiers. (a) With center-tapped transformer; (b) bridge-type. A filter like that of Fig. 26-11 is required.

an oscillator operating at perhaps a few hundred kilohertz. The output can be stepped up in voltage by an air-core radiofrequency transformer, followed by a half-wave rectifier or even a voltage doubler to give very high dc potentials. Filtering requirements are easily met because of the high ripple frequency. Modern trend is toward the use of an oscillator employing two power transistors, which with transformer and associated parts can be enclosed in a single shield can. This unit can be powered either from a single battery or from a line rectifier.

VOLTAGE REGULATION

In the rectifier power supplies just described, the output voltage and current are both subject to variation resulting from any change in load resistance or in ac line voltage. The effects of line-voltage fluctuations can usually be reduced to the point where they are negligible by the insertion of a *constant-voltage transformer* between the instrument and the line. The effects of changes in load are not so easily eliminated, as they depend on the effective internal resistance of the rectifier-filter combination. A choke-input filter gives a more stable voltage than does a capacitor-input, while a voltage-doubler shows less constancy. Two types of circuit for the regulation of voltage will be discussed; they are effective against variations either from line-voltage or load changes.

The first of these makes use of the constant-voltage characteristic of a zener diode. The diode is connected across the load, but with a series resistor between it and the rectifier, as in Fig. 26-14. The voltage across

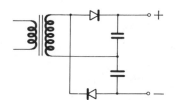

Fig. 26-13 Voltage-doubling rectifier. A filter is necessary.

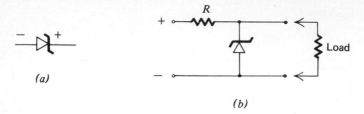

Fig. 26-14   Zener diode.   (a) Symbol; (b) a voltage regulator.

the diode will remain constant, no matter what current passes through it, within rather wide limits. Hence the current through resistor R will adjust itself to maintain whatever voltage drop is necessary between rectifier and diode. If the load resistance changes so that the load current is altered, the current through the diode will change by the same amount in the opposite direction, to maintain the constant voltage.

The regulation which can be achieved by this simple circuit is of the order of a 2 percent change in voltage for a load change from the maximum design current to zero current. This is adequate for some purposes, but quite inadequate for others.

Better regulation can be obtained by means of a more complex circuit, such as that in Fig. 26-15. Here a high-current transistor Q_1 is introduced in series with the load. Any change in either load current or line voltage will produce a change in the voltage supplied by the rectifier bridge to the collector of Q_1, but the zener voltage is continually compared with a portion of the load voltage by transistor Q_2, which supplies just enough base current to Q_1 to maintain a constant overall output. This circuit will keep the output voltage within about 0.5 percent for a change in

Fig. 26-15   A voltage-regulated power supply using a series-pass transistor (Q_1), an amplifier (Q_2), and a zener diode.

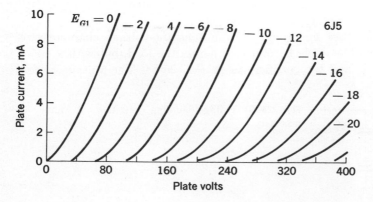

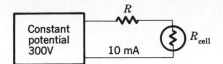

Fig. 26-16   A constant-current source.

load from 0 to 100 mA, and within 2 percent up to 400 mA. Other regulators can be designed which will give several orders of magnitude better constancy than this.

Regulator circuits generally respond in a time interval small compared with the period of the alternating source, so only a minimum of conventional filtering is required.

CURRENT REGULATION

A source of constant current is required for coulometric applications, and also as excitation for some types of light sources, such as the hydrogen-discharge lamp. One approach which is applicable to small currents in electrochemistry lies in the use of a constant high-voltage source with a large dropping resistor in series with the electrolysis cell (Fig. 26-16). Suppose 10 mA is desired and the drop across the cell is of the order of 1 V. Then for a 300-V supply the series resistor R must be $299/0.01 \approx 30$ kΩ. If the cell resistance should change by 100 percent, i.e., from 100 to 200 Ω, the current would only change by the factor $302/301$, or about 0.3 percent.

A more versatile device, which also eliminates any possible high-voltage hazard, consists of a regulating circuit comparable to Fig. 26-15, wherein the voltage supplied to the sensing transistor is the potential drop across a precision resistor in series with the load.

In the case of a power supply for a light source, it is desirable to initiate control by an auxiliary photocell which monitors the lamp.

THE EMITTER FOLLOWER

The output of an amplifier can be taken across the emitter resistor instead of at the collector. This is shown in Fig. 26-17. It can be shown that the voltage gain with this arrangement is just slightly less than unity—0.98 to 0.999. This slight loss in voltage is more than offset by the gain in

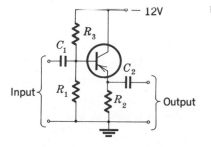

Fig. 26-17   An emitter-follower amplifier.

current which is available. This results from the fact that R_2 can be very much smaller than R_1 and, since the potentials across the two are essentially equal, much greater power is available at the output than in the input circuit. Thus we have a power amplifier but not a voltage amplifier. The name comes from the fact that the signal voltage on the emitter *follows* exactly whatever variations are presented to the base. The follower is utilized as an *impedance converter* to couple a high-impedance circuit to one of low impedance. The input impedance is high enough to receive signals from sources of moderately high resistance, though not usually high enough for measurements with the glass electrode or with ionization chambers.

DC AMPLIFIERS

An amplifier built around a single transistor will amplify signals at any frequency from 0 Hz (i.e., direct current) up to some maximum depending on the transistor type and on circuit constants. Special difficulties arise with dc signals when gain requirements dictate multiple stages. For one thing, the output potential of one stage may not be compatible with the input bias needed for the next stage. Another difficulty is concerned with *drift*. Any gradual change in the first stage, such as might be produced by a change in ambient temperature or by aging of some component, will be amplified by the succeeding stages and appear at the output indistinguishable from true signal. Neither of these problems occurs in ac amplifiers, because the capacitive (or inductive) coupling between stages prevents passage of dc potentials.

A highly successful method of eliminating drift is by *chopping* or *modulating* the signal, which has the effect of converting the dc signal to its ac equivalent. An ac amplifier, not subject to drift, can then be utilized.* Two examples of chopper amplifiers are shown in Fig. 26-18. The chopper unit consists of a vibrating metallic reed energized by an ac electromagnet. The points are contacted alternately by the grounded reed at the frequency of the magnetic field. A single chopper can be connected as in (*a*) so as to rectify synchronously the signal at the output of the amplifier. Less subject to noise pickup is a full-wave system (*b*) requiring two choppers and transformer coupling both into and out of the amplifier.

OPERATIONAL AMPLIFIERS

We will now consider a class of dc amplifiers especially designed to perform mathematical operations (hence the name) on signals presented to them.

---

* Note the similarity between this chopping of a dc electric signal and the chopping of a beam of radiation by means of a mechanical shutter, as is done in most recording spectrophotometers.

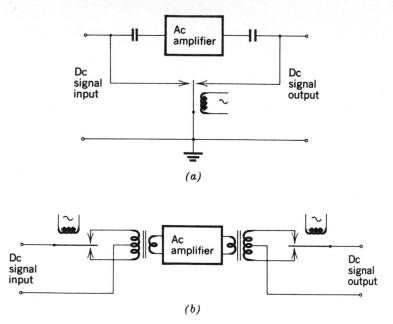

(a)

(b)

Fig. 26-18   The chopper amplifier.   *(a)* Half-wave circuit; *(b)* full-wave version.   Both circuits include synchronous rectification.

Appropriate external connections establish conditions such that the output voltage will be *(1)* the algebraic sum of two or more input voltages, *(2)* the product of an input voltage multiplied by a constant factor, *(3)* the time integral of the input, or *(4)* the derivative of the input with respect to time.   Other mathematical functions, such as taking logarithms or antilogarithms, squaring, or multiplying or dividing one variable quantity by another, can be implemented by the use of nonlinear components along with the amplifiers.

To be suitable for such applications an amplifier must have the following attributes: *(1)* It must have a large negative gain, at least -10^4, with many commercial units well beyond -10^6; the negative sign signifies inversion—the output is opposite in sign to the input; *(2)* it must have a large input impedance, not less than 10^5 Ω, often up to 10^{12} or even higher; *(3)* it must be capable of being nulled, that is, to give zero output for zero input; and *(4)* it must have only minimal drift.

Most operational amplifiers have two input terminals, only one of which produces an inversion of sign.   The terminals are conventionally marked $+$ and $-$, as in Fig. 26-19.   These designations do not mean that the terminals are to be connected only to potentials of the indicated sign, as would be the case with similar marks on a voltmeter, but rather that the one marked $-$ gives sign inversion and the other does not.   In

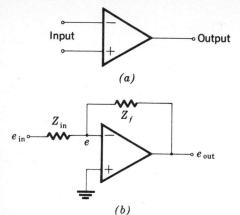

(a)

Fig. 26-19 Operational amplifiers. *(a)* General symbol; *(b)* connected as an inverting amplifier such that $e_{out} = -(Z_f/Z_{in})e_{in}$, provided the internal gain $A \gg 1$. The triangular symbol implies that a suitable power supply is provided; all potentials are referred to ground.

(b)

those operational amplifiers which have only a single input, it is always the noninverting one which is omitted. In case the noninverting input is not required in a particular application, it should be grounded to avoid instability.

The basic connections are shown in Fig. 26-19(*b*). Most circuits using operational amplifiers depend on negative feedback: A connection is made through a suitable impedance Z_f from the output to the inverting input. If the signal to be sensed by the amplifier is a voltage, then it must be applied through an impedance Z_{in}. Since the input to the amplifier proper draws negligible current, the current flowing in Z_{in}, namely, $(e_{in} - e)/Z_{in}$, must be equal to that in the feedback loop. But the current in the feedback loop, given by $(e - e_{out})/Z_f$, can come only from the output of the amplifier. Therefore, when an input signal is applied, the amplifier must adjust itself so that the feedback and input currents are precisely equal, or

$$\frac{e_{in} - e}{Z_{in}} = \frac{e - e_{out}}{Z_f} \tag{26-1}$$

which leads to

$$e_{out} = \frac{e(Z_f + Z_{in}) - e_{in}Z_f}{Z_{in}} \tag{26-2}$$

This relation can be greatly simplified by taking into consideration the high inherent gain of the amplifier (often called its *open-loop gain*). This means that potential e at the summing junction* must be very small compared to e_{out}. If the gain is 10^6, then an output of 10 V means that the

---

* The summing junction is a designation given to the inverting input connection of an operational amplifier for reasons that will appear shortly.

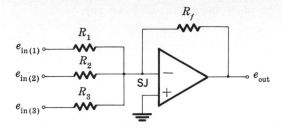

Fig. 26-20 Operational amplifier connected as a summer. SJ denotes the summing junction. For $A \gg 1$,

$$e_{\mathrm{out}} = -R_f \left[\frac{e_{\mathrm{in}(1)}}{R_1} + \frac{e_{\mathrm{in}(2)}}{R_2} + \frac{e_{\mathrm{in}(3)}}{R_3} \right]$$

input to the amplifier will be at a potential only 10 μV removed from ground. This is so close to ground that the summing junction is commonly said to be at *virtual ground*. Hence in Eq. (26-2), the term involving e can be neglected, giving

$$e_{\mathrm{out}} = -e_{\mathrm{in}} \frac{Z_f}{Z_{\mathrm{in}}} \qquad (26\text{-}3)$$

which is the basic working equation of an operational amplifier. In practice the ratio Z_f/Z_{in} is seldom made greater than 100, nor less than 0.01.

If Z_f and Z_{in} are purely resistive, they can be replaced by corresponding R's, and Eq. (26-3) shows that the output voltage will be the negative of the input multiplied by a constant, adjustable between 0.01 and 100.

Several inputs can be connected simultaneously to the summing junction, as in Fig. 26-20, in which case the output becomes the negative sum of the inputs, each multiplied by the appropriate ratio. It is because of this important property that the inverting input is called the *summing junction*. Any of these multiple inputs can be given a negative signal, resulting in subtraction.

To perform integration, the feedback element must be a capacitor, as in Fig. 26-21. Since negligible current can enter the amplifier proper,

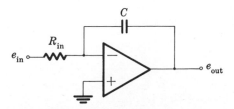

Fig. 26-21 Operational amplifier connected as an integrator. For $A \gg 1$,

$$e_{\mathrm{out}} = -\frac{1}{R_{\mathrm{in}}C} \int_0^t e_{\mathrm{in}} \, dt$$

the input and feedback currents must be equal, and we can write

$$\frac{e_{in}}{R_{in}} = -C \frac{de_{out}}{dt} \tag{26-4}$$

which is equivalent to

$$e_{out} = -\frac{1}{R_{in}C} \int_0^t e_{in}\, dt \tag{26-5}$$

To differentiate, the positions of R and C must be interchanged, and we obtain

$$e_{out} = -R_fC \frac{de_{in}}{dt} \tag{26-6}$$

This simple circuit is little used in practice, because it overemphasizes the effect of random noise at the input and may be unstable.

Another important function is the logarithm, which can be obtained by utilizing the exponential characteristic of a forward-biased p-n junction. Figure 26-22 shows the circuit. The equation is derivable, as before, by setting equal the currents through the input connection and the feedback. The current-voltage relation of a p-n diode is approximated by

$$\log I = k_1V \qquad \text{or} \qquad I = k_2 \text{ antilog } V \tag{26-7}$$

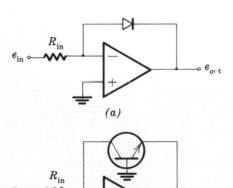

(a)

(b)

Fig. 26-22 Operational amplifier connected to give a logarithmic function. Feedback is (a) a silicon diode and (b) a silicon transistor with a grounded base, both shown oriented for positive e_{in}. For $A \gg 1$,

$$e_{out} = k \log e_{in} - k'$$

in which k and k' are numerical constants.

Hence we can write

$$\frac{e_{\text{in}}}{R_{\text{in}}} = k_2 \text{ antilog } e_{\text{out}}$$

which gives for e_{out}

$$e_{\text{out}} = k_3 \log e_{\text{in}} - k_4 \tag{26-8}$$

Experiment shows that a silicon diode is more satisfactory for this purpose than one of germanium, but that a silicon transistor with grounded base (Fig. 26-22b) is better yet; it can give a linear-log relation over at least four logarithmic decades.

Antilogarithms can be taken by interchanged connections, as in Fig. 26-23. It is possible to carry out multiplication or division of variables by taking their logarithms, adding or subtracting, and then extracting the antilogarithm. It is essential that all the functional transistors (or diodes) be held at the same temperature or compensated for temperature changes.

ERRORS IN OPERATIONAL AMPLIFIERS

One great advantage of operational amplifiers that makes them so convenient is that they follow closely the mathematical relations presented in the last few pages. There are, however, several sources of error that may sometimes become significant. The most important of these will be described here.

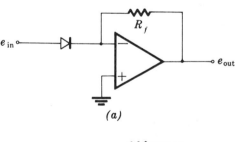

(a)

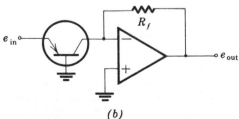

(b)

Fig. 26-23 Operational amplifier connected to give an exponential (or antilogarithmic) function. Input impedance is provided by (a) a silicon diode and (b) a silicon transistor, both shown for positive values of e_{in}. For $A \gg 1$,

$$e_{\text{out}} = k \exp{(-k'e_{\text{in}})}$$

in which k and k' are numerical constants.

Finite gain. The formulas given earlier are all based on the assumption that the inherent, or *open-loop*, gain of the amplifier is much larger than unity. A detailed analysis shows that the equation for the performance of a simple inverter (Fig. 26-19b) is

$$e_{\text{out}} = -e_{\text{in}} \frac{Z_f}{Z_{\text{in}}} \frac{A}{A - G} \tag{26-9}$$

where A is the open-loop gain, and G is the closed-loop gain (i.e., $-Z_f/Z_{\text{in}}$). For example, if $A = 10^5$ and $G = 10^3$, $e_{\text{out}}/e_{\text{in}}$ becomes 1.01×10^3, an error of 1 percent. This is the reason for limiting the Z_f/Z_{in} ratio to about 100 as previously mentioned.

Offset voltage. Ideally an amplifier should give zero output for zero input, but slight asymmetries always exist so that even with both inputs grounded a small voltage will appear at the output. A variable trimming resistor is usually provided to correct this condition by injecting a compensating potential. Slow drifts of the offset voltage may arise after the trimmer has been adjusted, and this is often the feature that limits the small-signal response of the amplifier. Offset is especially detrimental when the amplifier is connected as an integrator, since the effect is cumulative.

Bias current. Sometimes an amplifier fails to behave ideally, in that it draws some appreciable current at its inputs. This bias current is best dealt with by inserting a resistor of a few kilohms between the noninverting input and ground. The voltage drop produced by the bias current in this resistor tends to compensate the drop caused by the same current in the input and feedback resistors. As with offset voltage, this fault is particularly objectionable with integrators.

TRANSDUCER APPLICATIONS OF OPERATIONAL AMPLIFIERS

Transducers are devices by which chemical information is converted into electric signals. They can be classified according to the electrical quantity which represents the signal. There are only three major categories in which transducers can be considered to act as (*1*) a variable resistor, (*2*) a source of potential, or (*3*) a source of current.

The class of *resistive transducers* includes photoconductive cells, thermistors, metallic resistance thermometers, and cells for electrolytic conductivity (most operational amplifiers will work at audio frequencies as well as at direct current).

In principle, any of these can be used either as Z_{in} or Z_f in the circuit of Fig. 26-19. If e_{in} is replaced by a constant potential E_{in}, then observation of e_{out} will allow the unambiguous determination of the resistance of the transducer. If the quantity desired is actually *conductance*, then the transducer can conveniently be placed in the input so that the reciprocal will be obtained directly.

If a *ratio* of two resistances is desired, as might be the case in some dual-beam photometers, then it may be possible to use them as Z_{in} and Z_f with a single amplifier, Fig. 26-24a. If their *difference* is required, rather than their ratio, the circuits of Fig. 26-24b or c can be utilized. Circuit b requires two standard voltages of equal value but opposite sign, which may not always be available. Circuit c needs only a single reference voltage, but requires two amplifiers; it has an added feature which may be a disadvantage, that the output meter cannot be grounded.

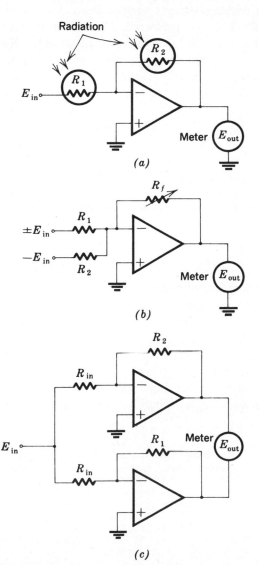

(a)

(b)

(c)

Fig. 26-24 Resistance measurement with operational amplifiers. (*a*) The ratio of two resistances (shown as light-sensitive): $E_{out} = -(R_2/R_1)E_{in}$; (*b*) the sum or difference of resistances; the two E_{in}'s must have the same absolute value, and then adjustment of R_f to make $E_{out} = E_{in}$ will give $R_f = (R_2 \mp R_1)$; (*c*) provided the two resistors marked R_{in} are equal, then $E_{ent} = (E_{in}/R_{in})(R_1 - R_2)$.

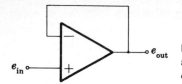

Fig. 26-25 Potential measurement with an operational amplifier, the voltage follower. If $A \gg 1$, then $e_{out} = e_{in}$.

Transducers which produce *potentials* include the many electrode combinations for potentiometry and chronopotentiometry, photovoltaic cells, and thermocouples. The general circuit of Fig. 26-19b is not suitable for potential measurement, as it requires current to flow from the source. Instead, the circuit of Fig. 26-25 is to be preferred. Whatever voltage is presented to the noninverting input will be reproduced at the output. No significant current is drawn from the transducer, but considerable current can be delivered by the amplifier to operate further equipment.

Currents produced by transducers are of importance in voltammetry and amperometry, with vacuum (or gas-filled) phototubes and photo-multipliers, and with flame-ionization GC detectors. These current sources can be connected directly to the summing junction of the amplifier, as in Fig. 26-26. In order to keep this junction at virtual ground, the amplifier forces an equal current through the feedback resistor, thereby building up a potential which is the output e_{out}. Examples of this current-to-voltage conversion appear in Figs. 3-17 and 3-18 in connection with photomultipliers and phototubes.

PHOTOCELLS

There are several types of semiconductor phototransducers. The most generally useful for light-measuring purposes is the *silicon p-i-n photodiode*. This is fabricated on a thin wafer of *intrinsic* silicon (*i*), i.e., the pure, undoped element, with two heavily doped regions, *p* type on the front surface, where the radiation is incident, and *n* type on the back. Boron and phosphorus are appropriate dopants. The *p-i-n* structure is preferable to the simpler *p-n*, in that it permits higher bias voltages to be applied without breakdown, and this favors a wide dynamic linear range.

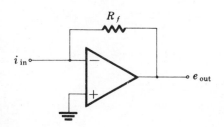

Fig. 26-26 Operational amplifier for current measurement. If $A \gg 1$, then $e_{out} = -R_f i_{in}$.

The operation of a photocell of this type can be described by the family of characteristic curves of Fig. 26-27. In the area to the right of the origin (first quadrant) the cell acts as a photoconductive device; the voltage shown is a bias potential applied across the cell with reverse polarity. In this mode the photocurrent is a linear function of illumination. The second quadrant corresponds to operation in the photovoltaic mode, which means that the cell acts as a generator of electrical energy. In this mode the output is no longer linear with illumination. The straight lines numbered 1, 2, and 3 correspond to three possible operating conditions: (*1*) with no applied-voltage bias, and with nearly zero series resistance—this mode gives good linearity and zero dark current; (*2*) with about —13 V bias to increase the response speed; and (*3*) with zero bias and a high resistance load, in which mode the voltage produced is proportional to the logarithm of the illumination.

Selenium photovoltaic cells, sometimes called *barrier-layer cells*, are less expensive than the silicon types just described, and are limited to service in the photovoltaic mode. They have been extensively employed in filter photometers and fluorometers, but are being displaced by silicon diodes in new designs. Their temperature coefficients are larger, and they suffer from "fatigue" effects, a lessening of sensitivity with time when exposed to light.

Photoconductive cells are made of sulfides or selenides of lead or cadmium. These show no photovoltaic properties. CdS and CdSe cells

Fig. 26-27 Current-voltage characteristic curves for a silicon photodiode; the load lines 1, 2, and 3 are discussed in the text. (*United Detector Technology.*)

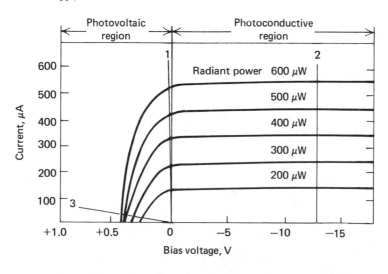

are widely used for all sorts of applications in control and measurement where the greater sensitivity and precision of silicon photodiodes is not required. PbS cells are the detectors of choice for the near infrared, and are widely used in spectrophotometers in this region.

A number of combination semiconductor units are available, such as *phototransistors* and *photo-FETs*, wherein the control function of the transistor is taken over or supplemented by radiation. These are used largely for punched-card and other data processing applications.

Figure 26-28 gives the wavelength dependence of representative photodetectors.

OTHER APPLICATIONS OF OPERATIONAL AMPLIFIERS

In addition to their functions in handling the output of transducers, operational amplifiers have wide areas of usefulness in auxiliary positions in connection with analytical systems.

An operational amplifier can serve as an adjustable constant-voltage source for comparison potentiometry, for controlling a potentiostat, or for other purposes. By this means the defined standard potential of a Weston cell can be utilized in the circuit of Fig. 26-25 without the possibility of damage to the cell from passage of current.

A constant-current source can be assembled by following Fig. 26-19b, where e_{in} and Z_{in} are held constant; the current through Z_f, which becomes the load, is determined by the ratio e_{in}/Z_{in} and is independent of Z_f. The difficulty with this is that the load cannot be grounded, which is sometimes

Fig. 26-28 Wavelength response of typical photocells of various kinds. The CdS, CdSe, and PbS cells are photoconductive, and the Se and Si are photovoltaic.

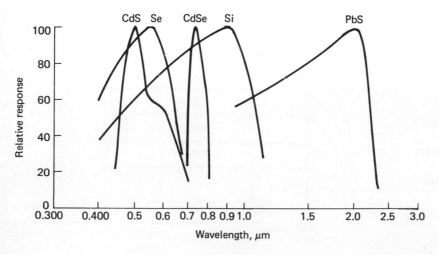

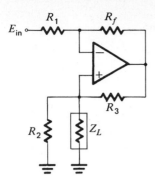

Fig. 26-29 Constant-current source for a grounded load, Z_L. If $R_3/R_2 = R_f/R_3$ (and $A \gg 1$), the load current will be $I_L = -E_{in}/R_2$, independent of variations in Z_L.

important. This restriction can be avoided by using the circuit of Fig. 26-29.

ANALOG COMPUTERS

An electronic analog computer consists of an assemblage of operational amplifiers together with the necessary power supplies and various items of supporting equipment. All input and output connections are brought out to a logical array of jacks on the panel. A selection of plug-in capacitors and resistors is made available, along with a supply of patch cords. Such a computer usually has a panel voltmeter with which to establish conditions and balance the amplifiers, but the ultimate readout is taken by oscilloscope or recorder. Lesser assemblies, known as manifolds rather than computers, may have only four to six amplifiers and fewer auxiliary components.

The original purpose of analog computers was the solution of algebraic or differential equations, especially in an engineering context. For such use a great many amplifiers might be needed, extending into the hundreds. Any of the smaller computers or manifolds is extremely useful in the analytical instrumentation laboratory, permitting the easy construction of any of the operational amplifier circuits discussed in this book.

As an example of more complex instruments that are easily assembled on an operational amplifier manifold, attention is directed to the three-electrode polarograph described in connection with Fig. 15-11. A still more involved piece of electrochemical equipment is an instrument for controlled potential coulometry (2), illustrated in Fig. 26-30, also readily assembled on a manifold. In this instrument, three electrical quantities are of interest: the potential of the working electrode against a reference, the current passing between working and auxiliary electrodes, and the number of coulombs required to carry out a chemical process. All these quantities can be controlled or observed with the analog equipment shown.

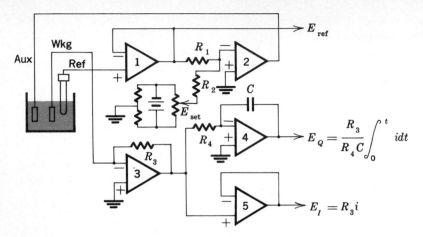

Fig. 26-30 Operational amplifier circuitry for controlled-potential coulometry. The reference potential, working current, and coulombs passed can be monitored simultaneously.

Notice first that the working electrode is connected directly to virtual ground, the summing junction of amplifier 3. Therefore, to maintain a desired voltage between the working and reference electrodes, the reference electrode must be established at a point away from ground. This is accomplished by amplifier 2, as the potentials of the reference electrode E_{ref} and of a manually settable voltage source E_{set} are summed at its summing junction. For convenience, R_1 may be made equal to R_2 so that, because of the virtual ground restriction, one may be sure that the reference electrode potential is equal to E_{set} but of opposite sign. Amplifier 2 delivers as much current to the auxiliary electrode as it needs to maintain the desired condition. Amplifier 1, connected as a voltage follower, is solely to prevent any current from being passed through the reference electrode.

The current flowing through the working electrode must also successively pass through R_3 and R_4, where it charges capacitor C of the integrator (amplifier 4) which measures coulombs. If there is likelihood of the integrator going off-scale, then provision can be made to discharge the capacitor when it reaches a particular voltage level (as 10.00 V) and to count on a mechanical register the number of times such discharging is required. The voltage follower (amplifier 5), connected between R_3 and R_4, gives a measure of the electrolysis current at any moment.

OSCILLATORS

A source of alternating current (other than line frequency) is often needed in laboratory instruments, and is most conveniently obtained electronically. In principle, any amplifier can be made into an oscillator by providing a positive feedback path with a suitable frequency characteristic. This means returning part of the output to the input in such a phase that an increase in output causes an increase in input, which in turn causes a decrease in output, and hence in input, and so on ad infinitum.

Figure 26-31 shows one of many circuits that will do this, the Hartley oscillator. The feedback from the output circuit to the input takes place through the mutual inductance of the two portions of the inductor L. The frequency of oscillation is $f = (2\pi)^{-1}(LC)^{-\frac{1}{2}}$ Hz, and is most conveniently altered by a change in the capacitance.

Another way in which frequency can be fixed is by a *twin-tee filter* (Fig. 26-32a). This network will pass all frequencies *except* that given by the formula $f = (2\pi RC)^{-1}$ Hz. A plot of impedance as a function of frequency shows a very sharp peak in (b); the sharpness depends on how accurately the resistors and capacitors are matched. If this network is connected as the feedback impedance of an amplifier (which may or may not be an operational amplifier as shown), then the net gain $e_{out}/e_{in} = -Z_f/Z_{in}$ will be low for all frequencies other than the characteristic frequency of the filter. We may suppose that an unshielded summing junction will pick up enough noise at random frequencies to constitute an input (shown dotted in the figure), and this will cause oscillations to start at the filter frequency, since this is *not* fed back negatively and hence is amplified by the full open-loop gain of the amplifier. To sustain oscillation, a small, untuned, positive feedback is injected at the noninverting input, via capacitor C_f. This oscillator is especially convenient for service at one or a few fixed frequencies.

A *square-wave generator* can be made by the simple process of clipping a sine wave with zener diodes, as in Fig. 26-33. If the amplitude of the sine wave is several times greater than the zener voltage, a good approximation to a square wave will result.

A square wave can be produced directly with the two-transistor circuit of Fig. 26-34, called a *multivibrator*. The two transistors are cross-connected in such a way that when one of them (say Q_1) turns on (i.e.,

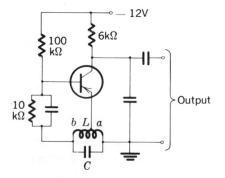

Fig. 26-31   A Hartley oscillator circuit.

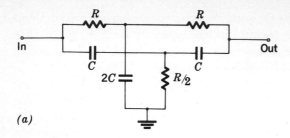

(a)

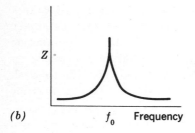

(b)

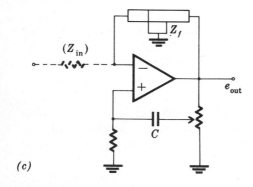

(c)

Fig. 26-32 The twin-tee rejection filter. *(a)* The filter circuit for which the characteristic frequency is $f_0 = 1/(2\pi RC)$ Hz. *(b)* Impedance of the filter as a function of frequency. *(c)* An operational amplifier as an oscillator, with twin-tee feedback.

starts to conduct), it sends a negative pulse to the other (Q_2), sufficient to prevent its immediate turn-on. However, as soon as the charge on capacitor C_1 leaks off through the resistor network, the base of Q_2 will become more positive, until Q_2 starts to conduct. This sends a negative pulse to turn off Q_1, and this cyclical action continues indefinitely. The square-wave output can be taken from the collector of either transistor. Its frequency is determined by the values of R_1C_1 and R_2C_2.

SERVOMECHANISMS

An instrument servo system in its most common form consists of a small motor, either dc or two-phase ac, controlled in speed and direction of rota-

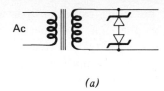

(a)

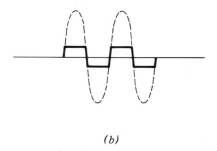

(b)

Fig. 26-33 Formation of a square wave by zener-clipping a sine wave. Note that, no matter which phase, one diode is forward-biased while the other operates in its breakdown region.

tion by an amplifier. The system is provided with some sort of feedback, so that the turning of the motor produces a signal, usually a voltage, which is automatically compared with a standard or reference potential. The difference, called the *error signal*, is returned to the amplifier to control the motor.

A servo may be included wherever a mechanical effect must be produced by a varying signal. An example is its use in a recording spectro-

Fig. 26-34 Multivibrator square-wave generator.

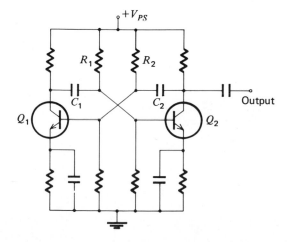

photometer to control the slit width. The beam of radiation passing through the reference cuvet (solvent) serves as the signal to operate the slit-control motor in such a manner as to maintain the transmitted energy constant as the spectrum is scanned. Another, even more widespread, application is in the self-balancing potentiometric recorder, described in the next section.

AUTOMATIC RECORDERS (3)

Electric recorders fall into two general classes: deflection and null instruments. Deflection types, which may be galvanometers, voltmeters, ammeters, wattmeters, etc., are generally less complex in design and can be much faster in response, following variations of input up to perhaps 100 Hz. However, their accuracy is generally inferior to null types, and they are restricted to narrow recording paper, sometimes with curvilinear coordinates.

A major drawback to deflection instruments is the loading effect on the system being measured. It requires a significant amount of power to operate the deflection system, and this must be provided by the circuit connected to the meter. Thus a deflecting millivoltmeter draws a significant current, and a deflecting ammeter causes a noticeable drop in the voltage being measured. Such errors can be eliminated by the addition of an operational amplifier between signal source and recorder movement.

Deflecting recorders designed for high speed of response are known as *oscillographs*. They are particularly useful in engineering measurements on dynamic systems, vibration studies, and the like, and are also applied to biological and clinical studies, including electrocardiography and encephalography. The principal application in chemistry is in mass spectrometry (see Fig. 21-12).

Null recorders may be typified by the self-balancing potentiometer shown schematically in Fig. 26-35. The unknown potential E_x is connected

Fig. 26-35   Self-balancing potentiometer, schematic.

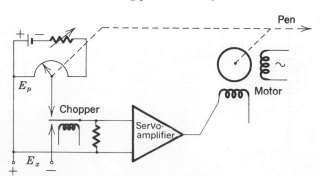

in series opposition with a potential E_p taken from a slide-wire potenti-ometer. A vibrating chopper is so connected that it throws the two poten-tials alternately onto the input of an ac amplifier, at the frequency of the power line. The output of the amplifier energizes one winding of a two-phase motor, the other winding of which is connected directly to the power line. The motor controls the moving contact on the potentiometer slide-wire. At the point of balance, $E_x = E_p$, and the amplifier receives no 60-Hz signal from the chopper, hence supplies no 60-Hz output, and the motor is idle. If E_x becomes larger than E_p, a proportionate signal is observed and amplified, and the motor turns in such a direction as to increase E_p to rebalance the circuit. If E_x becomes smaller than E_p, a similar action takes place, but the output of the amplifier is shifted in phase so that the motor turns in the opposite direction to attain balance. The motor shaft is mechanically linked to the recorder pen, causing it to move a distance proportional to the angular displacement of the sliding contact. This is therefore a measure of the unknown potential E_x. In some models, the amplifier output is rectified synchronously with the input chopping, so that a dc output is produced with its sign dependent on the sign of the input; the motor is then a dc, permanent-magnet type. In either case, the working current through the slide-wire must be adjusted so that the deflection of the pen agrees with the calibration of scale and paper. This may be accomplished by means of a standard cell and rheo-stat, or by a zener diode balancing circuit.

PROBLEMS

26-1 In the rectifier circuits of Figs. 26-11, 26-12a, and 26-13, predict what would happen if one of the diodes were accidentally reversed.

26-2 In the constant-current source of Fig. 26-16, what voltage would appear across the terminals of the cell if the cell resistance were 10 Ω? If the cell were open-circuited, as a result of removing the electrodes from the solution?

26-3 The mechanical chopper is sometimes replaced by one or more photoconductive cells closely coupled with small neon lamps. Show how these units could be used to implement the circuits of Fig. 26-18. The symbol for this electrooptic device is shown in Fig. 26-36.

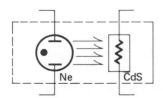

Fig. 26-36 Electrooptic coupler.

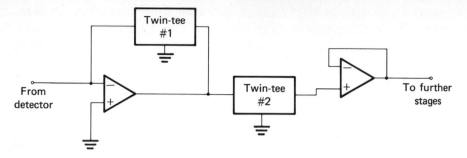

Fig. 26-37 Amplifier with two twin-tee filters.

26-4 In Fig. 26-24b, voltages E_1 and E_2 are required to be equal but of opposite sign. How could this be accomplished with a single voltage source?

26-5 The technique of biamperometric titration, discussed in Chap. 15, requires impressing a small constant voltage across a cell while monitoring the current produced. An analogous titration procedure calls for passing a small constant current through the cell while measuring the potential. Design operational amplifier circuits to implement these two techniques.

26-6 Design an operational amplifier circuit for implementing the equation

$$e_{\text{out}} = kXY^n$$

where X and Y are two variable voltages, n is a variable parameter, and k is a numerical constant. The circuit can be based on Figs. 26-22 and 26-23.

26-7 An infrared spectrophotometer has a rotating shutter that chops the radiation at 17 Hz. The signal amplifier system includes two twin-tee filters, tuned to frequencies of 17 and 60 Hz, arranged as in Fig. 26-37. In which positions should the two filters be placed? Explain their functions in the circuit.

REFERENCES

1. Gray, P. E., D. DeWitt, and A. R. Boothroyd: "SEEC Notes I; PEM: Physical Electronics and Circuit Models of Transistors," Education Development Center, Newton, Mass., 1962.
2. Propst, R. C.: A Multipurpose Instrument for Electrochemical Studies, *AEC Res. Dev. Rept.* DP-903 (1964).
3. Ewing, G. W., and H. A. Ashworth: "The Laboratory Recorder," Plenum, New York, 1974.

GENERAL REFERENCES

Diefenderfer, A. J.: "Principles of Electronic Instrumentation," W. B. Saunders, Philadelphia, 1972.

Malmstadt, H. V., C. G. Enke, S. R. Crouch, and G. Horlick: "Electronic Measurements for Scientists," W. A. Benjamin, New York, 1974.

Malmstadt, H. V., and C. G. Enke: "Digital Electronics for Scientists," W. A. Benjamin, New York, 1969.

Smith, J. I.: "Modern Operational Circuit Design," Wiley-Interscience, New York, 1971.

Graeme, J. G., "Applications of Operational Amplifiers: Third-Generation Techniques," McGraw-Hill, New York, 1973.

Graeme, J. G., G. E. Tobey, and L. P. Huelsman: "Operational Amplifiers: Design and Applications," McGraw-Hill, New York, 1971.

Vassos, B. H., and G. W. Ewing: "Analog and Digital Electronics for Scientists," Wiley-Interscience, New York, 1972.

Chapter 27

Instrument-Computer Interfacing

One of the most significant instrumental developments of recent times has only been hinted at in previous chapters: the interfacing of laboratory instruments with digital computers. This is far too large and complex a subject to treat in detail in this book; the most that can be done here is to summarize the fields of application and to indicate in general terms the most important features.

TYPES OF COMPUTERS*

For present purposes computers are best classified by size. Size means not only how much space is occupied in the laboratory, but also how many memory locations are available (often rated in thousands, as 4k or 8k). Large computers are not generally located in the laboratory, but are usually installed in a separate computer center, perhaps many miles distant. Their use in data processing and general problem solving is well known.

---

* Analog computers were mentioned briefly in Chap. 26. We are concerned here only with digital computers.

538

Of more immediate importance in connection with laboratory instrumentation are the *minicomputer,* and *microcomputer.* The minicomputer can well be located in or close to the laboratory. It may be a *dedicated* unit, which means that it is permanently connected to a particular instrument and is not available for other use even when that instrument is idle.* On the other hand, the minicomputer may be *time-shared,* through a multiplexing device, so that it can be used in several different ways by different users at the same time without interference. It can be used in three distinct functions, namely, to control the operation of an instrument, to acquire data from the instrument, and to process the data so obtained.

A microcomputer is always dedicated. It can be considered a stripped-down version of a minicomputer, retaining only those components that are actually needed for the specific application. It is not provided with its own separate control panel, but does its work as an integral part of the host instrument.

There are, of course, no sharp lines of distinction between these computer types, nor between a microcomputer and instrument electronics that would not ordinarily be considered a computer.

A digital computer is solely controlled by two-value switches (both manual and electronic) that can be on or off but can never stably occupy any intermediate position. Similarly, the computer can only act by establishing outputs at either of two voltage levels corresponding to this same duality. The state of logic at any point in the entire computer can be completely specified in terms of these two voltage levels. The two levels are designated by various names, as on-off, true-false, high-low, one-zero; we shall use the terms high and low. In many computers the corresponding voltage levels are high $= 5.0 \pm 2.0$ V and low $= 0.0 \pm 0.5$ V. The band of forbidden potentials between high and low (i.e., between $+0.5$ and $+3.0$ V) gives "noise immunity" to the system, which means that random noise pulses are most unlikely to be large enough to bridge this gap and cause false signals.

THE STRUCTURE OF A COMPUTER

Most digital computers, no matter what the size class, are built on the same fundamental plan, as diagrammed in Fig. 27-1. The component variously designated the *central processing unit* (CPU) or *arithmetic processor* is the heart of the computer, where the essential logical decisions are made. All other component parts are attached to the CPU. These include most of the storage area devoted to memory (the CPU may contain a small fraction of the memory as well), and the in/out (I/O) peripheral devices.

The bulk of the memory of all except perhaps the smallest computers is of the type called *random-access* (RAM). Data or instructions can

* An example is shown in Fig. 8-7.

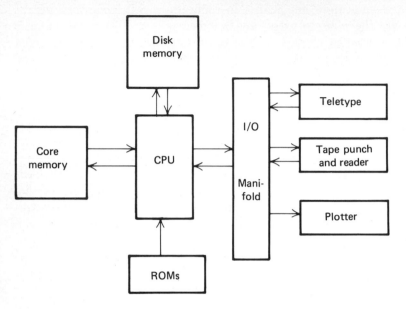

Fig. 27-1   Architecture of a digital computer.

be stored in it at any time, and recalled in whatever sequence is demanded by the program. Such memory units are often constructed of magnetizable *cores* made of ferrite material, in which the state of magnetization can be described by either of two oppositely directed vectors corresponding to the high and low logic levels. The tiny toroidal cores are arranged in square arrays, typically 32 rows of 32 cores for a total of 1024 positions, and there may be many such squares. The state of magnetization is induced and later sensed by fine wires threaded through the cores. Other memory units, less costly but slower of access, make use of magnetic tapes or magnetic disks. A single disk may hold as much information as 20 million cores. Disks are largely used for storage of programs for future use.

Another class of memory is called *read-only memory* (ROM). This may have a particular program or set of instructions built in at the time of manufacture. This program can be utilized whenever needed. An ROM, for example, could contain all the necessary instructions to extract square roots.

The I/O devices often connected to a computer include the electric typewriter or Teletype, that can both introduce data into the computer and print out analytical results. Digital plotters are available to give graphical outputs. Either magnetic or paper tape may be used to receive data from an experiment and subsequently introduce it into the computer.

COMPUTER INTERFACING

The laboratory instrument to be connected to a computer must provide its information in a form that can be utilized by the computer's system of switches. This means that the signals from the instrument must consist of a series of pulses of uniform height (nominally 5 V), but of variable duration and repetition rate.

A few types of analytical instruments normally give a digital output. A Geiger counter and a photon-counting photometer are two examples of instruments that give pulses all of the same size but at varying rates. From other instruments the signal must be translated to an acceptable form by an *analog-to-digital converter* (ADC). One type of ADC is a turns-counting device, applicable where the information from the instrument is carried by a rotating shaft. This might be the case, for example, in specifying the wavelength setting of a monochromator. If the desired resolution requires reading the position of the shaft with precision, a *digital shaft encoder* (Fig. 27-2) can be employed. This consists of a series of concentric rings with either opaque or transparent segments. A linear array of tiny photocells, usually photodiodes or phototransistors, is mounted radially behind the disk to detect the transparent areas. The sequence

Fig. 27-2   A binary shaft encoder.

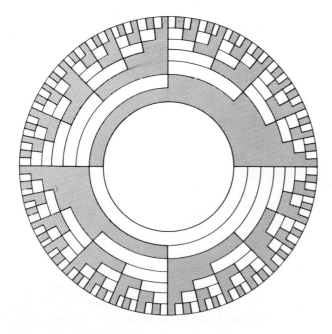

of pulses resulting from turning the disk can be interpreted digitally by the computer.

Other ADCs are purely electronic, usually based on counting pulses from a square-wave generator. In one type, the frequency of the generator is controlled by the input signal. In another, the time required for a ramp voltage to change from zero to the signal level is measured.

Several data inputs may be required from the same instrument. For example, an ultraviolet spectrophotometer might require signals from the scan drive, the slit control, and the beam-splitting chopper, as well as the photomultiplier. These can be handled through separate ADCs, or they can be *multiplexed*, which means that the several signals are sampled one at a time and digitized by a single ADC.

Once the signals are received in digital form, they are utilized by the computer in accordance with a *program* that has been entered previously into the computer's memory banks. The program then generates a new set of digital signals to serve as computer output. These output signals can drive such peripheral recording equipment as the electric typewriter or digital plotter. They can also, in many cases, be returned to the laboratory instrument to control its operating parameters. Thus the computer can examine the various signals coming from the instrument, evaluate them by performing arithmetic operations, make decisions as to the next step, and instruct the instrument accordingly. This is known as *closed-loop* control. Of course the capability to do all this must have been provided by the programmer.

Since most laboratory instruments are analog devices, i.e., operate with continuously variable quantities, the digital signals from the computer must usually be returned to the analog domain to be effective in controlling instrumental variables. This is generally accomplished by a *digital-to-analog converter* (DAC), the inverse of the ADC. Sometimes this can be avoided to advantage through the use of special components. An important example is the *stepping motor*, often used to drive the wavelength scan of a computer-controlled spectrophotometer. This motor operates on pulses, and will turn through an exactly reproducible fraction of a complete rotation for each pulse. Hence a computer-generated signal consisting of a series of pulses with controlled timing will drive the motor just as much or as little as required.

The entire set of ADCs, DACs, and multiplexors required to interconnect an instrument with a computer is termed the *interface*.

COMPUTER PROGRAMMING

Fundamentally, a computer can only respond to digital signals presented sequentially. The set of rules by which a program is established by direct setting of on-off switches is known as *machine language*. For any but the shortest programs this is exceedingly cumbersome. Instead the switch-setting steps are assembled into related groups, each designated by a code word, usually for convenience a mnemonic. Then the program can be written in terms of these mnemonic code words, a procedure known as *assembly language programming.* This saves much time over machine language, but is still too slow and laborious for many purposes, and higher-level languages such as Basic and Fortran have been devised for much actual problem-solving.

These high-level languages are designed to be easily comprehended by the programmer and user of the computer without the need for complete understanding of the internal workings of the machine. Fortran consists largely of symbolic sentences, called *statements,* which follow the customary patterns in which arithmetic and algebraic operations are written. Basic emphasizes conversational interaction between computer and operator. Either can be used successfully for data processing and calculations.

Perhaps the most powerful procedure is to utilize a combination of one of these high-level languages with subroutines written in assembly language. The control and data acquisition functions are handled most efficiently in assembly language, whereas the data processing and interaction with the operator are dealt with in Fortran or Basic. Minicomputers, applied to the control of instruments, are generally programmed in assembly language.

GENERAL REFERENCES

For a comprehensive treatment of the relation between computers and laboratory instruments the reader is referred to:

Perone, S. P., and D. O. Jones: "Digital Computers in Scientific Instrumentation: Applications to Chemistry," McGraw-Hill, New York, 1973.

A concise description of semiconductor devices that can be utilized in home-built interfacing is presented in:

Dessy, R. E., and J. A. Titus: *Anal. Chem.,* **45**:124A (1973).

Representative control applications are described in:

Mueller, K. A., and M. F. Burke: *Anal. Chem.,* **43**:641 (1971).

Eggert, A. A., G. P. Hicks, and J. E. Davis: *Anal. Chem.,* **43**:736 (1971).

Appendix

Standard electrode potentials*

Electrode	Reaction	$E°$, V versus NHE
F_2, F^-	$F_2 + 2e^- \rightarrow 2F^-$	$+2.65$
Co^{3+}, Co^{2+}, Pt	$Co^{3+} + e^- \rightarrow Co^{2+}$	$+1.82$
Au^+, Au	$Au^+ + e^- \rightarrow Au$	$+1.68$
Ce^{4+}, Ce^{3+}, Pt	$Ce^{4+} + e^- \rightarrow Ce^{3+}$	$+1.61$
MnO_4^-, Mn^{2+}, Pt	$MnO_4^- + 8H^+ + 5e^- \rightarrow Mn^{2+} + 4H_2O$	$+1.51$
Au^{3+}, Au	$Au^{3+} + 3e^- \rightarrow Au$	$+1.50$
Cl_2, Cl^-	$Cl_2 + 2e^- \rightarrow 2Cl^-$	$+1.360$
$Cr_2O_7^{2-}$, Cr^{3+}, Pt	$Cr_2O_7^{2-} + 14H^+ + 6e^- \rightarrow 2Cr^{3+} + 7H_2O$	$+1.33$
Tl^{3+}, Tl^+, Pt	$Tl^{3+} + 2e^- \rightarrow Tl^+$	$+1.25$
O_2, H_2O	$O_2 + 4H^+ + 4e^- \rightarrow 2H_2O$	$+1.229$
Pt^{2+}, Pt	$Pt^{2+} + 2e^- \rightarrow Pt$	$+1.2$
Br_2, Br^-	$Br_2(liq) + 2e^- \rightarrow 2Br^-$	$+1.065$
Hg^{2+}, Hg_2^{2+}, Pt	$2Hg^{2+} + 2e^- \rightarrow Hg_2^{2+}$	$+0.920$
Ag^+, Ag	$Ag^+ + e^- \rightarrow Ag$	$+0.799$
Hg_2^{2+}, Hg	$Hg_2^{2+} + 2e^- \rightarrow 2Hg$	$+0.789$
Fe^{3+}, Fe^{2+}, Pt	$Fe^{3+} + e^- \rightarrow Fe^{2+}$	$+0.771$
Ag_2SO_4, Ag	$Ag_2SO_4 + 2e^- \rightarrow Ag + SO_4^{2-}$	$+0.653$
$AgC_2H_3O_2$, Ag	$AgC_2H_3O_2 + e^- \rightarrow Ag + C_2H_3O_2^-$	$+0.643$
I_2, I^-	$I_2 + 2e^- \rightarrow 2I^-$	$+0.536$
Cu^+, Cu	$Cu^+ + e^- \rightarrow Cu$	$+0.521$
Ag_2CrO_4, Ag	$Ag_2CrO_4 + 2e^- \rightarrow 2Ag + CrO_4^{2-}$	$+0.446$
VO^{2+}, V^{3+}, Pt	$VO^{2+} + 2H^+ + e^- \rightarrow V^{3+} + H_2O$	$+0.361$
$Fe(CN)_6^{3-}$, $Fe(CN)_6^{4-}$, Pt	$Fe(CN)_6^{3-} + e^- \rightarrow Fe(CN)_6^{4-}$	$+0.36$
Cu^{2+}, Cu	$Cu^{2+} + 2e^- \rightarrow Cu$	$+0.337$
UO_2^{2+}, U^{4+}, Pt	$UO_2^{2+} + 4H^+ + 2e^- \rightarrow U^{4+} + H_2O$	$+0.334$
Hg_2Cl_2, Hg	$Hg_2Cl_2 + 2e^- \rightarrow 2Hg + 2Cl^-$	$+0.268$
AgCl, Ag	$AgCl + e^- \rightarrow Ag + Cl^-$	$+0.222$
$HgBr_4^{2-}$, Hg	$HgBr_4^{2-} + 2e^- \rightarrow Hg + 4Br^-$	$+0.21$
Cu^{2+}, Cu^+, Pt	$Cu^{2+} + e^- \rightarrow Cu^+$	$+0.153$
Sn^{4+}, Sn^{2+}, Pt	$Sn^{4+} + 2e^- \rightarrow Sn^{2+}$	$+0.15$
Hg_2Br_2, Hg	$Hg_2Br_2 + 2e^- \rightarrow 2Hg + 2Br^-$	$+0.140$
CuCl, Cu	$CuCl + e^- \rightarrow Cu + Cl^-$	$+0.137$
TiO^{2+}, Ti^{3+}, Pt	$TiO^{2+} + 2H^+ + e^- \rightarrow Ti^{3+} + H_2O$	$+0.1$
AgBr, Ag	$AgBr + e^- \rightarrow Ag + Br^-$	$+0.095$
UO_2^{2+}, UO_2^+, Pt	$UO_2^{2+} + e^- \rightarrow UO_2^+$	$+0.05$

Standard electrode potentials (Continued)

Electrode	Reaction	$E°, V$ versus NHE
CuBr, Cu	$CuBr + e^- \rightarrow Cu + Br^-$	$+0.033$
H^+, H_2	$2H^+ + 2e^- \rightarrow H_2$	0.000
$HgI_4{}^{2-}$, Hg	$HgI_4{}^{2-} + 2e^- \rightarrow Hg + 4I^-$	-0.04
Pb^{2+}, Pb	$Pb^{2+} + 2e^- \rightarrow Pb$	-0.126
Sn^{2+}, Sn	$Sn^{2+} + 2e^- \rightarrow Sn$	-0.136
AgI, Ag	$AgI + e^- \rightarrow Ag + I^-$	-0.151
CuI, Cu	$CuI + e^- \rightarrow Cu + I^-$	-0.185
Mo^{3+}, Mo	$Mo^{3+} + 3e^- \rightarrow Mo$	-0.2
Ni^{2+}, Ni	$Ni^{2+} + 2e^- \rightarrow Ni$	-0.250
V^{3+}, V^{2+}, Pt	$V^{3+} + e^- \rightarrow V^{2+}$	-0.255
$PbCl_2$, Pb	$PbCl_2 + 2e^- \rightarrow Pb + 2Cl^-$	-0.268
Co^{2+}, Co	$Co^{2+} + 2e^- \rightarrow Co$	-0.277
$PbBr_2$, Pb	$PbBr_2 + 2e^- \rightarrow Pb + 2Br^-$	-0.280
Tl^+, Tl	$Tl^+ + e^- \rightarrow Tl$	-0.336
$PbSO_4$, Pb	$PbSO_4 + 2e^- \rightarrow Pb + SO_4{}^{2-}$	-0.356
PbI_2, Pb	$PbI_2 + 2e^- \rightarrow Pb + 2I^-$	-0.365
Ti^{3+}, Ti^{2+}, Pt	$Ti^{3+} + e^- \rightarrow Ti^{2+}$	-0.37
Cd^{2+}, Cd	$Cd^{2+} + 2e^- \rightarrow Cd$	-0.403
Cr^{3+}, Cr^{2+}, Pt	$Cr^{3+} + e^- \rightarrow Cr^{2+}$	-0.41
Fe^{2+}, Fe	$Fe^{2+} + 2e^- \rightarrow Fe$	-0.440
Ga^{3+}, Ga	$Ga^{3+} + 3e^- \rightarrow Ga$	-0.53
TlCl, Tl	$TlCl + e^- \rightarrow Tl + Cl^-$	-0.557
U^{4+}, U^{3+}, Pt	$U^{4+} + e^- \rightarrow U^{3+}$	-0.61
TlBr, Tl	$TlBr + e^- \rightarrow Tl + Br^-$	-0.658
Cr^{3+}, Cr	$Cr^{3+} + 3e^- \rightarrow Cr$	-0.74
TlI, Tl	$TlI + e^- \rightarrow Tl + I^-$	-0.753
Zn^{2+}, Zn	$Zn^{2+} + 2e^- \rightarrow Zn$	-0.763
TiO^{2+}, Ti	$TiO^{2+} + 2H^+ + 4e^- \rightarrow Ti + H_2O$	-0.89
Mn^{2+}, Mn	$Mn^{2+} + 2e^- \rightarrow Mn$	-1.18
V^{2+}, V	$V^{2+} + 2e^- \rightarrow V$	-1.18
Ti^{2+}, Ti	$Ti^{2+} + 2e^- \rightarrow Ti$	-1.63
Al^{3+}, Al	$Al^{3+} + 3e^- \rightarrow Al$	-1.66
U^{3+}, U	$U^{3+} + 3e^- \rightarrow U$	-1.80
Be^{2+}, Be	$Be^{2+} + 2e^- \rightarrow Be$	-1.85
Np^{3+}, Np	$Np^{3+} + 3e^- \rightarrow Np$	-1.86
Th^{4+}, Th	$Th^{4+} + 4e^- \rightarrow Th$	-1.90
Pu^{3+}, Pu	$Pu^{3+} + 3e^- \rightarrow Pu$	-2.07
$AlF_6{}^{3-}$, Al	$AlF_6{}^{3-} + 3e^- \rightarrow Al + 6F^-$	-2.07
Mg^{2+}, Mg	$Mg^{2+} + 2e^- \rightarrow Mg$	-2.37
Ce^{3+}, Ce	$Ce^{3+} + 3e^- \rightarrow Ce$	-2.48
La^{3+}, La	$La^{3+} + 3e^- \rightarrow La$	-2.52
Na^+, Na	$Na^+ + e^- \rightarrow Na$	-2.714
Ca^{2+}, Ca	$Ca^{2+} + 2e^- \rightarrow Ca$	-2.87
Sr^{2+}, Sr	$Sr^{2+} + 2e^- \rightarrow Sr$	-2.89
Ba^{2+}, Ba	$Ba^{2+} + 2e^- \rightarrow Ba$	-2.90
K^+, K	$K^+ + e^- \rightarrow K$	-2.925
Li^+, Li	$Li^+ + e^- \rightarrow Li$	-3.045

* From W. M., Latimer, "Oxidation States of the Elements and Their Potentials in Aqueous Solution," 2d ed., Prentice-Hall, Englewood Cliffs, N.J., 1952.

Name Index

Subject Index